D1260124

NANOTECHNOLOGY

NANOTECHNOLOGY

Basic Calculations for Engineers and Scientists

Louis Theodore, Eng. Sc. D.

Consultant, Theodore Tutorials
East Williston, New York

WILEY-INTERSCIENCE

A JOHN WILEY & SONS, INC., PUBLICATION

Library of Congress Cataloging-in-Publication Data:

Theodore, Louis.
    Nanotechnology: basic calculations for engineers and scientists / Louis Theodore.
       p. cm.
    "Wiley-Interscience."
    Includes bibliographical references and index.
    ISBN-13: 978-0-471-73951-7
    ISBN-10: 0-471-73951-0
       1. Nanotechnology--Mathematics--Problems, exercises, etc. I. Title.

T174.7.T477 2006
620′.5--dc22

                                                                    2005047794

Printed in the United States of America

10 9 8 7 6 5 4 3 2

Nature is neutral. Man has wrestled from nature the power to make the world a desert or to make the deserts bloom. There is no evil in the atom; only in men's souls.

—Adlai Stevenson, 1952

Ill can he rule the great, that cannot reach the small.

—Edmund Spenser, 1596

The small and tiny shall become all-powerful

—L. Theodore, 2006

Contents

Preface

It is not a secret that the teaching of a nanotechnology course will soon be required in most engineering and science curricula. It is also generally accepted as one of the key state-of-the-art courses in applied science. The need to develop an understanding of this general subject matter for the practicing engineer and scientist of the future cannot be questioned.

One of the problems with nanotechnology is that its range of subject matter is so broad that nearly every engineering and science discipline falls under the nano umbrella; in effect, it is interdisciplinary. Adding to the confusion is that no clear-cut definition of nanotechnology has emerged since its infancy nearly a half century ago. The reader will soon note that the author has not laid claim to an end-all definition, but rather refers to nanotechnology simply as nanotechnology.

This project was a unique undertaking. Rather than prepare a textbook on nanotechnology, the author considered writing a problem-oriented book because of the dynamic nature of this emerging field. Ultimately, it was decided to prepare an overview of this subject through illustrative examples rather than to provide a comprehensive treatise. One of the key features of this book is that it could serve both academia (students) and industry. Thus, it offers material not only to individuals with limited technical background, but also to those with extensive industrial experience. As such, it can be used as a text in either a general engineering or science course and (perhaps primarily) as a training tool for industry.

As is usually the case in preparing a manuscript, the question of what to include and what to omit has been particularly difficult. However, the problems and solutions in this work attempt to address principles and basic calculations common to nanotechnology.

This basic calculations workbook is an outgrowth of the 2005 John Wiley & Sons book "Nanotechnology: Environmental Implications and Solutions". The desirability of publishing a workbook that focuses almost exclusively on nanotechnology calculations was obvious following the completion of that book.

This book contains nearly 300 problems related to a variety of topics of relevance to the nanotechnology field. These problems are organized into the following four Parts or Categories:

Chemistry Fundamentals and Principles
Particle Technology
Applications
Environmental Concerns

Each Part is divided into a number of problem Sections (or Chapters), with each set containing anywhere from 8 to 12 problems and solutions. The interrelationship between the problems is emphasized in all Parts.

The general approach employed involved the use of solved illustrative examples. However, introductory paragraphs are included in each Part and each Section. The remainder of the text consists of solved examples. In each Part, these have been chosen to emphasize the most important basic concepts, issues, and applications that arise in the topic covered by that Part.

Another feature of this work is that the solutions to the problems are presented in a stand-alone manner. Throughout the book, the problems are laid out in such a way as to develop the reader's technical understanding of the subject in question. Each problem contains a title, problem statement, data, and solution, with the more difficult problems located at or near the end of each problem set (Section). Although some of the topics are somewhat segmented and compartmentalized (relative to each other), every attempt was made to present and arrange each subject in a logical order.

The author cannot claim sole authorship to all the problems and material in this book. The main sources that were employed in preparing the problems included numerous Theodore Tutorials (plus those concerned with the professional engineering exam) and the Reynolds, Jeris and Theodore 2004 Wiley-Interscience text, "Handbook of Chemical and Environmental Engineering Calculations". Finally, the author wishes to acknowledge the National Science Foundation for supporting several faculty workshops that produced a number of problems appearing in this work.

The author also wishes to thank Dr. Albert Swertka, Professor Emeritus of Physicis, U.S. Merchant Marine Academy, for contributing an outstanding write-up in layman terms on "Quantum Mechanics". It can be found in the Appendix. This material was included for those readers interested in obtaining a (better) understanding of how quantum mechanics is related to nanotechnology.

Somehow, the editor usually escapes acknowledgement. I was particularly fortunate to have Bob Esposito ("Espo" to us) of John Wiley & Sons serve as my editor. His advice, support, and encouragement is appreciated.

It is the hope of the editor and author that this basic calculations text provides support in developing an understanding of nanotechnology, and that it will become a useful resource for the training of engineers and scientists in mastering this critical topic area.

Louis Theodore
January 2006

Introduction

Technical individuals have traditionally conducted calculation-related studies using one of a combination of the following approaches (see Figure A):

1. Macroscopic level
2. Microscopic level
3. Molecular level

These studies generally involve the application of a conservation law, e.g., mass, energy, and momentum. For example, if one were interested in determining changes occuring at the inlet and outlet of a system under study, the conservation law is applied on a "macrocopic" level to the entire system. The resultant equation describes the overall changes occuring to the system without regard for internal variations *within* the system. This approach is usually employed in a Unit Operations (for chemical engineers) course. The microscopic approach is employed when detailed information concerning the behavior *within* the system is required, and this is often requested of and by technical personnel. The conservation law is then applied to a differential element within the system, which is large compared to an individual molecule, but small compared to the entire system. The resultant equation is then expanded, via an integration, to describe the behavior of the entire system. This has come to be defined by some as the *transport phenomena approach*. The molecular approach involves the application of the conservation law to individual molecules. This leads to a study of statistical and quantum mechanics – both of which are beyond the scope of this text.

Approaches (1) and (2) are normally in the domain of the engineer, while (2) and (3) are employed by the scientist, particularly the physicist. In a very real sense, this

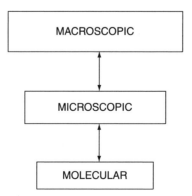

Figure A   Engineering and Science Approaches

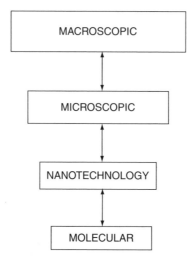

Figure B   Nanotechnology Approach

text emphasizes (1), since it has been written for the practicing engineer and scientist, and attempts to provide solutions to real-world nanotechnology applications. Not withstanding this, material in Part 1 delves into some science principles and fundamentals, and an abbreviated introduction to quantum mechanics can be found in the Appendix.

However, nanotechnology has disrupted the above classical approach to the describing behavior of systems. The nanotechnology field today belongs somewhere between (2) and (3), i.e., between the microscopic and molecular approaches (Figure B).

Nanoparticles cannot be correctly described by applying either the microscopic or molecular method of analysis. This new, so-called, in-between field gives rise to some very unusual physics.

This unusual behavior results because the (physical, chemical, and so on) properties are a strong function of the size of the substance. At microscopic or macroscopic sizes, one chunk of iron (an element) has the exact same properties of another chunk of iron. At the molecular level, an atom of iron has the exact same properties of another atom of iron. However, something happened on the way to the forum ... when the size of the iron particle is in the nano range. The chemical, physical, mechanical, electrical, etc., properties of these bulk materials are different in the nanometer range. Further, a 10 nanometer particle has different properties than a particle of different size, e.g., 20 nanometers. (Note: A nanometer is one billionth of a meter; thus, one nanometer equals 10^{-9} meters.) The same phenomena is experienced with iron oxide or any other solid particle. What does all of this mean? It permits a new way to vary and control the properties of materials. In effect, one need only change the size of the particle rather than its composition.

PART 1
Chemistry Fundamentals and Principles

One of the most used definitions of chemistry is that it is concerned with the study of the properties of materials and the changes that materials undergo. It has also been said that chemistry deals with the combination of atoms, and physics with the forces between atoms. One of the objects of physical chemistry is to interpret the relationship between atoms and molecules by examining the forces that exist at the atomic level.

As indicated above, chemistry involves studying the properties and behavior of matter. *Matter* is the physical material that composes the universe; it represents anything that has mass and occupies space. It can take any and many forms. Over the years, the chemist learned that the tremendous variety of matter is due to special combinations of 112 very "elementary" substances, called *elements*. The properties of matter can be related to *atoms*, the special or infinitesimally small building blocks of matter. The atoms can be combined to form molecules. Molecular properties are a function of the number of different atoms and how their component atoms connect to each other. For example, consider the two molecules methane and methanol. Methane molecules contain one atom of carbon and four atoms of hydrogen. Methane is also referred to as natural gas, since the principle component in natural gas is methane. The addition of one oxygen atom to a molecule of methane converts it to methanol. Methanol is a liquid alcohol. This demonstrates how simple changes in the atomic structure of matter can cause significant changes in its properties.

One of the future challenges facing nanotechnology is to change molecules in a controlled way, creating new substances with very special properties. These changes can lead to improvement of healthcare, conservation of natural resources, protection of the environment, and provision of everyday needs for food, clothing, and protective armaments.

When familiar materials such as metals, metal oxides, ceramics, and polymers, and novel forms of carbon are converted into infinitesimally small particle sizes, the resulting particles have orders of magnitude increases in available surface

Nanotechnology: Basic Calculations for Engineers and Scientists, by Louis Theodore
Copyright © 2006 John Wiley & Sons, Inc.

area. It is this remarkable surface of particles in the nanometer range (1.0 nanometer $= 10^{-9}$ meter) that confers upon hem unique properties, especially when compared to macroscopic particles of the same material [1].

This first Part of the book is specifically devoted to chemistry principles and fundamentals. An understanding of this subject is a prerequisite for understanding the basics of nanotechnology. Atoms, the modern theory of atomic structure, elements, the periodic table, molecules, conversion constants and dimensional analysis, concentration terms, surface area determination, crystal structure, and physical/chemical property estimation all can be factored into the mix when studying nanotechnology.

The objective of Part 1 is to introduce the reader to some simple science principles and fundamentals. Most of this material can be directly or indirectly related to the nanotechnology field. A more basic analysis of subatomic particles can be found in the Appendix (see Quantum Mechanics).

The first Part contains a host of solved problems that concentrate on seven important areas, arranged in Chapters:

1. Units, Conversion Constants, and Dimensional Analysis
2. Atoms, Elements, and the Periodic Table
3. Molecular Rearrangements
4. Concentration Terms
5. Particle Size, Surface Area, and Volume
6. Crystal Structure
7. Physical and Chemical Property Estimation

1 Units, Conversion Constants, and Dimensional Analysis

This first Chapter is primarily concerned with units, conversion constants, and dimensional analysis. Each of these receives treatment in the problems that follow; emphasis is placed on the conversion of units using a dimensional analysis approach. The Chapter concludes with two problems on significant figures.

Many engineering and scientific terms are *quantitative;* i.e., they are associated with numbers. When a number represents quantity, the units (unless dimensionless) of that quantity should be specified, i.e., both a number and unit need to be provided. To say that the diameter of an atom is 1.5 is meaningless. To say that it is 1.5 nanometers (nm) correctly specifies the length. The units used for scientific applications, particularly in nanotechnology, are those of the *metric* and *SI systems.* These two systems are reviewed in the first two Problems.

Converting a measurement from one unit to another can conveniently be accomplished by using *unit conversion factors*; these factors are obtained from the simple equation that relates the two units numerically. For example, from

$$1 \ \text{foot(ft)} = 12 \ \text{inches(in)}$$

the following conversion factor can be obtained

$$12 \ \text{in}/1 \ \text{ft} = 1$$

Since this factor is equal to unity, multiplying some quantity (e.g., 18 ft) by the factor cannot alter its value. Hence

$$18 \ \text{ft}(12 \ \text{in}/1 \ \text{ft}) = 216 \ \text{in}$$

Note that the old units of *feet* on the left-hand side cancel out leaving only the desired units of *inches.*

Similarly

$$1 \ \text{meter (m)} = 10^9 \ \text{nanometer (nm)}$$

Nanotechnology: Basic Calculations for Engineers and Scientists, by Louis Theodore
Copyright © 2006 John Wiley & Sons, Inc.

and the corresponding conversion constant or factor is

$$10^9 \text{ nm}/1 \text{ m} = 1$$

Physical equations must be dimensionally consistent. For the equality to hold, each term in the equation must have the same dimensions. This condition can be and should be checked when solving engineering problems. Throughout the text, and in particular in this Chapter, great care is exercised in maintaining the dimensional formulas of all terms and dimensional homogeneity of each equation. Note that equations are generally provided or developed in term of specific units rather than general dimensions (e.g., nanometers, rather than length). This approach should help the reader to more easily attach physical significance to the terms and equations presented in this text.

Since conversion constants are equal to unity with no units, more than one conversion factor can be employed in the solution of a problem. Further modification of an equation can be accomplished if the units for one or more of the terms of the equation are altered through the use of conversion factors. This is discussed in the next paragraph.

One of the properties of equations that has a rational basis and is deduced from general relations is that they must be dimensionally homogeneous, or consistent. This can be demonstrated theoretically. If two sides of an equation should have different dimensions, the equation is in error. Thus, one test for the consistency of an equation is whether all the terms in the equation contain the same units. If this condition is satisfied, the equation is said to be dimensionally consistent (correct); if not, it is incorrect. However, this aforementioned property of dimensional homogeneity does not apply to empirical (not based on theoretical or physical principles) equations. Also note that the application of the principle of consistency not only must be applied to the magnitude of the terms in an equation, i.e., one term cannot be finite in magnitude while another term is a differential or is differentially small, but also to significant figures.

1.1 BACKGROUND ON THE METRIC SYSTEM

Provide background information on the metric system.

SOLUTION

The need for a single worldwide coordinated measurement system was recognized over 300 years ago. Gabriel Mouton, Vicar of St. Paul in Lyons, proposed a comprehensive decimal measurement system in 1670 based on the length of one minute of arc of a great circle of the earth. In 1671, Jean Picard, a French astronomer, proposed the length of a pendulum beating seconds as the unit of length. (Such a pendulum would have been fairly easily reproducible, thus facilitating the widespread distribution of uniform standard.) Other proposals were made, but over a century elapsed before any action was taken.

In 1790, in the midst of the French Revolution, the National Assembly of France requested the French Academy of Sciences to "deduce an invariable standard for all the measures and weights." The commission appointed by the Academy created a system that was, at once, simple and scientific. The unit of length was to be a portion of the Earth's circumference. Measures for capacity (volume) and mass (weight) were to be derived from the unit of length, thus relating the basic units of the system to each other and to nature. Furthermore, the larger and smaller versions of each unit were to be created by multiplying or dividing the basic units by 10 and its multiples. This feature provided a great convenience to users of the system by eliminating the need for calculating and dividing by 16 (to convert ounces to pounds) or by 12 (to convert inches to feet). Similar calculations in the metric system could be performed simply by shifting the decimal point. Thus, the metric system is a *base-10* or *decimal* system.

The commission assigned the name *metre* (or *meter*) to the unit of length. This name was derived from the Greek word *metron*, meaning "a measure." The physical standard representing the meter was to be constructed so that it would equal one ten-millionth of the distance from the North Pole to the equator along the meridian of the Earth running near Dunkirk in France and Barcelona in Spain.

The metric unit of mass, called the *gram*, was defined as the mass of one cubic centimeter (a cube that is $1/100$ of a meter on each side) of water as its temperature of maximum density. The cubic decimeter (a cube $1/10$ of a meter on each side) was chosen as the unit of fluid capacity. This measure was given the name *litre* (*liter*).

Although the metric system was not accepted with enthusiasm at first, adoption by other nations occurred steadily after France made its use compulsory in 1840. The standardized character and decimal features of the metric system made it well suited to scientific and engineering work. Consequently, it is not surprising that the rapid spread of the system coincided with an age of rapid technological development. In the United States, by Act of congress in 1866, it was made "lawful throughout the United States of America to employ the weights and measures of the metric system in all contracts, dealings, or court proceedings."

By the late 1860s, even better metric standards were needed to keep pace with scientific advances. In 1875, an international treaty, the "Treaty of the Meter," set up well-defined metric standards for length and mass, and established permanent machinery to recommend and adopt further refinements in the metric system. This treaty, known as the *Metric Convention*, was signed by 17 countries, including the United States.

As a result of the Treaty, metric standards were constructed and distributed to each nation that ratified the convention. Since 1893, the internationally agreed metric standards have served as the fundamental weights and measures standards of the United States.

By 1900 a total of 35 nations – including the major nations of continental Europe and most of South America – had officially accepted the metric system. Today, with the exception of the United States and a few small countries, the entire world is predominantly using the metric system or is committed to such use. In 1971 the Secretary of Commerce, in transmitting to Congress the results of a three-year study authorized by the Metric Study Act of 1968, recommended that the United States change to predominant use of the metric system through a coordinated national program.

The International Bureau of Weights and Measures located at Sevres, France, serves as a permanent secretariat for the Metric Convention, coordinating the exchange of information about the use and refinement of the metric system. As measurement science develops more precise and easily reproducible ways of defining the measurement units, the General Conference of Weights and Measures – the diplomatic organization made up of adherents to the Convention – meets periodically to ratify improvements in the system and the standards.

1.2   THE SI SYSTEM OF UNITS

Describe the SI system of units.

SOLUTION

In 1960, the General Conference adopted an extensive revision and simplification of the system. The name *Le Systeme International d'Unites* (International System of Units), with the International abbreviation *SI*, was adopted for this modernized metric system. Further improvements in and additions to SI were made by the General Conference in 1964, 1968, and 1971.

The basic units in the SI system are the *kilogram* (mass), *meter* (length), *second* (time), *Kelvin* (temperature), *ampere* (electric current), *candela* (the unit of luminous intensity), and *radian* (angular measure). All are commonly used by the scientist engineer. The Celsius scale of temperature ($0°C-273.15$ K) is commonly used with the absolute Kelvin scale. The important derived units are the *newton* (SI unit of force), the *joule* (SI unit of energy), the *watt* (SI unit of power), the *pascal* (SI unit of pressure), and the *hertz* (unit of frequency). There are a number of electrical units: *coulomb* (charge), *farad* (capacitance), *henry* (inductance), *volt* (potential), and *weber* (magnetic flux). One of the major advantages of the metric system is that larger and smaller units are given in powers of ten. In the SI system, a further simplification is introduced by recommending only those units with multipliers of 10^3 be employed, e.g., there are 10^9 nanometers in a meter. Thus, for lengths in engineering, the nanometer *micrometer* (previously *micron*), *millimeter*, and *kilometer* are recommended, and the *centimeter* is generally avoided. A further simplification is that the decimal point may be substituted by a comma (as in France, Germany, and South Africa), while the other number, before and after the comma, will be separated by spaces between groups of three, i.e., one million dollars will be $1 000 000,00. More details are provided below.

1.2.1 Seven Base Units

Length, Meter (m) The meter (common international spelling, *metre*) is defined as 1 650 763.00 wavelengths in vacuum of the orange-red line of the spectrum of krypton-86. The SI unit of area is the *square meter* (m^2). The SI unit of volume is the *cubic meter* (m^3). The *liter* (0.001 cubic meter), although not an SI unit, is commonly used to measure fluid volume.

Mass, Kilogram (kg) The standard for the unit of mass, the *kilogram*, is a cylinder of platinum–iridium alloy kept by the International Bureau of Weights and Measures at Paris. A duplicate in the custody of the National Bureau of Standards serves as the mass standard for the United States. This is the only base unit still defined by an artifact. The SI unit of force is the *newton* (N). It is the force which, when applied to a 1 kilogram mass, will give the kilogram mass an acceleration of 1 (meter per second) per second: $1 \text{ N} = 1 \text{ kg-m/s}^2$. The SI unit for pressure is the *pascal* (Pa), where $1 \text{ Pa} = 1 \text{ N/m}^2$. The SI unit for work and energy of any kind is the *joule* (J): $1 \text{ J} = 1 \text{ N-m}$. The SI unit for power of any kind is the *watt* (W), where $1 \text{ w} = 1 \text{ J/s}$.

Time, Second (s) The *second* is defined as the duration of 9 192 632 770 cycles of the radiation associated with a specified transition of the cesium-133 atom. It is realized by tuning an oscillator to the resonance frequency of cesium-133 atoms as they pass through a system of magnets and a resonant cavity into a detector. The number of periods or cycles per second is called *frequency*. The SI unit for frequency is the *hertz* (Hz). One hertz equals one cycle per second. The SI unit for speed is the *meter per second* (m/s). The SI unit for acceleration is the (*meter per second*) *per second* (m/s^2).

Electric Current, Ampere (A) The *ampere* is defined as that current which, if maintained in each of two long parallel wires separated by one meter in free space, would produce a force between the two wires (due to their magnetic fields) of 2×10^{-7} newtons for each meter of length. The SI unit of voltage is the *volt* (V), where $1 \text{ V} = 1 \text{ W/A}$. The SI unit of electrical resistance is the *ohm* (Ω), where $1 \text{ }\Omega = 1 \text{ V/A}$.

Temperature, Kelvin (K) The *Kelvin* is defined as the fraction 1/273.16 of the thermodynamic temperature of the triple point of water. The temperature 0 K is *absolute zero*. Water freezes at about 0°C and boils at about 100°C. The °C is defined as an interval of 1 K, and the Celsius temperature 0°C is defined as 273.15 K. 1.8 Fahrenheit scale degrees are equal to 1.0°C or 1.0 K; the Fahrenheit scale uses 32°F as a temperature corresponding to 0°C.

Amount of Substance, Mole (mol) The *mole* is the amount of substance of a system that contains as many elementary entities as there are atoms in 0.012 kilograms of carbon-12. When the mole is used, the elementary entities must be specified and may be atoms, molecules, ions, electrons, other particles, or specified groups of such particles. The SI unit of concentration (of amount of substance) is the *mole per cubic meter* (mol/m³).

Luminous Intensity, Candela (cd)   The *candela* is defined as the luminous intensity of 1/600 000 of a square meter of a blackbody at the temperature of freezing platinum (2045 K). The SI unit of light flux is the *lumen* (lm). A source having an intensity of 1 candela in all directions radiates a light flux of 4π lumens.

1.2.2   Two Supplementary Units

Phase Angle, Radian (rad)   The *radian* is the plane angle with its vertex at the center of a circle that is subtended by an arc equal in length to the radius.

Solid Angle, Steradian (sr)   The *steradian* is the solid angle with its vertex at the center of a sphere that is subtended by an area of the spherical surface equal to that of a square with sides equal in length to the radius.

1.2.3   SI Multiples and Prefixes

These are provided in Table 1.1

TABLE 1.1   SI Multiples and Prefixes

Multiples and Submultiples		Prefixes	Symbols
100 000 000 000	10^{12}	tera (ter'a)	T
100 000 000	10^{9}	giga (ji'ga)	G
100 000	10^{6}	mega (meg'a)	M
1 000	10^{3}	kilo (kil'o)	K
100	10^{2}	hecto (hek'to)	h
10	10^{1}	deka (dek'a)	da
Base unit 1	10^{0}		
0.1	10^{-1}	deci (des'i)	d
0.01	10^{-2}	centi (sen'ti)	c
0.001	10^{-3}	milli (mil'i)	m
0.000 001	10^{-6}	micro (mi'kro)	μ
0.000 000 001	10^{-9}	nano (nan'o)	n
0.000 000 000 001	10^{-12}	pico (pe'ko)	p
0.000 000 000 000 001	10^{-15}	femto (fem'to)	f
0.000 000 000 000 000 001	10^{-18}	atto (at'to)	a

1.3   THE CONVERSION CONSTANT g_c

Define the "gravitation" conversion constant g_c.

SOLUTION

The momentum of a system is defined as the product of the mass and velocity of the system.

$$\text{Momentum} = (\text{mass} \times \text{velocity})$$

One set of units for momentum are, therefore, (lb)(ft)/s. The units of the time rate of change of momentum (hereafter referred to as rate of momentum) are simply the units of momentum divided by time, that is

$$\text{Rate of momentum} \equiv \frac{\text{lb} - \text{ft}}{\text{s}^2}$$

The above units can be converted to lb_f if multiplied by an appropriate constant. A defining equation from Newton's Law is

$$\text{Force} = 1\,\text{lb}_f = 32.2\frac{(\text{lb})(\text{ft})}{(\text{s}^2)}$$

If this equation is divided by lb_f, one obtains

$$1.0 = 32.2\frac{(\text{lb})(\text{ft})}{(\text{lb}_f)(\text{s}^2)}$$

This serves to define conversion constant g_c. If the rate of momentum is divided by g_c as 32.2 (lb)(ft)/(lb$_f$)(s^2) – this operation being equivalent to dividing by 1.0 – the following units result:

$$\text{Rate of momentum} \equiv \left(\frac{\text{lb} - \text{ft}}{\text{s}^2}\right)\left(\frac{\text{lb}_f - \text{s}^2}{\text{lb} - \text{ft}}\right)$$

$$\equiv \text{lb}_f$$

Thus a force is equivalent to a rate of momentum from the above dimensional analysis.

Similarly, in the SI system in which the unit of force is defined to be the newton (N), then when 1 kg is accelerated at 1 m/s², it will experience a force of 1 N.

$$\text{Force} = 1\,\text{N} = (1\,\text{kg})\left(\frac{1\,\text{m}}{\text{s}^2}\right)$$

$$g_c = \frac{1\,\text{N}}{\text{kg} - \text{m/s}^2} = 1$$

Thus, this form of the conversion constant in SI units is unity with no units (dimensionless).

1.4 UNIT CONVERSION FACTORS: GENERAL APPROACH

Convert the following:

1. 8.03 yr to seconds (s)
2. 150 mile/h to yard/h
3. 100.0 m/s² to ft/min²

SOLUTION

The following conversion factors are needed:

365 day/yr
24 hr/day
60 min/hr
60 s/min
5280 ft/mile
30.48 cm/ft
3 ft/yd

1. Arranging the conversion factors so that units cancel to leave only the desired units, the following is obtained:

$$(8.03\,\text{yr})\left(\frac{365\,\text{day}}{yr}\right)\left(\frac{24\,\text{h}}{\text{day}}\right)\left(\frac{60\,\text{min}}{\text{h}}\right)\left(\frac{60\,\text{s}}{\text{min}}\right) = 2.53 \times 10^8\,\text{s}$$

2. In similar fashion, $\left(\frac{150\,\text{mile}}{\text{h}}\right)\left(\frac{5280\,\text{ft}}{\text{mile}}\right)\left(\frac{\text{yd}}{3\text{ft}}\right) = 2.6 \times 10^5\ \text{yd/h}$

3. $(100.0\,\text{m/s}^2)\left(\frac{100\,\text{cm}}{\text{m}}\right)\left(\frac{\text{ft}}{30.48\,\text{cm}}\right)\left(\frac{60\,\text{s}}{\text{min}}\right)^2 = 1.181 \times 10^6\,\text{ft/min}^2$

1.5   TEMPERATURE CONVERSIONS

Convert the following temperatures:

1. 20°C to °F, K, and °R
2. 20°F to °C, K, and °R

SOLUTION

The following key equations are employed:

$$T\,(°F) = 1.8T\,(°C) + 32$$
$$T\,(K) = T\,(°C) + 273$$
$$T\,(°R) = T\,(°F) + 460$$
$$T\,(°R) = 1.8T\,(K)$$

1. $T\,(°F) = 1.8(20°C) + 32 = 68°F$
   $T\,(K) = (20°C) + 273 = 293\ K$
   $T\,(°R) = 1.8(293\ K) = 527°R$
2. $T\,(°C) = (20°F\text{-}32)/1.8 = -6.7°C$
   $T\,(K) = -6.7°C + 273 = 266\ K$
   $T\,(°R) = 20°F + 460 = 480°R$

1.6   PRESSURE CALCULATIONS

The height of a liquid column of mercury is 2.493 ft. Assume the density of mercury is $848.7\ lb/ft^3$ and atmospheric pressure is $2116\ lb_f/ft^2$ absolute. Calculate the gage pressure in lb_f/ft^2 and the absolute pressure in lb_f/ft^2, psia, mm Hg, and in. H_2O [2].

SOLUTION

Expressed in various units, the standard atmosphere is equal to:

1.0	Atmospheres (atm)
33.91	Feet of water (ft H_2O)
14.7	Pounds-force per square inch absolute (psia)
2116	Pounds-force per square foot absolute (psfa)
29.92	Inches of mercury (in. Hg)
760.0	Millimeters of mercury (mm Hg)
1.013×10^5	Newtons per square meter (N/m²)

The equation describing the gage pressure in terms of the column height and liquid density is

$$P_g = \rho g h / g_c$$

where P_g = gauge pressure, ρ = liquid density, h = height of column, g = acceleration of gravity, and g_c = conversion constant.
Thus,

$$P_g = (848.7 \, lb/ft^3)\left(1 \frac{lb_f}{lb}\right)(2.493 \, ft)$$

$$= 2116 lb_f/ft^2 \text{ gauge}$$

The pressure in lb_f/ft^2 absolute is

$$P_{absolute} = P_g + P_{atmospheric}$$

$$= 2116 \, lb_f/ft^2 + 2116 \, lb_f/ft^2$$

$$= 4232 \, lb_f/ft^2 \text{ absolute}$$

The pressure in psia is

$$P(psia) = (4232 \, psfa)\left(\frac{1 \, ft^2}{144 \, in.^2}\right) = 29.4 \, psia$$

The corresponding gage pressure in psig is

$$P(psig) = 29.4 - 14.7 = 14.7 \, psig$$

The pressure in mm Hg is

$$P(mmHg) = (29.4 \, psia)\left(\frac{760 \, mm \, Hg}{14.7 \, psia}\right) = 1520 \, mm \, Hg$$

Note that 760 mm Hg is equal to 14.7 psia, which in turn is equal to 1.0 atm.
   Finally, the pressure in in. H_2O is

$$P_{(in. H_2O)} = \left(\frac{29.4 \, psia}{14.7 \, psia/atm}\right)\left(\frac{33.91 \, ft \, H_2O}{atm}\right)\left(\frac{12 \, in}{ft}\right)$$

$$= 813.8 \text{ in. } H_2O$$

The reader should note that absolute and gage pressures are usually expressed with units of atm, psi, or mm Hg. This statement also applies to partial pressures. One of the most common units employed in industry to describe pressure drop is inches of H_2O, with the notation in. H_2O or IWC (inches of water column).

1.7 DENSITY AND THERMAL CONDUCTIVITY

Convert a value of density for copper from lb/ft^3 to g/cc, and a value of thermal conductivity for methanol from cal/m-s-°C to Btu/ft-h-°F. Data are provided as follows:

Density of copper $= 557 \, lb/ft^3$

Thermal conductivity of methanol at 60°F $= 0.1512 \, cal/m-s-°C$

SOLUTION

Density (to be revisited in earnest in Chapter 4) and thermal conductivity are two physical properties of importance in nanotechnology. Density is the ratio of a material's mass to its volume; typical units are g/m^3, g/cm^3 (or g/cc), and lb/ft^3. Thermal conductivity provides a measure of how fast (or how easily) heat flows through a substance; it is defined as the amount of heat that flows in unit time through a unit area of unit thickness as a result of a unit difference in temperature. Typical units are cal/m-s-°C and Btu/ft-h-°F.

The conversion factors for g/lb and cm/in are 454 g/lb and 2.54 cm/in, respectively. Convert the density of copper from lb/ft^3 to g/cc.

$$\left(\frac{557 \, lb}{ft^3}\right)\left(\frac{454 \, g}{lb}\right)\left(\frac{1 \, ft^3}{12^3 \, in^3}\right)\left(\frac{1 \, in^3}{2.54^3 \, cm^3}\right) = 8.93 \, g/cm^3$$

$$= 8.93 \, g/cc$$

The conversion factors for Btu/cal and ft/m are 3.974×10^{-3} Btu/cal and 3.281 ft/m, respectively. Convert the thermal conductivity of methanol from cal/m-s-°C to Btu/ft-h-°F.

$$\left(\frac{0.1512 \, cal}{m-s-°C}\right)\left(\frac{3.974 \times 10^{-3} \, Btu}{cal}\right)\left(\frac{1 \, m}{3.281 \, ft}\right)\left(\frac{3600 \, s}{h}\right)\left(\frac{1°C}{1.8°F}\right)$$

$$= 0.366 \, Btu/ft-h-°F$$

1.8 VISCOSITY CONVERSION

What is the kinematic viscosity of a gas in ft^2/s? The specific gravity and absolute viscosity of a gas are 0.84 and 0.019 cP, respectively.

SOLUTION

This problem requires the conversion of cP to lb/ft-s. Therefore, the viscosity μ, is

$$\mu = \left(\frac{0.019\,\text{cP}}{1}\right)\left(\frac{6.720 \times 10^{-4}\ \text{lb/ft-s}}{1\ \text{cP}}\right) = 1.277 \times 10^{-5}\text{lb/ft-s}$$

The density ρ is given as

$$\rho = (SG)(\rho_{\text{ref}}) = (0.84)(62.43\ \text{lb/ft}^3) = 52.44\,\text{lb/ft}^3$$

Since the kinematic viscosity, v, is given by μ/ρ,

$$v = \mu/\rho = (1.277 \times 10^{-5}\ \text{lb/ft-s})/(52.44\ \text{lb/ft}^3) = 2.435 \times 10^{-7}\,\text{ft}^2/s$$

1.9 AIR QUALITY STANDARD

Convert the air quality standard of 9.0 ppmv for carbon monoxide at 25°C and 1 atm to mg/m^3.

SOLUTION

The parts per million of volume standard for carbon monoxide (CO) may be written as

$$9.0\,\text{ppmv} = 9.0\ \frac{\text{mL(CO)}}{\text{m}^3(\text{air})}$$

From the ideal gas law, at 25°C and 1 atm, 1 gmol CO occupies 24.5 L. Since the molecular weight of CO is 28,

$$\text{CO (STD)} = \left(9.0 \ \frac{mL_{CO}}{m^3_{air}}\right)\left(\frac{L_{CO}}{10^3 \ mL_{CO}}\right)\left(\frac{mole_{CO}}{24.5 \ L_{CO}}\right)\left(28 \ \frac{g_{CO}}{mole_{CO}}\right)\left(\frac{10^3 mg_{CO}}{g_{CO}}\right)$$

$$= 10.3 \ mg(CO)/m^3(air)$$

1.10 CONVERSION FACTORS FOR PARTICULATE MEASUREMENTS

Provide a list of common conversion factors for particulates.

SOLUTION

Particulate conversion factors are provided in Table 1.2.

1.11 SIGNIFICANT FIGURES AND SCIENTIFIC NOTATION

Discuss significant figures and scientific notations.

SOLUTION

Significant figures provide an indication of the precision with which a quantity is measured or known. The last digit represents in a qualitative sense, some degree of doubt. For example, a measurement of 8.32 nm implies that the actual quantity is somewhere between 8.315 and 8.325 nm. This applies to calculated and measured quantities; quantities that are known exactly (e.g., pure integers) have an infinite number of significant figures.

The significant digits of a number are the digits from the first nonzero digit on the left to either (a) the last digit (whether it is nonzero or zero) on the right if there is a decimal point, or (b) the last nonzero digit of the number if there is no decimal point. For example:

370	has 2 significant figures
370.	has 3 significant figures
370.0	has 4 significant figures
28 070	has 4 significant figures
0.037	has 2 significant figures
0.0370	has 3 significant figures
0.02807	has 4 significant figures

TABLE 1.2 Particulate Conversion Factors

From	To	Multiply By
mg/m^3	g/ft^3	283.2×10^{-6}
	g/m^3	0.001
	$\mu g/m^3$	1.0×10^3
	ng/m^3	1.0×10^6
	$\mu g/ft^3$	28.32
	ng/ft^3	28.32×10^3
	$lb/1000ft^3$	62.43×10^{-6}
g/ft^3	mg/m^3	35.3145×10^3
	g/m^3	35.314
	$\mu g/m^3$	35.314×10^6
	ng/m^3	35.314×10^9
	$\mu g/ft^3$	1.0×10^6
	ng/ft^3	1.0×10^9
	$lb/1000ft^3$	2.2046
g/m^3	mg/m^3	1.0×10^3
	g/ft^3	0.02832
	$\mu g/m^3$	1.0×10^6
	ng/m^3	1.0×10^9
	$\mu g/ft^3$	28.317×10^3
	$lb/1000ft^3$	0.06243
$\mu g/m^3$	mg/m^3	0.001
	ng/m^3	1.0×10^3
	g/ft^3	28.317×10^{-9}
	g/m^3	1.0×10^{-6}
	$\mu g/ft^3$	0.02832
	ng/m^3	28.32
	$lb/1000ft^3$	62.43×10^{-9}
$\mu g/ft^3$	mg/m^3	35.314×10^{-3}
	g/ft^3	1.0×10^{-6}
	g/m^3	35.314×10^{-6}
	$\mu g/m^3$	35.314
	ng/m^3	3.531×10^3
	$lb/1000ft^3$	2.2046×10^{-6}
$lb/10^3 ft^3$	mg/m^3	16.018×10^3
	g/ft^3	0.34314
	$\mu g/m^3$	16.018×10^6
	ng/m^3	16×10^9
	g/m^3	16.018
	$\mu g/ft^3$	353.14×10^3
No. of Particles/ft^3	$no./m^3$	35.314
	$no./l$	35.314×10^{-3}
	$no./cm^3$	35.314×10^{-6}
ton/mi^2	$lb/acre$	3.125

	$lb/1000ft^2$	0.07174
	g/m^2	0.3503
	kg/km^2	350.3
	mg/m^2	350.3
	mg/cm^2	0.03503
	g/ft^2	0.03254
lb	gr	7000.0
μm	in	39.37×10^{-5}
	mm	1.0×10^{-3}
	nm	1.0×10^3

Whenever quantities are combined by multiplication and/or division, the number of significant figures in the result should equal the lowest number of significant figures of any of the quantities. In long calculations, the final result should be rounded off to the correct number of significant figures. When quantities are combined by addition and/or subtraction, the final result cannot be more precise than any of the quantities added or subtracted. Therefore, the position (relative to the decimal point) of the last significant digit in the number that has the lowest degree of precision is the position of the last permissible significant digit in the result. For example, the sum of 3702, 370, 0.037, 4, and 37, should be reported as 4110 (without a decimal). The least precise of the five numbers is 370, which has its last significant digit in the *tens* position. The answer should also have its last significant digit in the *tens* position.

Unfortunately, engineers and scientists rarely concern themselves with significant figures in their calculations. However, it is recommended that the reader attempt to follow the calculational procedure set forth in this and the next Problem.

In the process of performing scientific calculations, very large and very small numbers are often encountered. A convenient way to represent these numbers is to use *scientific notation*. Generally, a number represented in scientific notation is the product of a number and 10 raised to an integer power. For example,

$$28\,070\,000\,000 = 2.807 \times 10^{10}$$

$$0.000\,002\,807 = 2.807 \times 10^{-6}$$

A nice feature of using scientific notation is that only the significant figures need appear in the number.

1.12 UNCERTAINTY IN MEASUREMENT

The size of a nanoparticle has been reported in source X as 9.5 Å, while source Y reports its value as 9.523 Å. What is the difference?

SOLUTION:

Two kinds of numbers are employed in engineering and scientific practice: *exact numbers* (those values that are known exactly) and *inexact numbers* (those values that have some uncertainty). Exact numbers are those that have defined values or are integers. Numbers obtained by measurement are always *inexact*, since *uncertainties always exist in measured quantities.*

Regarding the size of the nanoparticle, most lay people would say there is no difference. However, there is a difference, and the difference resides in the number of significant figures employed. Source X has two significant figures, while source Y has four significant figures. This difference indicates that Y is more precise.

As noted in the previous Problem, a size measurement of 9.5 Å suggests that the "true" measured value is somewhere between 9.45 and 9.55. For source Y, the measurement indicates that the actual value is in the 9.5225–9.5235 range.

2 Atoms, Elements, and the Periodic Table

This Second Chapter of Part 1 is primarily concerned with atoms and elements. The Periodic Table is also included in the presentation since it is a natural extension of these two key subjects. A brief introduction to each topic follows.

All atoms are made of three classes of particles, often referred to as subatomic particles. The three particles are the electron, proton, and neutron. The electron (negative) and the proton (positive) have equal but opposite charges. Since an atom is neutral in charge, the number of electrons must equal the number of protons. Further, these particles are the same in all atoms. The difference between atoms of distinct elements, for example, hydrogen and lead, is due entirely to the difference in the number of subatomic particles in each atom. Thus, an atom can be viewed as the smallest form of an element since an atom loses its identity when reduced to these subatomic particles.

At the turn of this century, 112 elements were known. These elements vary widely in location and abundance on planet Earth. For example, over 75 percent of the Earth's crust consists of oxygen and silicon. Interestingly, approximately 65 percent by mass of the human body is oxygen. As the number and information on elements increased, chemists attempted to find similarities in elemental as well as chemical behavior. These efforts ultimately resulted in the development of the Periodic Table. This has gone through significant changes over the years, and its latest "form" is presented below. This arrangement of elements in order of increasing atomic number and with elements having similar properties placed in vertical columns (both of which are discussed in the Problems that follow) is known as the *Periodic Table* (Figure 2.1).

Regarding nanotechnology, some scientists and engineers are pursuing ambitious – and some would say fantastic or futuristic – applications of this powerful new technology. For instance, the research community is presently working toward being able to design and manipulate nanoscaled objects, devices, and systems by the manipulation of individual atoms and molecules (see next Chapter). Such forward-looking researchers hope that by using atom-by-atom construction techniques, they will someday be able to create not just substances with remarkable functionality, but also tiny devices and machines (thus far dubbed nanobots or assemblers by some) that could be programmed to perform tasks that were previously considered impossible [1].

Nanotechnology: Basic Calculations for Engineers and Scientists, by Louis Theodore
Copyright © 2006 John Wiley & Sons, Inc.

PERIODIC TABLE

Main groups

	1A 1	2A 2	3B 3	4B 4	5B 5	6B 6	7B 7	8B 8	8B 9	8B 10	1B 11	2B 12	3A 13	4A 14	5A 15	6A 16	7A 17	8A 18
1	1 H 1.00794																	2 He 4.002602
2	3 Li 6.941	4 Be 9.012182											5 B 10.811	6 C 12.0107	7 N 14.00674	8 O 15.9994	9 F 18.998403	10 Ne 20.1797
3	11 Na 22.989770	12 Mg 24.3050											13 Al 26.981538	14 Si 28.0855	15 P 30.973762	16 S 32.066	17 Cl 35.4527	18 Ar 39.948
4	19 K 39.0983	20 Ca 40.078	21 Sc 44.95591	22 Ti 47.897	23 V 50.9415	24 Cr 51.9961	25 Mn 54.938049	26 Fe 55.845	27 Co 58.933200	28 Ni 58.6934	29 Cu 63.546	30 Zn 65.39	31 Ga 69.723	32 Ge 72.61	33 As 74.92160	34 Se 78.96	35 Br 79.904	36 Kr 83.80
5	37 Rb 85.4678	38 Sr 87.62	39 Y 88.90585	40 Zr 91.224	41 Nb 92.90638	42 Mo 95.94	43 Tc [98]	44 Ru 101.07	45 Rh 102.90550	46 Pd 106.42	47 Ag 107.8682	48 Cd 112.411	49 In 114.818	50 Sn 118.710	51 Sb 121.760	52 Te 127.60	53 I 126.90447	54 Xe 131.29
6	55 Cs 132.90545	56 Ba 137.327	57 *La 138.9055	72 Hf 178.49	73 Ta 180.9479	74 W 183.84	75 Re 186.207	76 Os 190.23	77 Ir 192.217	78 Pt 195.078	79 Au 196.96655	80 Hg 200.59	81 Tl 204.3833	82 Pb 207.2	83 Bi 208.98038	84 Po [210]	85 At [210]	86 Rn [222]
7	87 Fr [223]	88 Ra [226]	89 +Ac [227]	104 Rf [261]	105 Db [262]	106 Sg [266]	107 Bh [264]	108 Hs [265]	109 Mt [268]	110 [269]	111 [272]	112 [277]	113	114 [285]	115	116 [289]	117	118 [293]

Transition metals

*Lanthanide series	58 Ce 140.116	59 Pr 140.90765	60 Nd 144.24	61 Pm [145]	62 Sm 150.36	63 Eu 151.964	64 Gd 157.25	65 Tb 158.92534	66 Dy 162.50	67 Ho 164.93032	68 Er 167.26	69 Tm 168.93421	70 Yb 173.04	71 Lu 174.967
+Actinide series	90 Th 232.0381	91 Pa 231.03588	92 U 238.0289	93 Np [237]	94 Pu [244]	95 Am [243]	96 Cm [247]	97 Bk [247]	98 Cf [251]	99 Es [252]	100 Fm [257]	101 Md [258]	102 No [259]	103 Lr [262]

The names and symbols for elements 110 and above have yet to be decided.
Atomic weights are provided each element symbol.
Atomic weights in brackets are the masses of the longest-lived or most important isotope of radioactive elements.

Figure 2.1 Periodic table. The labels on top (1A,2A,3B etc.) are common usage in the United States.

2.1   ATOMIC THEORY

Briefly discuss present-day atomic theory.

SOLUTION

In ancient Greek philosophy, the word atomos was used to describe the smallest bit of matter that could be conceived. This "fundamental particle" was thought of as indestructible; in fact, atomos means "not divisible." Knowledge about the size and nature of the atom grew slowly throughout the centuries.

As discussed in the Introduction to this Section, the atom consists of three subatomic particles: the *proton, neutron*, and *electron*. The charge of an electron is -1.602×10^{-19} C (coulombs), and that of a proton is $+1.602 \times 10^{-19}$ C. The quantity -1.602×10^{-19} C is defined as the *electronic charge*. Note that the charges of these subatomic particles are expressed as multiples of this charge rather than in coulombs. Thus, the charge of an electron is 1−, and that of a proton is 1+. Neutrons carry no charge, i.e., they are electrically neutral. Since an atom has an equal number of electrons and protons, it has zero or no net electric charge.

Both the protons and neutrons reside in the nucleus of the atom, which is extremely small. Most of the atom's volume is the space in which the electrons reside. The external electrons are attracted to the protons in the nucleus because of their opposite electrical charge.

2.2   THE AVOGADRO NUMBER

Define the Avogadro Number.

SOLUTION

A *mole* (sometimes also called a *gmol* or a *gmole*) of a compound is an Avogadro number, or 6.023×10^{23} molecules of that compound; a mole of an element is 6.023×10^{23} atoms of that element. Equivalently, a *lbmol* of an element is (454) (6.023×10^{23}) or 2.734×10^{26} atoms of the element. The number 454 is the number of grams in a pound. The atomic weights used in the Periodic Table may be given any of the following sets of units: amu/atom, g/mol, or lb/lbmol. The following conversion factors follow from these definitions and will be useful in solving the Problems.

1 lbmol = 454 mol
1 g = 6.023×10^{23} amu
1 lb = 2.734×10^{26} amu

2.3   MASS AND SIZE OF ATOMS

The following information on a penny (one cent) is provided. Assume the penny to be pure elemental copper.

Diameter of a penny $= 0.019$ m
Copper atom diameter $= 0.26$ nm

Approximately how many copper atoms can be arranged, centerline-to-centerline, around the diameter of one penny? Also, calculate the mass of these copper atoms.

SOLUTION

The mass of an atom is extremely small. In fact, it is nearly differentially small, with the mass of the heaviest known atom approximately 4×10^{-25} kg (or 4×10^{-22} g). These small masses are often expressed with a unit defined as the *atomic mass unit*, or amu. An amu equals 1.66054×10^{-24} g. For reference, a proton has a mass of 1.0073 amu, a neutron 1.0087 amu, and an electron 5.486×10^{-4} amu. The nucleus, consisting of protons and neutrons, therefore contains almost all of the mass of an atom.

Atoms are also extremely small in size; most have diameters between 0.1 and 50 nm. A convenient unit of length occasionally used is called the *angstrom* (Å), where the angstrom equals 10^{-10} m or 10^{-1} nm. Thus, most atoms have diameters on the order of 1–5 Å. The lightest of all atoms, hydrogen, has a diameter of 1 Å or 0.1 nm and weighs 1.7×10^{-24} grams.

For the problem at hand, the circumference (C) of the penny is

$$C = \pi D = (3.14)(0.019\,\text{m})$$
$$= 0.0597\,\text{m}$$

The number (N) of copper atoms "circumscribing" the penny is then

$$N = \frac{(0.0597\,\text{m})(10^9\,\text{mm/m})}{0.26\,\text{mm}/1\,\text{atom Cu}}$$
$$= 2.3 \times 10^8\,\text{atoms}$$

The mass (M) of these atoms (MW $= 63.546$) is therefore

$$M = (2.3 \times 10^8\,\text{atoms})\left(63.546\,\frac{\text{amu}}{\text{atoms}}\right)\left(\frac{1\,\text{g}}{6.023 \times 10^{23}\,\text{amu}}\right)$$
$$= 2.4 \times 10^{-14}\,\text{g}$$

The reader should note the magnitude of this mass.

2.4 ATOMIC CONVERSIONS

Find the mass of

1. a single uranium atom,
2. a *mole* of uranium and
3. a *lbmole* (or *lbmol*) of uranium

in atomic mass units (amu), grams and pounds. The atomic weight of uranium is 238.

SOLUTION

Calculate the mass of a single uranium atom in amu. The answer is 238 amu/atom, since this is an acceptable unit for atomic weight. The mass of a mole of uranium in amu is therefore

$$(238 \text{ g/mol})(6.023 \times 10^{23} \text{ amu/g}) = 1.43 \times 10^{26} \text{ amu/mol}$$

The mass of a lbmol of uranium in amu is

$$(238 \text{ lb/lbmol})(2.734 \times 10^{26} \text{ amu/lb}) = 6.51 \times 10^{28} \text{ amu/lbmol}$$

The mass of a single uranium atom in grams is

$$(238 \text{ amu/atom})(1 \text{ g}/6.023 \times 10^{23} \text{ amu}) = 3.95 \times 10^{-22} \text{ g/atom}$$

Alternatively the mass of a mole of uranium in grams is 238 g/mol, since this is an acceptable unit for atomic weight.

The mass of a lbmol of uranium in grams is

$$(238 \text{ g/mol})(454 \text{ mol/lbmol}) = 1.08 \times 10^{5} \text{ g/lbmol}$$

Alternatively, the mass of a single uranium atom in lb is

$$(238 \text{ amu/atom}) \times (1 \text{ lb}/2.734 \times 10^{26} \text{ amu}) = 8.71 \times 10^{-25} \text{ lb/atom}$$

while the mass of a mole of uranium in lb is

$$(238 \text{ lb/lbmol})(1 \text{ lbmol}/454 \text{ mol}) = 0.524 \text{ lb/mol}$$

Finally the mass of a lbmol of uranium in lb is 238 lb/lbmol, since this is also an acceptable unit for atomic weight.

2.5   ATOMIC NUMBER, ATOMIC WEIGHT, AND MASS NUMBER

Discuss atomic number, atomic weight, and mass number.

SOLUTION

What differentiates an atom of one element from an atom of another element? The answer is based on the number of protons in the nucleus of the atom in question. Further, all atoms of an element have the same number of protons in the nucleus with the specific number of protons different for different elements. As noted earlier, since an atom has no net electrical charge, the number of electrons in it must equal its number of protons. For example, all atoms of the element carbon have six protons and six electrons. Note that it is customary to speak of "atomic weights," although "atomic masses" would perhaps be more accurate. Mass is a measure of the quantity of matter in a body, whereas weight is the force exerted on the body under the influence of gravity. Furthermore, "atomic weight" is measured in amu.

The standard used for the calculation of atomic weights was recently changed. The new standard completely replaced the two earlier standards. The new standard is particularly appropriate because carbon-12 is often used as a reference standard in computations of atomic masses. The mass number (M) of an atom is defined as the integer closest to the atomic weight of the atom.

To summarize, the average mass of each element (expressed in amu) is also known as the atomic weight. Although the term *average atomic mass* or atomic mass may be more appropriate, the term *atomic weight* has become common. The atomic weights of the elements are listed both in the Periodic Table in the Introduction to this Part, and in the table of elements (Table 2.1) provided in Problem 2.7.

2.6   BISMUTH APPLICATION

How many protons, neutrons, and electrons are in an atom of ^{209}Bi (Bismuth)?

SOLUTION

The superscript 209 is the mass number, the sum of the numbers of protons and neutrons. According to the list of elements in the Periodic Table, Bismuth has an atomic number of 83. Consequently, a ^{209}Bi atom has 83 protons, 83 electrons, and $209 - 83 = 126$ neutrons.

2.7   ELEMENTS

Define and describe elements.

TABLE 2.1 Atomic Weights of the Elements[a,b]

Element	Symbol	Atomic Number	Atomic Weight
Actinium	Ac	89	227.028
Aluminum	Al	13	26.9815
Americium	Am	95	(243)
Antimony	Sb	51	121.75
Argon	Ar	18	39.948
Arsenic	As	33	74.9216
Astatine	At	85	(210)
Barium	Ba	56	137.34
Berkelium	Bk	97	(247)
Beryllium	Be	4	9.0122
Bismuth	Bi	83	208.98
Boron	B	5	10.811
Bromine	Br	35	79.904
Cadmium	Cd	48	112.4
Calcium	Ca	20	40.08
Californium	Cf	98	(251)
Carbon	C	6	12.0112
Cerium	Ce	58	140.12
Cesium	Cs	55	132.905
Chlorine	Cl	17	35.453
Chromium	Cr	24	51.996
Cobalt	Co	27	58.9332
Copper	Cu	29	63.546
Curium	Cm	96	(247)
Dysprosium	Dy	66	18.9984
Einsteinium	Es	99	(252)
Erbium	Er	68	167.26
Europium	Eu	63	151.96
Fermium	Fm	100	(257)
Fluorine	F	9	18.9984
Francium	Fr	87	(223)
Gadolinium	Gd	64	157.25
Gallium	Ga	31	69.72
Germanium	Ge	32	72.59
Gold	Au	79	196.967
Hafnium	Hf	72	178.49
Helium	He	2	4.0026
Holmium	Ho	67	164.93
Hydrogen	H	1	1.00797
Indium	In	49	114.82
Iodine	I	53	126.904
Iridium	Ir	77	192.2
Iron	Fe	26	55.847

(*Continued*)

TABLE 2.1   *Continued*

Element	Symbol	Atomic Number	Atomic Weight
Krypton	Kr	36	83.8
Lanthanum	La	57	138.91
Lawrencium	Lr	103	(260)
Lead	Pb	82	207.19
Lithium	Li	3	6.939
Lutetium	Lu	71	174.97
Magnesium	Mg	12	24.312
Manganese	Mn	25	54.938
Mendelevium	Md	109	(258)
Mercury	Hg	80	200.59
Molybdenum	Mo	42	95.94
Neodymium	Nd	60	144.24
Neon	Nc	10	20.183
Neodymium	Np	93	237.048
Nickel	Ni	28	58.71
Niobium	Nb	41	92.906
Nitrogen	N	7	14.0067
Nobelium	No	102	(259)
Osmium	Os	76	190.2
Oxygen	O	8	15.9994
Pallasium	Pd	46	106.4
Phosphorus	P	15	30.9738
Platinum	Pt	78	195.09
Plutonium	Pu	94	(244)
Potassium	Po	84	(209)
Praseodymium	K	19	39.102
Promethium	Pr	59	140.907
Radium	Pm	61	(145)
Radon	Pa	91	231.036
Rhenium	Ra	88	226.025
Radon	Rn	86	(222)
Rhenium	Re	75	186.2
Rhodium	Rh	45	102.905
Rubidium	Rb	37	84.57
Rutherium	Ru	44	101.07
Samarium	Sm	62	150.35
Scandium	Sc	21	44.956
Selenium	Se	34	78.96
Silicon	Si	14	28.086
Silver	Ag	47	107.868
Sodium	Na	11	22.9898
Strontium	Sr	38	87.62
Sulfur	S	16	32.064
Tantalum	Ta	73	180.948

(Continued)

TABLE 2.1 *Continued*

Element	Symbol	Atomic Number	Atomic Weight
Technetium	Tc	43	(98)
Tellurium	Te	52	127.6
Terbium	Tb	65	158.924
Thallium	Tl	81	204.37
Thorium	Th	90	232.038
Thulium	Tm	69	168.934
Tin	Sn	50	118.69
Titanium	Ti	22	47.9
Tungsten	W	74	183.85
Uranium	U	92	238.09
Vanadium	V	23	50.942
Xenon	Xe	54	137.04
Ytterbium	Yb	70	173.04
Yttrium	Y	39	88.905
Zinc	Zn	30	65.37
Zirconium	Zr	40	91.22

[a]Atomic weights apply to naturally occurring isotopic compositions and are based on an atomic mass of $^{12}C = 12$.

[b]For radioactive elements, a value given in parentheses is the atomic mass number of the isotope of the longest known half-life. Isotopes are discussed in Problem 2.10.

[c]Geologically exceptional samples are known in which the element has an isotopic composition outside the limits for normal material.

SOLUTION

Elements may be viewed as substances that cannot be decomposed, or broken into more elementary substances, by ordinary chemical means. Elements were believed at one time to be the fundamental substances but have, as described earlier, been separated into their constituent fundamental particles.

Each element is enclosed in a box in the aforementioned Periodic Table. Notice that the symbol for each element consists of one or two letters, with the first letter capitalized. These symbols are often derived from the English name for the element (first and second columns in Table 2.1), but have also sometimes been derived from a foreign name instead. A formal alphabetical listing of these elements, with atomic weights and atomic numbers, is provided in Table 2.1.

2.8 SYMBOLS FOR ELEMENTS

Answer the following three questions

1. Provide the elements represented by the symbols C, Cd, and F.

2. Provide the symbol for the elements chromium, arsenic, and oxygen.

3. Provide the name of the elements with symbols Si, S and Ca.

SOLUTION

Refer to either Table 2.1 in Problem 2.7 and/or the Periodic Table. The answers are

1. Carbon, cadmium, and fluorine

2. Cr, As and O

3. Silicon, sulfur, and calcium.

2.9 PERIODIC TABLE APPLICATION

Which of the following elements have, relatively speaking, similar chemical and physical properties: Na, He, Cl, K, S, and Br?

SOLUTION

Many elements show strong similarities to each other. This unique property is demonstrated in the Periodic Table. As indicated earlier, elements are arranged by atomic number (number of protons). Elements in the vertical columns, known as *groups*, exhibit similar but gradually changing properties as one proceeds from top to bottom. For example, Group 1, occupying the extreme left-hand column of the Periodic Table, consists of hydrogen and the alkali metals – a highly reactive set of elements. Except for hydrogen, all Group 1 elements are soft metals that are good conductors of heat and electricity.

By contrast, occupying the extreme right-hand column of the Periodic Table are the inert, rare, or noble gases as they are called, appearing in Group 8A. As such, these elements are relatively unreactive and rarely react or combine with other elements. To their left lie the halogens, a highly reactive set of nonmetals that are poor conductors of heat and electricity. Each of the groups occupying other vertical columns between these extremes shows a progressive change in metallic character and exhibits its own unique sets of properties.

Meanwhile, each horizontal row in the Periodic Table is called a period. The first period contains only hydrogen and helium. Within a given group (vertical column), the size of the atom increases from the top period to the bottom, and with this changing atom size come changes in other properties that are related to atomic size. For example, the halogens proceed from gas to liquid to solid as one moves down the table.

In general, size increases from right to left in a given period. Ionization energy, electron affinity, and electronegativity increase from the lower left-hand corner of the table to the upper right; atomic and ionic radii increase from the upper right to the lower left.

With reference to the Problem statement, and employing the Periodic Table, one concludes that

1. The elements sodium (Na) and potassium (K) appear in group 1A – Alkali Metals. Both are soft metals and reactive.
2. The elements chlorine (Cl) and bromine (Br) appear in Group 7A – Halogens. Both are gases.

2.10 ISOTOPES

Discuss isotopes.

SOLUTION

An *isotope* is defined as one of two or more species of an atom having the same atomic number, hence constituting the same element, but differing in mass number. Since the atomic number is equivalent to the number of protons in the nucleus, and the mass number is the sum total of the protons plus the neutrons in the nucleus, isotopes of the same element differ from one another only in the number of neutrons in their nuclei.

Thus, atoms of the same element that differ in weight are known as isotopes. In the case of chlorine two isotopes occur in nature. Experiments show that chlorine is a mixture of three parts of chlorine-35 for every one part of the heavier chlorine-37 isotope. This proportion accounts for the observed atomic weight of chlorine.

Summarizing, one can conclude that atoms of a particular atomic number but having different mass numbers are defined as isotopes of the element of that atomic number. Since the number of protons is the same, it is the differences in the number of neutrons that results in different mass numbers.

3 Molecular Rearrangements

There are a near infinite number of compounds known today, all made up from or drawn from the elements discussed in Chapter 2. There are 112 elements from which these compounds may be derived. The development of new nanochemicals in the future will be derived by combining these elements in a way once thought impossible (see the Problem 3.2). This is similar to the 26 letters in the alphabet, which can be combined in various ways to form words. Examples include the name OTTO, the month APRIL, and the element SODIUM. Similarly, the 10 basic numeric digits, that is 0 to 9, can be combined to form the numbers 13, 475, or 1 303 928. License plates for most states are drawn from combinations of letters and digits (see the Problem Problem 3.1).

As discussed earlier, elements are "substances" that cannot be broken down or decomposed into simpler substances. Further, each element is made up of only one kind of atom. Some elements can be combined, for example O to form O_2; here, O_2 is consider a molecule. On the other hand compounds are defined as consisting of two or more different elements and therefore contain two or more kinds of atoms.

Combinations of "substances" can react and undergo a molecular rearrangement. A chemical equation is a short-hand method for describing a chemical transformation. The substances on the left-hand side of the equation are called *reactants*, while those on the right-hand side are called *products*. The equation

$$NaOH + HCl \rightarrow NaCl + H_2O$$

indicates that sodium hydroxide reacts with hydrogen chloride to form sodium chloride (salt), and water. This general topic leads to the subject of stoichiometry and is treated in more detail later in this Chapter.

3.1 LICENSE PLATE SETS

Determine the number of possible license plates, drawn from 26 letters and 10 integers, that consist of 7 symbols. An example is the author's current license plate CCY9126. Note that this calculation involves *sets*, where the order of the symbol matters, and the symbol may be repeated.

Nanotechnology: Basic Calculations for Engineers and Scientists, by Louis Theodore
Copyright © 2006 John Wiley & Sons, Inc.

SOLUTION

There are 26 letters and 10 integers. Therefore, 36 symbols are available from which to choose. The *set*(s) of r symbols that can be drawn from a pool of n symbols is given by

$$N(n, r) = (n)^r$$

For this case, $r = 7$ and $n = 36$. The number of sets is therefore

$$N(n, r) = (36)^7$$

$$= 7.84 \times 10^{10}$$

Obviously, there are a rather large number of sets.

3.2   CHEMICAL PERMUTATIONS AND COMBINATIONS

Determine the number of four-element chemical compounds that can theoretically be generated from a pool of 112 elements. Assume each element counts only once in the chemical formula. An example of a three-element compound is H_2SO_4 (sulfuric acid), or CH_3OH (methanol). An example of a four-element compound is $NaHCO_3$.

SOLUTION

By definition, each different ordering or arrangement of all or part of a set of symbols (or objects) where the order matters is defined as a permutation. The number of different permutations of n symbols taken r at a time is given by

$$P(n, r) = \frac{n!}{(n - r)!}$$

There are 112 elements to choose from in this application. A modified form of the equation presented in Section 3.1 applies; however, in this case, $r = 4$ and $n = 112$, and the equation involves a permutation (P) calculation. For this application, the describing equation is given by

$$P = \frac{n!}{(n - r)!}$$

such that

$$n! = n(n - 1) \ldots (3)(2)(1)$$

and

$$n! = n(n - 1)!$$

with

$$O! = 1$$

Based on the problem statement, $n = 112$ and $r = 4$. Therefore,

$$P(112, 4) = \frac{112!}{(112 - 4)!}$$

$$= \frac{112!}{108!}$$

$$= (112)(111)(110)(109)$$

$$= 1.49 \times 10^8$$

As with the previous problem, this too is a large number. This number would be further increased if the number of a particular element was greater than one, for example HCN (hydrogenyamide) compared with C_3H_3N (acrylonitrate)

The above can be extended to combinations where the arrangement of symbols is such that the order does not matter, as with a chemical compound. For example, HCN is the same as CNH. In this case,

$$C(n, r) = \frac{n!}{r!(n - r)!}$$

substituting once again gives

$$C(112, 4) = \frac{112!}{(4!)(112 - 4)!}$$

$$= \frac{112!}{(4!)(108!)}$$

$$= \frac{(112)(111)(110)(109)}{(4)(3)(2)(1)}$$

$$= 6.21 \times 10^6$$

Two points need to be made – one concerning the calculation and the other concerning the chemistry of the compound.

1. For calculations involving large factorals, it is often convenient to use an approximation known as Stirling's formula:

$$n! \approx (2\pi)^{0.5} e^{-n} n^{n+0.5}$$

2. For a real-word "viable" compound, the elements involved must be capable of bonding.

3.3 FORMULA WEIGHT AND MOLECULAR WEIGHT

Answer the following two problems

1. Calculate the formula weight of methanol.
2. Calculate the molecular weight of calcium nitrate.

SOLUTION

The *molecular weight* (MW) of a compound is the sum of the atomic weights of the atoms that make up the molecule. Units of atomic mass units per molecule (amu/molecule) are employed for formula weight, while grams per gram-mole (g/gmol) are used for molecular weight. One gram-mole (gmol) contains an Avogadro number of molecules, i.e., 6.023×10^{23} molecules. The reader should note that formula weight and molecular weight are the same.

1. The formula weight of CH_3OH (methanol) is given by

$$
\begin{aligned}
1\,C\,\text{atom} &= (1)(12.0\,\text{amu}) = 12.0\,\text{amu} \\
4\,H\,\text{atom} &= (4)(1.0\,\text{amu}) = 4.0\,\text{amu} \\
1\,O\,\text{atom} &= (1)(16.0\,\text{amu}) = \underline{16.0\,\text{amu}} \\
\text{Total} &= 32.0\,\text{amu}
\end{aligned}
$$

The formula weight of methanol is 32.0.

2. The molecular weight of calcium nitrate is obtained from the pertinent atomic weights.

$$
\begin{aligned}
1\,Ca &= (1)(40.1) = 40.1 \\
2\,N &= (2)(14.0) = 28.0 \\
6\,O &= (6)(16.0) = \underline{96.0} \\
\text{Total} &= 164.1
\end{aligned}
$$

The molecular weight of calcium nitrate is 164.1.

3.4 MOLE/MOLECULE RELATIONSHIP

Determine the number of molecules contained in 50 g of nitrobenzene ($C_6H_5O_2N$ or $C_6H_5NO_2$).

SOLUTION

The molecular formula of nitrobenzene is normally written as $C_6H_5NO_2$. Pertinent atomic weights are listed below:

Carbon = 12
Hydrogen = 1
Oxygen = 16
Nitrogen = 14

The molecular weight of nitrobenzene is then

$$(6)(12) + (5)(1) + (2)(16) + (1)(14) = 123 \text{ g/gmol}$$

To convert a mass to moles, divide by the molecular weight:

$$(50.0 \text{ g})\left(\frac{\text{gmol}}{123 \text{ g}}\right) = 0.407 \text{ gmol}$$

There are 6.02×10^{23} (Avogadro's number) molecule/gmol. Therefore,

$$(0.407 \text{ gmol})(6.02 \times 10^{23} \text{ molecules/gmol}) = 2.45 \times 10^{23} \text{ molecules}$$

3.5 POLLUTANT CHEMICAL FORMULAS

Provide the chemical formula and molecular weight (where applicable) for the following five pollutants.

Asbestos
Fine Mineral Fiber
Tetrachlorodibenzo-p-dioxin
Trichlorophenol
2,4-Toluene Diamine

SOLUTION

The chemical formula and molecular weight (MW) for each pollutant is provided below.

Asbestos: $MG_6 (SI_4O_{10})(OH)$; most common MW not applicable
Fine mineral fibers: Not applicable
Tetrachlorodibenzo-p-dioxin: $C_{12}H_4Cl_4O_2$; 321.96

Trichlorophenol: $C_6H_3Cl_3O$; 197.45

2,4-Toluene Diamine: $C_7H_{10}N_2$; 122.19

Additional details on all the various classes of pollutants are provided in the literature [3].

3.6 STOICHIOMETRY

Define stoichiometry [4].

SOLUTION

The term *stoichiometry* has come to mean different things to different people. In a loose sense, stoichiometry involves the balancing of an equation for a chemical reaction that provides a quantitative relationship among the reactants and products. In the simplest stoichiometric situation, exact quantities of pure reactants are available, and these quantities react completely to give the desired product(s). In an industrial process, the reactants usually are not pure. One reactant is usually in excess of what is needed for the reaction, and the desired reaction may not go to completion because of a host of other considerations. For example, complete combustion of pure hydrocarbons yields carbon dioxide and water as the reaction products.

Consider the combustion of methane in oxygen:

$$CH_4 + O_2 \rightarrow CO_2 + H_2O$$

In order to balance this reaction, two molecules of oxygen are needed. This requires that there be four oxygen atoms on the right side of the reaction. This is satisfied by introducing two molecules of water as product. The final balanced reaction becomes

$$CH_4 + 2O_2 \rightarrow CO_2 + 2H_2O$$

Thus, two molecules (or moles) of oxygen are required to completely combust one molecule (or mole) of methane to yield one molecule (or mole) of carbon dioxide and two molecules (or mole) of water. Note that the numbers of carbon, oxygen, and hydrogen atoms on the right-hand side of this reaction are equal to those on the left-hand side. The reader should verify that the total mass (obtained by multiplying the number of each molecule by its molecular weight and summing) on each side of the reaction is the same. (The term *moles* here may refer to either gram-moles (gmol) or pound-moles (lbmol); it makes no difference.)

3.7 LIMITING AND EXCESS REACTANTS

Discuss limiting and excess reactants.

SOLUTION

The terms used to describe a reaction that does not involve stoichiometric ratios of reactants must be carefully defined in order to avoid confusion. If the reactants are not present in formula or stoichiometric ratio, one reactant is said to be *limiting*; the others are said to be in *excess*. Consider the following reaction:

$$CO + \frac{1}{2}O_2 \rightarrow CO_2$$

If the starting amounts are 1 mol of CO and 3 mol of oxygen, CO is the limiting reactant, with O_2 present in excess. There are 2.5 mol of excess O_2, because only 0.5 mol is required to combine with the CO. The percentage of excess must be defined in relation to the amount of the reactant necessary to react completely with the limiting reactant. Thus, there is 500% excess oxygen present. If for some reason only part of the CO actually reacts, this does not alter the fact that the oxygen is in excess by 500%. However there are often several possible products. For instance, the reactions

$$CO + O_2 \rightarrow CO_2 \quad \text{and} \quad C + \frac{1}{2}O_2 \rightarrow CO$$

can occur simultaneously. In this case, if there are 3 mol of oxygen present per mole of carbon, the oxygen is in excess. The extent of this excess, however, cannot be definitely fixed. It is customary to choose one product (e.g., the desired one) and specify the excess reactant in terms of this product. For this case, there is 200% excess oxygen for the reaction going to CO_2, and there is 500% excess oxygen for the reaction going to CO. The discussion on excess oxygen can be extended to excess air using the same approach. Stoichiometric or theoretical oxygen (or air) is defined as 0% excess oxygen (or air). This is an important concept since many reaction systems operate with excess air.

3.8 COMBUSTION OF CHLOROBENZENE

Generate a complete mole and mass balance for the stoichiometric combustion of chlorobenzene in air [4]. Employ English units.

SOLUTION

For the combustion of 1.0 lbmol of chlorobenzene (C_6H_5Cl) in stoichiometric oxygen, the balanced reaction is

$$C_6H_5Cl + 7O_2 \rightarrow 6CO_2 + 2H_2O + HCl$$

Since air, not oxygen, is employed in the process this reaction with air becomes,

$$C_6H_5Cl + 7O_2 + 26.3N_2 \rightarrow 6CO_2 + 2H_2O + HCl + 26.3N_2$$

where the nitrogen in the air has been retained on both sides of the equation since it does not participate in the reaction. The moles and masses involved in this reaction, based on the stoichiometric combustion of 1.0 lbmol of C_6H_5Cl are given in Table 3.1.

Note that, in accordance with the conservation law for mass, the initial and final masses balance. The number of moles, as is typically the case in many calculations, do not balance. The concentrations of the various species may also be calculated. For example.

$$\%CO_2 \text{ by weight} = (264/1072.9)100\% = 24.61\%$$
$$\%CO_2 \text{ by mol (or volume)} = (6/35.3)100\% = 17.0\%$$
$$\%CO_2 \text{ by weight (dry basis)} = (264/1036.9)100\% = 25.46\%$$
$$\%CO_2 \text{ by mol (dry basis)} = (6/33.3)100\% = 18.0\%$$

The air requirement for this reaction is 33.3 lbmol. This is stoichiometric or 0% excess air (EA). For 100% EA (100% above stoichiometric), one would use (33.3)(2.0) or 66.6 lbmol of air. For 50% EA one would use (33.3)(1.5) or 50 lbmol of air; for this condition 16.7 lbmol excess or additional air is employed.

If 100% excess air is employed in the combustion of 1.0 lbmol of C_6H_5Cl, the combustion reaction would become

$$C_6H_5Cl + 14O_2 + 52.6N_2 \rightarrow 6CO_2 + 2H_2O + HCl + 7O_2 + 52.6N_2$$

The reader is left with the exercise of verifying these results:

$$\text{Total final number of moles} = 68.6$$
$$\%O_2 \text{ by mol (or volume)} = (7/68.6)100\% = 10.2\%$$

TABLE 3.1   Moles and Masses in Stoichiometric Combustion of 1.0 lbmol of C_6H_5Cl

	C_6H_5Cl	+	$7O_2$	+	$26.3N_2$	→	$6CO_2$	+	$2H_2O$	+	HCl	+	$26.3N_2$
Moles	1		7		26.3		6		2		1		26.3
MW	112.5		32		28		44		18		36.5		28
Mass	112.5		224		736.4		264		36		36.5		736.4

$$\text{Initial mass} = 1072.9, \quad \text{initial number of moles} = 34.3$$
$$\text{Final mass} = 1072.9, \quad \text{final number of moles} = 35.3$$

$$\%\text{HCl by mol} = (1/68.6)100\% = 1.46\%$$
$$\%\text{H}_2\text{O by mol} = (2/68.6)100\% = 2.92\%$$

Assuming atmospheric conditions (1 atm),

$$\text{Partial pressure O}_2 = 0.102 \text{ atm}$$
$$\text{Partial pressure HCl} = 0.0146 \text{ atm}$$
$$\text{Partial pressure H}_2\text{O} = 0.0292 \text{ atm}$$

If the C_6H_5Cl waste contains 0.5% sulfur (S) by mass, then

$$\text{Weight of S} = (0.005)(112.5) = 0.5625 \text{ lb}$$
$$\text{Number of lbmol of S} = 0.5625/32 = 0.0176 \text{ lbmol}$$
$$\text{Number of lbmol of SO}_2 \text{ formed} = \text{number of lbmol of S} = 0.0176 \text{ lbmol}$$

For this condition (approximately),

$$\%\text{SO}_2 \text{ by mol} = (0.0176/68.6)100\% = 0.0257\%$$
$$\text{Partial pressure SO}_2 = 2.57 \times 10^{-4} \text{ atm}$$

3.9 METAL ALLOY CALCULATION

A metal alloy consists of 25 w/o (weight percent of) nickel and 75 w/o aluminum. Determine the atom percent of the components.

SOLUTION

By definition:

$$\text{w/o of A} = \text{weight of A/total weight}$$
$$\text{a/o of A} = \text{atoms of A/total number of atoms}$$

Note that for all Problems in this text, the terms weight and mass are both being used simply to quantify amounts of materials and hence are used interchangeably. The usual distinction between these two terms is not necessary since it is being assumed in all problems that whatever is occurring is on the Earth's surface and the gravitational acceleration is constant.

Choose a convenient weight or mass basis for the alloy calculation. The term basis is used to indicate that amount of material on which a calculation is being performed. In some problems (such as this one) the amount of material is not specified, in which case the choice of a basis is arbitrary. Since the weight percents of the two components are given, a basis such as 100 amu or 100 g should be chosen to simplify the next step.

In this case, select as a basis 100 amu of the alloy. Determine the weight of each component

$$\text{Ni: } (0.25)(100\,\text{amu}) = 25\,\text{amu}$$
$$\text{Al: } (0.75)(100\,\text{amu}) = 75\,\text{amu}$$

Find the atomic weight (AW) of each component; use the Periodic Table.

$$AW\,(\text{Ni}) = 58.7$$
$$AW\,(\text{Al}) = 27.0$$

The number of atoms of each component and the total number of atoms is the

$$\text{Ni: } (25\,\text{amu})/(58.7\,\text{amu/atom}) = 0.426\,\text{atoms}$$
$$\text{Al: } (75\,\text{amu})/(27.0\,\text{amu/atom}) = 2.778\,\text{atoms}$$
$$\text{Total} \quad = 3.204\,\text{atoms}$$

Finally, the atom fraction and atomic percent of each component may be calculated:

$$\text{Ni: } 0.426/3.204 = 0.133$$
$$= 13.3\,\text{a/o Ni}$$
$$\text{Al: } 2.778/3.204 = 0.867$$
$$= 86.7\,\text{a/o Al}$$

Note that the percent of aluminum jumped from 75.0 to 86.7 during the conversion from w/o to a/o. This is predictable; the aluminum is the lighter atom and therefore makes a greater contribution on an atom number basis than on a weight basis.

3.10   CHEMICAL PRODUCTION

Prepare a list of the top 20 chemicals produced by the chemical industry. Also include information on the annual quantity produced.

SOLUTION

The data in Table 3.2 has been compiled from Chemical and Engineering News, June 25, 2001 [5].

The top chemical is sulfuric acid. About 40 million tons are produced annually. There are two major processes used in the production of H_2SO_4, the lead chamber

TABLE 3.2 Top 20 Chemicals Produced by the Chemical Industry

Rank	Chemical	Production, 10^{-9} kg
1	Sulfuric acid	39.62
2	Ethylene	25.15
3	Lime	20.12
4	Phosphoric acid	16.16
5	Ammonia	15.03
6	Propylene	14.45
7	Chlorine	12.01
8	Sodium hydroxide	10.99
9	Sodium carbonate	10.21
10	Ethylene chloride	9.92
11	Nitric acid	7.99
12	Ammonium nitrate	7.49
13	Urea	6.96
14	Ethylbenzene	5.97
15	Styrene	5.41
16	Hydrogen chloride	4.34
17	Ethylene oxide	3.87
18	Cumene	3.74
19	Ammonium sulfate	2.60
20	1,3-Butadiene	2.01

process and contact process. The lead chamber process is the older of the two processes, and its product is aqueous sulfuric acid containing 62% to 78% H_2SO_4. The contact process yields pure sulfuric acid. In both processes, sulfur dioxide, SO_2, is obtained by burning sulfur and is oxidized to sulfur trioxide, SO_3, and the SO_3 is dissolved in water.

The reader is left the exercise of writing the formula of each chemical in a "fourth" column and then calculating the number of moles produced. Finally, rank the chemicals in order of moles produced.

4 Concentration Terms

Describing the concentration of a nanoparticle/nanochemical/nanomixture can take on many different meanings. The usual methods employed in expressing concentration and concentration-related terms needs to be clearly understood. This fourth Chapter of Part 1 attempts to define and relate the various concentration terms currently used by practicing engineers and scientists. Many of these terms and approaches are and will continue to be employed in the nanotechnology field.

Few substances encountered in practice are chemically pure; most substances are *not* chemically pure. Theoretically and inpractically, impure substances can be separated into pure substances. For example, seawater consists of salt (NaCl) and water (H_2O). Thus, mixtures consist of combinations of different substances, with the substances in question being either an element or a compound, or both. The substances that make up the mixture are usually defined as the components of the mixture; the terms solute and solvent are often applied to the components of a solution.

Pure substances can either be gaseous, liquid, or solid. The same can be said for mixtures. For example, oxygen is a pure gas, while air is a mixture of gases. Alternately, water is a pure liquid while a solution of water and alcohol is a liquid mixture.

Mixtures can be homogeneous or heterogeneous. An example of a homogeneous mixture is the aforementioned air, consisting primarily of oxygen and nitrogen. Heterogeneous mixtures possess varying properties throughout the mixture. Thus, and generally speaking, matter may be classified either as pure substances or as mixtures on the basis that a pure substance has a definite composition whereas the composition of a mixture can be variable.

4.1 DENSITY, SPECIFIC GRAVITY, AND BULK DENSITY

Calculate the density and specific gravity of a 200 nm diameter nanoparticle whose weight has been approximately determined to be 3.7×10^{-8} µg. Also estimate the particle bulk density.

SOLUTION

This first Problem in Chapter 4 is concerned with density, and its accompanying term, specific gravity. These two terms are generally applied to pure substances

Nanotechnology: Basic Calculations for Engineers and Scientists, by Louis Theodore
Copyright © 2006 John Wiley & Sons, Inc.

and most tabulated values in the literature are available for pure substances [6]. However, they can also be applied to mixtures. In fact, industry often maintains a check on the quality of raw materials, intermediates, end-products, and by-products through density or specific gravity measurements. This is referred to as "finger printing." Normally either the density or the specific gravity can be determined by plant personnel with relatively simple and inexpensive instrument methods.

The *density* (ρ) of a substance is the ratio of its mass to its volume and may be expressed in units of pounds per cubic foot (lb/ft^3), kilograms per cubic meter (kg/m^3), and so on. For solids, density can be easily determined by placing a known mass of the substance in a liquid and determining the displaced volume. The density of a liquid can be measured by weighing a known volume of the liquid in a volumetric flask. For gases, the ideal gas law can be used to calculate the density from the pressure, temperature, and molecular weight of the gas.

Densities of pure solids and liquids are relatively independent of temperature and pressure and can be found in standard reference books. The *specific volume (v)* of a substance is its volume per unit mass (ft^3/lb, m^3/kg, and so on) and is, therefore, the inverse of its density.

The *specific gravity* (SG) is the ratio of the density of a substance to the density of a reference substance at a specific condition:

$$SG = \rho/\rho_{ref}$$

The reference most commonly used for *solids* and *liquids* is water at its maximum density, which occurs at 4°C; this reference density is 1.000 g/cm^3, 1000 kg/m^3, or 62.43 lb/ft^3. Note that, since the specific gravity is a ratio of two densities, it is dimensionless. Therefore, any set of units may be employed for the two densities as long as they are consistent. The specific gravity of *gases* is used only rarely; when it is, air is usually employed as the reference substance at the same conditions of temperature and pressure as the gas.

Another dimensionless quantity related to density is the API (American Petroleum Institute) gravity, which is often used to indicate densities of fuel oils and some liquid hazardous wastes. The relationship between the API scale and the specific gravity is

$$\text{degrees API} = \frac{141.5}{SG(60/60°F)} - 131.5$$

where $SG(60/60°F)$ = specific gravity of the liquid at 60°F using water at 60°F as the reference.

For particles, a distinction should always be made between *true* density and *bulk* density. The true density is the actual density of a discrete particle or solid, whereas the bulk density is the density of the crushed or powdered form. In lieu of data, the bulk density may be assumed $\rho\theta$; for purposes of engineering calculations to be approximately 60% of the true density ref 2.

For this application, the volume of the particle is

$$V = (4/3)\pi r^3; \qquad r = d_p/2$$
$$= (4/3)(\pi)(100)^3$$
$$= 4.19 \times 10^6 \, (\text{mm})^3$$

The density is therefore given by

$$\rho = \frac{3.7 \times 10^{-8} \, \mu g}{4.19 \times 10^6 \, (\text{mm}^3)}$$
$$= 0.883 \times 10^{-14} \, \mu g/\text{mm}^3$$
$$= (0.883 \times 10^{-14})(10^{-6})(10^{+7})^3$$
$$= 8.83 \, \text{g/cm}^3 = 551 \, \text{lb/ft}^3$$

The specific gravity is

$$SG = 8.83$$

The bulk density, ρ_B is approximately

$$\rho_B = (0.6)(8.83)$$
$$= 5.30 \, \text{g/cm}^3$$
$$= 33.1 \, \text{lb/ft}^3$$

4.2 CLASSES OF SOLUTION

Describe the various classes of solution.

SOLUTION

Physically speaking, there are three general classes of state: gaseous, liquid, and solid. Therefore, there are nine possible types of solutions of mixtures of these states. These are shown in Table 4.1 with an example of each [7].

4.3 MOLALITY VERSUS MOLARITY

Discuss the difference between molality and molarity.

SOLUTIONS

Two basic methods are employed to describe the concentration and/or the composition of a system consisting of more than one pure substance:

1. Mass volume, or mole ratio.
2. Mass volume, or mole fraction (or percent).

TABLE 4.1   The Nine possible Types of Solution

Solution Type	Application
Gaseous solution	
Gas in a gas	Air (nitrogen and oxygen)
Liquid in a gas	Water mist in air
Solid in a gas	Particulate in air
Liquid solutions	
Gas in a liquid	Carbon dioxide in water
Liquid in a liquid	Alcohol in water
Solid in a liquid	Suspended particulate in water
Solid solutions	
Gas in a solid	Sulfur dioxide in palladium
Liquid in a solid	Mercury in gold
Solid in a solid	Nichel in lead

One method, on a mole basis, is called the *molality*. It is defined as the number of gram mols of solute dissolved in a given mass (equivalent to a given number of mols) of solvent, usually taken as 1000 grams. Sometimes, the ratio of the number of mols of solute to the number of mols of solvent, or its reciprocal ratio, is employed.

A second method is more generally used, and on a mole basis is termed the *molarity*. It is defined as the number of gram mols of solute in a given volume of solution (equivalent to a given number of mols of solute plus solvent) taken as 1000 milliliters. For multicomponent systems in general, the mol fraction, expressed as the number of mols of the given component divided by the total number of mols of all components, is preferred. It has the decided advantage in expressing composition, since the sum of the mol fractions of all components in the system is unity. In addition, the mol fraction is constrained to values between zero and unity.

In general, the second option [2], that is, mass, volume or mole fraction is normally employed in most engineering and scientific calculations. Fractional terms can be converted to percentages simply by multiplying by 100.

4.4   MOLAR RELATIONSHIPS

A mixture contains 20 lb of O_2, 2 lb of SO_2, and 3 lb of SO_3. Determine the weight fraction and mole fraction of each component.

SOLUTION

By definition:

Weight fraction = weight of A/total weight

Moles of A = weight of A/molecular weight of A

Mole fraction = moles of A/total moles

First, calculate the weight fraction of each component.

Compound	Weight (lb)	Weight Fraction
O_2	20	$20/25 = 0.8$
SO_2	2	0.08
SO_3	3	0.12
Total	25	1.00

Calculate the mole fraction of each component, noting that moles = weight/molecular weight.

The molecular weights of O_2, SO_2, and SO_3 are 32, 64, and 80, respectively. The following table can be completed.

Compound	Weight	Molecular Weight	Moles	Mole Fraction
O_2	20	32	$20/32 = 0.6250$	0.901
SO_2	2	64	0.0313	0.045
SO_3	3	80	0.0375	0.054
Total			0.6938	1.000

The reader should note that, in general, weight fraction (or percent) is *not* equal to mole fraction (or percent).

4.5 CONCENTRATION CONVERSION

Express the concentration of 72 g of HCl in 128 cm^3 of water into terms of fraction and percent by weight, parts per million (by weight), and molarity.

SOLUTION

The fraction by weight (in solution) can be calculated as follows:

$$72\,g/(72\,g + 128\,g) = 0.36$$

The percent by weight can be calculated from the fraction by weight:

$$(0.36)(100\%) = 36\%$$

The ppm (parts per million in water by weight) can be calculated as follows:

$$(72 \text{ g}/128 \text{ g})(10^6) = 562\ 500 \text{ ppm}$$

The molarity (M) is defined as follows:

$$M = \text{moles of solute/volume of solution (L)}$$

Using atomic weights,

$$\text{MW of HCl} = 1.0079 + 35.453 = 36.4609$$

$$M = \left[(72 \text{ g HCl}) \left(\frac{1 \text{ mol HCl}}{36.4609 \text{ g HCl}} \right) \right] \Big/ \left(\frac{128 \text{ cm}^3}{1000 \text{ cm}^3/\text{L}} \right) = 15.43 \text{ mol/L}$$

To compare data collected at different conditions, actual concentrations are often converted to standard temperature and pressure (STP). According to the U.S. Environmental Protection Agency, STP conditions for atmospheric or ambient sampling are usually 25°C and 1 atm.

4.6 CHLORINE CONCENTRATION

The concentration of chlorine vapor is measured to be 15 mg/m^3 at a pressure of 600 mm Hg and at a temperature of 10°C.

1. Convert the concentration units to ppmv.
2. Calculate the concentration in units of mg/m^3 at STP (2).

SOLUTION

Atmospheric concentrations of chemical agents are usually reported using two classes of units:

1. Mass of pollutant per volume of air, that is mg/m^3, μg/m^3, ng/m^3, and so on;
2. Parts of pollutant per parts of air (by volume), that is ppmv, ppbv, and so on

1. Choosing a basis of 1 m^3 of air, the number of moles of Cl$_2$ (MW = 71) is

$$n_{Cl_2} = 0.015 \text{ g}/(71 \text{ g/gmol}) = 2.11 \times 10^{-4} \text{ gmol}$$

The volume contribution of chlorine to the total volume, V_{Cl_2}, often referred to as the *pure component volume*, is

$$V_{Cl_2} = \frac{(2.11 \times 10^{-4}\ \text{gmol})(0.082\ \text{atm} \cdot \text{L/gmol} \cdot \text{K})(10 + 273\ \text{K})}{(600/760)\ \text{atm}}$$

$$= 0.00620\ \text{L} = 6.20\ \text{mL}$$

Since there are 10^6 mL in a m^3, the concentration, C, in ppmv can be expressed as mL/m^3 or mL/10^6 mL. Thus,

$$C = 6.20\ \text{mL/m}^3 = 6.20\ \text{ppmv}$$

2. Applying the ideal gas law to adjust the air volume (1 m^3) to STP,

$$V = (1\ \text{m}^3)\left(\frac{273\ \text{K}}{283\ \text{K}}\right)\left(\frac{600\ \text{mmHg}}{760\ \text{mmHg}}\right)$$

$$= 0.762\ \text{m}^3$$

The concentration, C, in mg/m^3, is then

$$C = \frac{15\ \text{mg}}{0.762\ \text{m}^3}$$

$$= 19.7\ \text{mg/m}^3$$

4.7   TRACE CONCENTRATION

Some wastewater and water standards and regulations are based on a term defined as *parts per million*, ppm, or *parts per billion*, ppb. Define the two major classes of these terms and describe the interrelationship from a calculational point of view. Also, convert 10 calcium parts per million parts of water on a mass basis to parts per million on a mole basis (2).

SOLUTION

Water streams seldom consist of a single component. They may also contain two or more phases (a dissolved gas and/or suspended solids), or a mixture of one or more solutes. For mixtures of substances, it is convenient to express compositions in mole fractions or mass fractions. As indicated earlier, the following definitions are often used to represent the composition of component A in a mixture of components:

$$w_A = \frac{\text{Mass of A}}{\text{Total mass of water stream}} = \text{Mass fraction of A}$$

$$y_A = \frac{\text{Moles of A}}{\text{Total moles of water stream}} = \text{Mole fraction of A}$$

Trace quantities of substances in water streams are often expressed in parts per million (ppmw) or as parts per billion (ppbw) on a mass basis. These concentrations can also be provided on a mass per volume basis for liquids and on a mass per mass basis for solids. Gas concentrations are usually represented on a mole or volume basis (e.g., ppmm or ppmv). The following equations apply:

$$\text{ppmw} = 10^6 w_A$$

$$= 10^3 \text{ ppbw}$$

$$\text{ppmv} = 10^6 y_A$$

$$= 10^3 \text{ ppbv}$$

The two terms ppmv and ppmm are related through the molecular weight.

To convert 10 ppmv Ca to ppmm, select a basis of 10^6 g of solution. The mass fraction of Ca is first obtained by the following equation:

$$\text{Mass of Ca} = 10 \text{ g}$$

$$\text{Moles Ca} = \frac{10 \text{ g}}{40 \text{ g/mol}} = 0.25 \text{ mol}$$

$$\text{Moles H}_2\text{O} = \frac{10^6 \text{ g} - 10 \text{ g}}{18 \text{ g/mol}} = 55\,555 \text{ mol}$$

$$\text{Mole fraction Ca} = y_{Ca} = \frac{0.25 \text{ mol}}{0.25 \text{ mol} + 55\,555 \text{ mol}} = 4.5 \times 10^{-6}$$

$$\text{ppmm of Ca} = 10^6 \, y_{Ca}$$

$$= (10^6)(4.5 \times 10^{-6})$$

$$= 4.5$$

4.8 ASH EMISSION

A small quantity of a waste mixture is incinerated at 2000°F using 5% ash coal to provide energy necessary for combustion. If 300 ft^3 of the flue gas (measured at 2000°F) is produced for every pound of coal burned, what is the maximum effluent particulate loading in gr/ft^3 (at 2000°F). Assume no contribution to the particulates from the waste.

SOLUTION

Based on the problem statement, assume the volume of gas produced on incinerating the waste is negligible. Select as a basis 1.0 lb of coal burned.

$$\text{Mass of particulates} = (1.0 \text{ lb})(0.05)$$
$$= 0.05 \text{ lb}$$

The maximum particulate loading is given by:

$$W = \frac{0.05}{300}$$
$$= 1.667 \times 10^{-4} \text{ lb/ft}^3$$

where W = particulate loading (mass of particulate/total volume of flue gas).
   To convert to gr/ft^3, note that there are 7000 gr/lb. Therefore,

$$W = (1.667 \times 10^{-4})(7000)$$
$$= 1.167 \text{ gr/ft}^3$$

Note that gr represents grains.

4.9   DILUTION FACTOR

With reference to Problem 4.8 what dilution factor is required to achieve an ambient air quality standard of 75 $\mu g/m^3$ for the particulates?

SOLUTION

First convert W_{in} to micrograms per cubic meter ($\mu g/m^3$).

$$W_{in} = 1.667 \times 10^{-4} \text{ lb/ft}^3$$
$$= 2.6727 \times 10^6 \text{ } \mu g/m^3$$

where W_{in} = inlet loading (mass particulate/volume gas).
   The dilution factor (DF) may now be calculated:

$$DF = \frac{W_{in}}{W_{std}} = \frac{2.6727 \times 10^6}{75} = 3.564 \times 10^4$$

where W_{std} = ambient air quality standard.

4.10   NANO EXHAUST TO ATMOSPHERE

The exhaust to the atmosphere from a new nanoprocess has a chemical agent partial measure (p) of 0.15 mmHg. Calculate the ppm of the agent in the exhaust.

SOLUTION

First calculate the mole fraction (y). By definition.

$$y = p/P$$

Since the exhaust is discharged to the atmosphere, the atmospheric pressure (760 mm Hg) is the total pressure (P). Thus,

$$y = (0.15)/(760) = 1.97 \times 10^{-4}$$

$$\text{ppm} = (y)(10^6) = (1.97 \times 10^{-4})(10^6)$$
$$= 197 \text{ ppm}$$

Note: Since the concentration of *gases* is involved, it is understood that ppm is on a volume basis. When ppm is applied to liquids or solids, it is almost always based on mass.

4.11   FLUE GAS ANALYSIS

The mole percent (gas analysis) of a flue gas is given below:

$N_2 = 79\%$
$O_2 = 5\%$
$CO_2 = 10\%$
$CO = 6\%$

Calculate the average molecular weight of the mixture.

SOLUTION

Employ the procedure set forth in Chapter 2. First write the molecular weight of each component:

$MW(N_2) = 28$
$MW(O_2) = 32$

$$MW(CO_2) = 44$$
$$MW(CO) = 28$$

The following table can be completed by multiplying the molecular weight of each component by its mole percent.

Compound	Molecular Weight	Mole Fraction	Weight (lb)
N_2	28	0.79	22.1
O_2	32	0.05	1.6
CO_2	44	0.10	4.4
CO	28	0.06	1.7
Total		1.00	

Finally, calculate the average molecular weight of the gas mixture:

$$\text{Average molecular weight} = 22.1 + 1.6 + 4.4 + 1.7 = 29.8$$

The sum of the weights in pounds represents the average molecular weight because the calculation above is based on 1.0 mol of the gas mixture.

The reader should also note once again that in a gas, molar percent (or fraction) equals volume percent and vice versa. Therefore, a volume percent can be used to determine weight fraction as illustrated in the table. The term y is used in engineering practice to represent mole (or volume) fraction of gases; the term x is often used for liquids and solids. Also, note that for liquid and solid mixtures, the molar percent almost always does not equal the volume percent.

4.12   pH

Classify whether the following three solutions are acidic or basic.

1. pH = 2.0
2. pH = 10.5
3. pH = 7.0

Also determine the pH of a solution with a hydrogen concentration of 1.05×10^{-6}.

SOLUTION

An important chemical property of an aqueous solution is its pH. The pH measures the acidity or basicity of the solution. In a neutral solution, such as pure water, the hydrogen (H^+) and hydroxyl (OH^-) ion concentrations are equal. At ordinary temperatures, this concentration is

$$C_{H^+} = C_{OH^-} = 10^{-7} \text{ g-ion/L}$$

where C_{H^+} is the hydrogen ion concentration and C_{OH^-} is the hydroxyl ion concentration.

The unit g-ion stands for gram-ion, which represent an Avogadro number of ions. In all aqueous solutions, whether neutral, basic, or acidic, a chemical equilibrium or balance is established between these two concentrations, so that

$$K_{eq} = C_{H^+} C_{OH^-} = 10^{-14}$$

where $K_{eq} =$ equilibrium constant.

The numerical value for K_{eq} given above holds for room temperature and only when the concentrations are expressed in gram-ion per liter (g-ion/L). In acid solutions, C_{H^+} is $> C_{OH^-}$; in basic solutions, C_{OH^-} predominates. The pH is a direct measure of the hydrogen ion concentration and is defined by

$$pH = -\log C_{H^+}$$

Thus, an acidic solution is characterized by a pH below 7 (the lower the pH, the higher the acidity); a basic solution, by a pH above 7; and, a neutral solution by a pH of 7.

It should be pointed out that this last equation is not the exact definition of pH but is a close approximation to it. Strictly speaking, the *activity* of the hydrogen ion a_{H^+}, and not the ion concentration, belongs in the above equation. For a discussion of chemical activities, the reader is directed to the literature [8].

Based on the above discussion, one can conclude the following:

1. a pH of 2.0 is a acidic solution.
2. a pH of 10.5 is a basic solution.
3. a pH of 7.0 is a neutral solution.

If the hydrogen ion concentration is $1.05 \times 10^{-6.0}$, the pH is given by

$$pH = -\log(1.05 \times 10^{-6.0})$$
$$= 5.98$$

One concludes that this solution is mildly acidic.

5 Particle Size, Surface Area, and Volume

As described earlier, it is not only the chemical composition, but also the size, shape and surface characteristics of ultrafine particles in the nanoscale range that determine the properties of various substances. For instance, when produced at infinitesimally small particle sizes, materials have an extraordinary ratio of surface area to either particle size or particle volume. And at these minute sizes, these particles are small enough to subsequently affect light scattering behavior, they show quantum effects, and their optical, electrical, and magnetic properties, as well as hardness, toughness, and melting point can differ markedly from the properties exhibited by macrosized particles of the same materials.

There is an inverse relationship between particle size and surface area, and one of the hallmarks of nanotechnology is the desire to produce and use nanometer-sized particles of various materials in order to take advantage of the remarkable characteristics and performance attributes discussed above that many materials exhibit at these infinitesimally small particle sizes. To illustrate the relationship between shrinking particle size and increasing surface area, envision a child's alphabet block that starts out being just the right size to fit into the chubby hand of a curious toddler. Now run an imaginary knife through the block, along its x-, y-, and z-axes, to divide the original playing piece into 8 smaller blocks of equal size. While the original block had just enough surface area to hold six colorful pictures, three quick swipes of the imaginary knife produce eight smaller blocks, which now have additional – previously unavailable – surface area for picture display. With each of the eight smaller blocks now having six sides of its own, the newly size-reduced blocks can display 48 little pictures of circus animals, letters or numbers – much to the delight of the appreciative child.

Continue to divide each of these smaller blocks with three quick swipes of the imaginary knife, and one can see the exponential relationship between particle size and surface area. This inverse relationship between particle size and surface area is a key underpinning in the field of nanotechnology. This phenomenon is illustrated in some of the later Problems in this Chapter.

One particular type of nanometer-scaled structure with a unique geometry that is generating considerable interest is the carbon nanotube. These nanotubes are essentially seamless cylinders composed of carbon atoms in a regular hexagonal

Nanotechnology: Basic Calculations for Engineers and Scientists, by Louis Theodore
Copyright © 2006 John Wiley & Sons, Inc.

arrangement, closed on both ends by hemispherical endcaps. They can be produced as single-wall nanotubes (SWNT) or multiwall nanotubes (MWNT). Interestingly, SWNT can have diameters ranging from 0.7 to 2 nm, while MWNT can have diameters ranging from 10 to 200 nm; individual nanotube lengths can be up to 20 cm. They have a surface area of up to 11,500 m^2/g, a density of 1.33 to 1.40 g/cm^3, and as few as 2000 molecules residing in its space.

Finally, to illustrate and possibly explain the unique behavior and properties of nanosized particles, one can show that a nanoparticle micrometer (1 micron or 1000 nanometers) in diameter has approximately 1.5×10^{-3} percent (1.5×10^{-5} fractional basis) of all its atoms located on its surface. However, a particle with a diameter of just 10 nanometers has approximately 15% of its atoms on the surface [9].

5.1   SPHERE, CUBE, RECTANGULAR PARALLELEPIPED, AND CYLINDER

Calculate the area and volume of the following shaped particles:

1. Sphere of radius R;
2. Cube of side A;
3. Rectangular parallelepiped of sides A and B;
4. Cylinder of radius R and height H.

SOLUTION

1. For a sphere of radius R:

$$\text{Volume} = (4/3)\pi R^3$$
$$\text{Area} = 4\pi R^2$$

2. For a cube of side A:

$$\text{Volume} = A^3$$
$$\text{Area} = 6A^2$$

3. For a rectangular parallelepiped of depth C:

$$\text{Volume} = ABC$$
$$\text{Area} = 2(AB + AC + BC)$$

4. For a cylinder of radial R and height H:

$$\text{Volume} = \pi R^2 H$$
$$\text{Area} = 2\pi RH$$

Some of the material from this and later Problems in this Chapter will come into play in the next Chapter.

5.2 PARALLELOGRAM, TRIANGLE, AND TRAPEZOID

Calculate the area and perimeter of the following shaped particles.

1. Parallelogram of height H, side A, and base B;
2. Triangle of height H, base B, and left side A. The angle between side A and base B is ϕ;
3. Trapezoid of height H and parallel sides A and B. The lower left angle is ϕ and the lower right angle is θ.

SOLUTION

1. For the parallelogram:

$$\text{Area} = (H)(B) = (A)(B)\sin\phi$$
$$\text{Perimeter} = 2A + 2B$$

2. For the triangle:

$$\text{Area} = \frac{1}{2}(B)(H) = \frac{1}{2}(A)(B)\sin\phi$$
$$\text{Perimeter} = A + B + C$$

3. For the trapezoid:

$$\text{Area} = \frac{1}{2}H(A + B)$$
$$\text{Perimeter} = A + B + H\left(\frac{1}{\sin\phi} + \frac{1}{\sin\theta}\right)$$
$$= A + B + H(\csc\phi + \csc\theta)$$

5.3 POLYGONS

Calculate the area and perimeter of the following shaped particles.

1. Regular polygon of N sides, each of length B;
2. Regular polygon of N sides inscribed in circle of radius R.

SOLUTION

1. $\text{Area} = \dfrac{1}{4}NB^2 \cot\dfrac{\pi}{N} = \dfrac{1}{4}NB^2 \dfrac{\cos(\pi/N)}{\sin(\pi/N)}$

$\text{Perimeter} = (N)(B)$

2. $\text{Area} = \dfrac{1}{2}NR^2 \sin\dfrac{2\pi}{N} = \dfrac{1}{2}NR^2 \sin\dfrac{360°}{N}$

$\text{Perimeter} = 2(N)(R)\sin(\pi/N) = 2(N)(R)\sin(180/N)$

5.4 ELLIPSE AND ELLIPSOID

Calculate the area, perimeter, or volume of the following shaped particles.

1. Ellipse of semimajor axis A and semiminor axis B;
2. Ellipsoid of semi-axes A, B, C.

SOLUTION

1. $\text{Area} = \pi(A)(B)$

$\text{Perimeter} = 4A \displaystyle\int \sqrt{1 - k^2 \sin^2 \theta}\, d\theta$

$= 2\pi\sqrt{\dfrac{1}{2}(A^2 + B^2)} \text{ (approximately)}$

where $k = \sqrt{(A^2 + B^2)/A}$.

2. $\text{Volume} = \dfrac{1}{2}\pi B^2 A$

5.5 CONES

Calculate the volume and surface area of the following shaped particles.

1. Right circular cone of radius R, lateral length L, and height H;
2. Frustrum of right circular cone of radii A, B, lateral length L, and height H.

SOLUTION

1.   Volume $= \dfrac{1}{3} \pi R^2 H$

    Surface area $= \pi R \sqrt{R^2 + H^2} = \pi R L$

2.   Volume $= \dfrac{1}{3} \pi H (A^2 + AB + B^2)$

    Surface area $= \pi (A + B) \sqrt{H^2 + (B - A)^2} = \pi (A + B) L$

5.6 TORUS

Calculate the volume and surface area of a torus of inner radius A and outer radius B.

SOLUTION

$$\text{Volume} = \frac{1}{4} \pi^2 (A + B)(B - A)^2$$

$$\text{Surface area} = \pi^2 (B^2 - A^2)$$

5.7 AREA TO VOLUME RATIOS

Develop the area to volume ratio equations for the shaped particles from Problems 5.1, 5.5, and 5.6

SOLUTION

For 5.1

1. Area/Volume $= 3/R$
2. Area/Volume $= 6/A$
3. Area/Volume $= 2 \left[(1/A) + (1/B) + (1/C) \right]$
4. Area/Volume $= 2/R$

For 5.5

1. Area/Volume $= 3L/RH$
2. Area/Volume $= [3(A + B)L]/[H(A^2 + AB + B^2)]$

For 5.6

1. Area/Volume $= 4/(B - A)$

5.8   AREA TO VOLUME CALCULATION

Refer to Problem 5.7 and calculate the area to volume ratio given the following data: $A = 1, B = 2, C = 2, L = 1, H = 1$, and $R = 1$.

SOLUTION

For 5.1

1. Area/Volume = 3
2. Area/Volume = 6
3. Area/Volume = 4
4. Area/Volume = 2

For 5.5

1. Area/Volume = 3
2. Area/Volume = 9/7

For 5.6

1. Area/Volume = 4

5.9   INCREASE IN SPHERE SURFACE AREA

Consider a sphere with a diameter of 1.0 μm. If this same mass of sphere is converted (through a size reduction process) to spheres with a diameter of 1.0 nm, calculate the increase in surface area of the smaller sized spheres.

SOLUTION

From Problem 5.1 the volume of a sphere is given by

$$V = \frac{4}{3}\pi R^3; \qquad R = D/2$$
$$= \pi D^3/6$$

The volume of a 1.0 μm sphere is therefore

$$V (1.0\,\mu\text{m}) = \pi(1.0)^3/6$$

$$= 0.5236\,(\mu\text{m})^3; \quad 1.0\,\mu\text{m} = 10^3\,\text{nm}$$

$$= 0.5236 \times 10^{+9}\,(\text{nm})^3$$

The volume of a 1.0 nm sphere is correspondingly

$$V (1.0\,\text{nm}) = \pi(1.0)^3/6$$

$$= 0.5236\,(\text{nm})^3$$

Since the mass – as well as the total volume – remain the same, the smaller number of particles is

$$N = \frac{0.5236 \times 10^9}{0.5236} = 10^9$$

The surface area of the 1.0 μm particle is

$$\text{SA} (1.0\,\mu\text{m}) = 4\pi R^2 = \pi D^2$$

$$= (3.14)(1)^2$$

$$= 3.14\,(\mu\text{m})^2$$

$$= 3.14 \times 10^6\,(\text{nm})^2$$

and

$$\text{SA} (1.0\,\text{nm}) = (3.14)(1)^2$$

$$= 3.14\,(\text{nm})^2$$

However, there are 10^9 of the 1.0 nm particles. Therefore, the 10^9 smaller sized particles have a total surface area of $3.14 \times 10^9\,(\text{nm})^2$. This reduction in size has increased the surface area by a factor of 1000 or 10^3. The reader should note that the particle diameter has decreased by the same factor.

5.10   INCREASE IN CUBE SURFACE AREA

Consider a cube with side of 1.0 μm. If the same mass of cube is converted to cubes with sides of 1.0 nm, calculate the increase in surface area of the smaller sized cubes.

SOLUTION

From Problem 5.1, the volume of a cube is given by

$$V = A^3$$

Therefore,

$$V(1.0 \ \mu m) = (1)^3$$
$$= 1.0 \, (\mu m)^3$$
$$= 1.0 \times 10^9 \, (nm)^3$$

The volume of the 1.0 nm sided cube is

$$V(1.0 \ nm) = 1.0 \, (nm)^3$$

Since the volume remains the same, the number of smaller cubes is

$$N = 10^9/1.0$$
$$= 10^9$$

The surface area of a 1.0 μm sided cubes is

$$SA(1.0 \ \mu m) = 6A^2$$
$$= 6 \, (\mu m)^2$$
$$= 6 \times 10^6 \, (nm)^2$$

and

$$SA(1.0 \ nm) = 6A^2$$
$$= 6 \, (nm)^2$$

Since there are 10^9 of the 1.0 nm sized cubes, their total surface area is $6 \times 10^9 \, (nm)^2$. Once again, the reduction in size has increased the surface area by a factor of 1000 due to a similar corresponding decrease in the side length of the cube.

6 Materials Science Principles

The topic *Materials Science and Engineering* implies a double focus – one geared toward a fundamental study of materials and their properties, and the other toward the production and use of materials for the benefit of society. This Chapter concentrates solely on the former subject that is Materials Science.

The term *Materials* denotes a vast area of complied knowledge. There is very little in all of engineering and science that does not involve materials. Obviously, the first task in preparing problems in the study of materials must be the application of limits on the subject matter to be covered – a focus on specific types of materials. It is generally understood that *Materials* covers only the solid–state of matter; liquids are considered only in certain cases where solid–liquid equilibrium is involved. There are many types of solids, however, and further focusing is required. Most solids can be categorized into one of three types: metals, polymers/plastics, or ceramics. (Ceramics are compounds of metallic and nonmetallic elements such as ferrous oxide.) In this Chapter, emphasis has been placed on metals because, in the opinion of the author, this class of materials has the widest impact on all of the major fields of science and engineering, and in particular on nanotechnology.

The major topic covered in this chapter is the *Crystallography of Perfect Crystals* (CPC). As discussed earlier, all matter is ultimately composed of atomic particles. How these particles are put together plays an extremely important role in determining a material's properties and the various uses of that material. The purpose of the Chapter is to provide the reader some insights into how solids (mainly metals and ionic materials) are organized at the atomic level and how this organization is reflected in some of the properties of the solids. Almost all information about the structure of crystals comes from x-ray diffraction measurements.

There are 10 problems in this Chapter. The first two problems are qualitative in nature. The remaining eight problems primarily address crystal structure and have been adopted from the work of Reynolds [10]. The reader is referred to the literature for additional details on *materials science* [11, 12].

6.1 METALS, POLYMERS AND CERAMICS

Briefly describe metals, polymers, and ceramics.

Nanotechnology: Basic Calculations for Engineers and Scientists, by Louis Theodore
Copyright © 2006 John Wiley & Sons, Inc.

SOLUTION

Metals and alloys, which include steel, aluminum, magnesium, zinc, cast iron, titanium, copper, nickel, and so on, have properties that include good electrical and thermal conductivity, high strength, and stiffness. They are also ductile and shock-resistant. Although pure metals are occasionally used, combinations of metals called *alloys* are employed to provide improvement in a particular property or properties. They are extensively used in structural applications.

Ceramics are compounds that may be classified between metallic and nonmetallic elements; they are most frequently oxides, nitrides, and carbides. The materials include clay minerals, cement, and glass. These materials typically resist the passage of electricity and heat. They are also more resistant to high temperatures and corrosive environments than most metals and some polymers. Although ceramics have excellent strength and hardness, they are very brittle.

Polymers include the typical plastic and rubber materials. Many of them are organic compounds that are produced by forming larger molecules from organic molecules by a process defined as polymerization. These materials are usually lightweight, flexible, and resistant to corrosion.

6.2   COMPOSITES, SEMICONDUCTORS AND BIOMATERIALS

Materials of interest – other than those discussed in Problem 6.1 – also include composites, semiconductors and biomaterials. Provide a brief overview of each of these materials.

SOLUTION

Composites consist of two or more materials, possessing properties that cannot be obtained by any single material. They generally display some of the better characteristics of each of the component materials.

Semiconductors have electrical properties that are intermediate (thus the word semi) between the electrical conductors and insulators. Their properties are very sensitive to extremely low concentrations of peripheral atoms that are strategically located in very small spatial regions. Semiconductors have revolutionized the electronics and computer industries over the past three decades.

Biomaterials are used for replacement of diseased or damaged human body parts. These materials must not cause adverse biological reactions into the body. All of the materials discussed in this and the previous Problem – metals, ceramics, polymers, composites, and semiconductors – can be employed as biomaterials.

6.3   CRYSTAL COORDINATION NUMBERS

Determine the minimum radius ratios, $(C/R)_{min}$ (a) for a coordination number of 3, and (b) for a coordination number of 8.

SOLUTION

All matter is ultimately composed of atomic particles. How these particles are put together plays an extremely important role in determining a material's properties and in the various uses of that material. The purpose of the remainder of this Chapter is to give the reader some insights into how solids (mainly metals and ionic materials) are organized on the atomic level and how this organization is reflected in some of the properties of the solids.

Solid materials may be either *amorphous* or *crystalline*. The word *amorphous* literally means "without form" and the atoms or molecules of solids in this category have little organization. The word *crystalline* implies that the component atoms, ions, or molecules that make up the material are arranged spatially in an ordered pattern, often referred to as a *crystal lattice*. In the solid state, metallic and ionic materials are almost universally found as crystals in nature; many covalent materials are crystalline as well.

There are many different types of crystal patterns or structures. For ionic and metallic materials, the main factor that determines the pattern or crystal type is the packing efficiency or the *packing factor*. Both metallic and ionic bonds are electrostatic in nature. As a result, the closer the bonded atoms are, the stronger are the bonding forces and the more stable is the crystal. For the purposes of this explanation, assume the atoms of a crystal to be small hard spheres tightly packed together in an ordered pattern. The metallic crystal is composed of spheres all having the same size and the ionic crystal is constructed of spheres of at least two different sizes. The *packing factor* is defined as the fraction of space occupied by the spheres.

The *coordination* number (CN) (Figure 6.1) of an atom in a crystal is defined as the number of nearest neighbors that atom possesses. All "nearest neighbors" must be equidistant from the atom in question, which shall be referred to as the *central atom*. In the case of ionic crystals, electrical stability requires that the central and neighboring atoms be oppositely charged. In a metal, all atoms of the crystal are positively charged and are held together by an electron cloud which pervades the entire crystal.

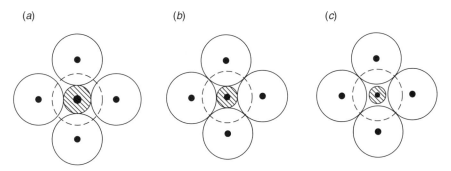

Figure 6.1   Coordination number.

There are five coordination numbers that occur in nature; these are 3, 4, 6, 8 and 12 (Figure 6.2). The crystal pattern that a given pair of ions form depends mainly on the relative sizes of the atoms or, equivalently, on the radius ratio (r/R). In this ratio, r represents the radius of the central atom and R the radius of the neighboring atoms. The central atom is always chosen as the smaller of the two ions.

Figure 6.1 shows arrangements for a CN of six. Note that the solid circles represent atoms whose centers are in the plane of the page; the dotted circle represents two atoms whose centers are above and below the plane of the page. In Figure 1(a), the central atom is in contact with all six neighbors simultaneously, a fact that is critical for ionic bonding. As the (r/R) ratio is decreased, the spacing between the neighboring atoms becomes smaller until the situation depicted in Figure 1(b) is achieved. In this diagram, the r/R ratio is 0.414, which is the minimum radius ratio that is capable of supporting a CN of six. For a ratio below this minimum, it is impossible to have the central atom contacting all six neighbors at the same time. In Table 6.1 the minimum radius ratios for the five coordination numbers are presented and in Figure 6.2, the atomic arrangements for the five CNs are depicted. In each of the five diagrams of Figure 6.2 the (r/R) ratio is at the minimum for that coordination number.

The minimum radius ratios in Table 6.1 can be calculated using simple geometric and trigonometric principles. Taking the coordination number of six as an example (Figure 6.2), the three dimensional figure obtained by joining the centers of the neighboring atoms is an *octahedron*. The two-dimensional figure formed by connecting atom centers A, B, and D is an isosceles right triangle. Since each leg of the triangle is $2R$ and the hypothenuse is $2R + 2r$, the application of the Pythagorean theorem yields

$$(2R + 2r)^2 = (2R)^2 + (2R)^2$$

or

$$2R + 2r = \sqrt{8}R$$

which can be rearranged to give

$$(r/R) = (r/R)_{min} = 0.414$$

TABLE 6.1 Minimum CN Radius Ratios

CN	$(r/R)_{min}$
3	0.155
4	0.225
6	0.414
8	0.732
12	1.000

The significance of the information contained in Table 6.1 lies in the fact that it can be used to help explain why ionic and metallic materials form the types of crystals that they do. The radius ratio of sodium chloride, for example, has been determined by x-ray diffraction to be 0.54. Since this ratio is less than 0.732, sodium chloride cannot crystallize in a pattern that requires a CN of 8; the sodium ion is simply too small to fit eight chloride ions around its periphery. This leaves coordination numbers of 6, 4, and 3 as possibilities, with 6 as the most likely prospect, since of the three, it would result in the highest packing factor.

Returning to the problem statement, for a coordination of 3, determine two distances in Figure 6.2(a) that can be represented directly in terms of the radii, r and R. The two distances are $\overline{AB}$ (which is equal to $r + R$) and either $\overline{AC}$ (which is equal to R) or, equivalently, the side of the equilateral triangle (which is equal to $2R$).

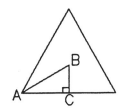

Find an equation that relates the two distances using the name figure. Since $\overline{AB}$ bisects the 60° angle of the equilateral triangle,

$$\angle BAC = 30°$$

and therefore

$$\cos 30° = \overline{AC}/\overline{AB}$$

Substituting from above,

$$\cos 30° = R/(r + R)$$

This equation may be solved for r/R. The result is the answer to (a), that is, the $(r/R)_{min}$ for a coordination number of 3.

$$r + R = R/\cos 30°$$

or

$$r/R + 1 = 1/\cos 30°$$
$$r/R = (r/R)_{min} = 0.155; \qquad \text{answer to (a)}$$

For a coordination number (CN) of 8, determine two distances in Figure 2d that can be represented directly in terms of the radii, r and R. The two distances are $\overline{HE}$ (which is equal to $2R$) and $\overline{CH}$ (which is equal to $2r + 2R$). Using the geometry of Figure 2d an equation that relates the two distances can be formed. Applying the

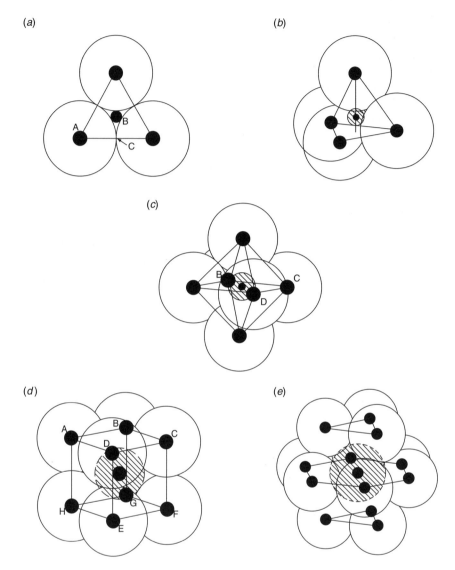

Figure 6.2   Cooridination numbers of (*a*) 3, (*b*) 4, (*c*) 6, (*d*) 8, and (*e*) 12.

Pythagorean theorem to triangle HEF,

$$\overline{HF}^2 = \overline{HE}^2 + \overline{EF}$$

or since $\overline{HE} = \overline{EF}$,

$$\overline{GE} = \sqrt{2}\,\overline{GH}$$

Applying the Pythagorean theorem to triangle CHF,

$$\overline{CH}^2 = \overline{HF}^2 + \overline{CF}^2$$

or since

$$\overline{HF} = \sqrt{2}\,\overline{HE}$$

and

$$\overline{CF} = \overline{HE},$$

$$\overline{CH} = \sqrt{3}\,\overline{HE}$$

Substituting yields

$$2r + 2R = \sqrt{3}\,(2R)$$

This equation may be solved for r/R. The result is the answer to (b), that is the $(r/R)_{min}$ for a coordination number of 8.

$$2(r/R) + 2 = \sqrt{3}\,(2)$$
$$r/R = (r/R)_{min} = 0.732; \qquad \text{answer to (b)}$$

The reader should be aware of the fact that an assumption has been made in the derivation of the information in Table 6.1. In obtaining the minimum radius ratios, the atoms of the crystal have been assumed to be incompressible spheres. Atoms, in fact, are neither incompressible nor spherical; as a result, if rigorously applied, the information in Table 6.1 can sometimes lead to erroneous results. For example, the (r/R) ratio for all metals is obviously 1.0, since all atoms of a metal crystal are the same size. This would lead to the conclusion that all metals should crystallize with a coordination number of 12 (for which the minimum radius ratio is 1.0). In fact, many metals, sodium and potassium among them, form crystals with a CN of 8.

In solving some of the subsequent Problems, the reader might find the following relationships among the edge, a (or, using triangle CHF in Figure 6.2d as a reference, $\overline{CF}$), the face diagonal, f (or $\overline{HF}$), and the body diagonal, b (or $\overline{CH}$), of a cube useful. They are taken from the above solution and can be applied to any cube:

$$f = \sqrt{2}a$$

and

$$b = \sqrt{3}a$$

6.4   GEOMETRY OF METALLIC UNIT CELLS

Answer the following:

1. Determine the numbers of atoms in a bcc, an fcc, and an hcp metal unit cell.
2. Find an expression for the edge of a bcc cell in terms of the atomic radius.
3. Find an expression for the edge of an fcc cell in terms of the atomic radius.
4. Given that the altitude of a tetrahedron is $\sqrt{2/3}$ of its edge, find an expression for the height of the unit cell in terms of the atomic radius.

SOLUTION

The term *Bravais lattice* refers to one of 14 different patterns employed in the structure of crystals. Three of these (the only three in which the points are arranged to form a cubic pattern) are shown in Figure 6.3. These three Bravais lattices are called *simple cubic* (sc), *body centered cubic* (bcc), and *face centered cubic* (fcc).

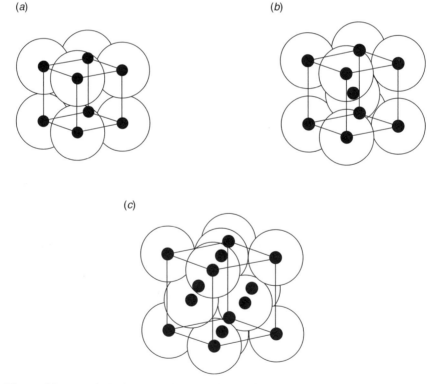

(a)

(b)

(c)

Figure 6.3   Bravais lattices: (a) simple cubic, (b) body-centered cubic, (c) face-centered cubic.

Although the points of the Bravais lattices are depicted as spheres in these diagrams, the points do not (at least for now) represent atoms. The *Bravais lattice* is an abstraction consisting only of a six- or eight-sided box with points placed either inside or on the box surface. The diagram in Figure 6.4 would show the *hexagonal* Bravais lattice if the three points that fall completely inside the eight-sided box were absent.

In order to transform the abstraction of the Bravais lattice into a real crystal structure, each point of the lattice is allowed to represent a single atom or set of atoms (like a pair of ions or a molecule). If each point of the fcc Bravais lattice shown in Figure 6.3(c) is replaced by a methane molecule, e.g. the unit cell of the methane crystal (which, since methane is a gas at ambient conditions, is found only at very low temperatures) results. A unit cell can be considered the basic "building block" for the construction of the crystal, and as such must be representative of the entire crystalline material. By repeating the methane unit cell over and over again, the solid methane crystal would result. As another example, if each point of the sc Bravais lattice is replaced by one Cs^+ and on Cl^- ion, the unit cell of cesium chloride is formed (Figure 6.6 in Problem 6.5). The simplest examples of this transformation from the Bravais lattice to the unit cell of a crystal are found in many metals where each point is replaced by a single atom. Substituting sodium atoms for the points in Figure 6.3(b) yields the bcc unit cell of sodium, and substituting copper atoms for the points in Figure 6.3(c) produces the fcc unit cell of copper. The unit cell structure shown in Figure 6.4, called the *hexagonal close-packed* (hcp) structure, was generated by replacing each point of the hexagonal Bravais lattice by two atoms instead of one. (This may not be obvious from Figure 6.4, however.) The metal zinc has an hcp structure.

To solve several of the problems in this Chapter, information contained in Table 6.2 should prove helpful. In this table, the structures of many of the elements are given along with their "atomic" and "ionic" radii. These radii were measured using x-ray diffraction and, once again, the assumption that the atom is a small incompressible sphere was made. By definition, the "atomic radius" represents the size of the atom in a crystal of the pure element. For example, the atomic radius of sodium (1.857Å) was measured using a pure sodium crystal. On the other hand, the "ionic radius" represents the size of the ion as it occurs in an ionic

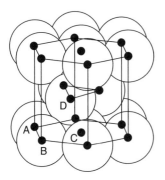
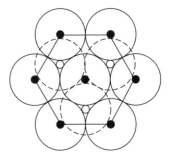

Figure 6.4   The hexagonal close-packed (hcp) unit cell.

TABLE 6.2   Table of Selected Elements

Element	Symbol	Atomic Number	Atomic Weight	Density (g/cc)	Structure	Atomic radius (Å)	Ionic radius (Å)
Lithium	Li	3	6.94	0.534	bcc	1.519	0.78
Berylium	Be	4	9.01	1.85	hcp	1.12	0.34
Carbon	C	6	12.01	2.25	Hex	0.71	
Nitrogen	N	7	14.01			0.71	
Oxygen	O	8	16.00			0.5	1.32
Fluorine	F	9	19.00			0.6	1.33
Sodium	Na	11	22.99	0.97	bcc	1.857	0.98
Magnesium	Mg	12	24.31	1.74	hcp	1.594	0.78
Aluminum	Al	13	26.98	2.699	Fcc	1.431	0.57
Silicon	Si	14	28.09	2.4	Cubic	1.176	0.42
Phosphorus	P	15	30.97	1.8		0.11	0.35
Sulfur	S	16	32.06	2.07	fc ortho	1.06	1.74
Chlorine	Cl	17	35.45			0.905	1.81
Potassium	K	19	39.10	0.86	bcc	2.312	1.33
Calcium	Ca	20	40.08	1.55	fcc	1.969	1.06
Scandium	Sc	21	44.96	2.5	fcc	1.605	0.83
Titanium	Ti	22	47.90	4.54	fcc	1.458	0.64
Vanadium	V	23	50.94	6.0	hcp	1.316	0.65
Chromium	Cr	24	52.00	7.19	bcc	1.249	0.64
Manganese	Mn	25	54.94	7.43	Cubic	1.12	0.91
Iron	Fe	26	55.85	7.87	bcc	1.241	0.83
Cobalt	Co	27	58.93	8.9	hcp	1.248	0.82
Nickel	Ni	28	58.71	8.90	fcc	1.245	0.78
Copper	Cu	29	63.54	8.96	fcc	1.278	0.96
Zinc	Zn	30	65.37	7.133	hcp	1.332	0.83
Gallium	Ga	31	69.72	5.91	fc ortho	1.218	0.62
Germanium	Ge	32	72.59	5.36	Cubic	1.224	0.44
Arsenic	As	33	74.92	5.73	Rhombic	1.25	0.69
Selenium	Se	34	78.96	4.81	Hex	1.16	1.91
Bromine	Br	35	79.91			1.13	1.96
Rubidium	Rb	37	85.47	1.53	bcc	2.44	1.49
Strontium	Sr	38	87.62	2.6	fcc	2.15	1.27
Yttrium	Y	39	88.91	5.51	hcp	1.79	1.06
Zirconium	Zr	40	91.22	6.5	hcp	1.58	0.87
Niobium	Nb	41	92.91	8.57	bcc	1.429	0.69
Molybdenum	Mo	42	95.94	10.2	bcc	1.36	0.68
Ruthenium	Ru	44	101.07	12.2	hcp	1.352	0.65
Rhodium	Rh	45	102.91	12.44	fcc	1.344	0.68
Palladium	Pd	46	106.4	12.0	fcc	1.375	
Silver	Ag	47	107.87	10.49	fcc	1.444	1.13
Cadmium	Cd	48	112.40	8.65	hcp	1.489	1.03

(*Continued*)

TABLE 6.2 *Continued*

Element	Symbol	Atomic Number	Atomic Weight	Density (g/cc)	Structure	Atomic radius (Å)	Ionic radius (Å)
Tin	Sn	50	118.69	7.298	bc tetra	1.509	0.74
Antimony	Sb	51	121.75	6.62	Rhombic	1.452	0.90
Tellurium	Te	52	127.6	6.24	Hex	1.43	2.11
Iodine	I	53	126.9	4.93	Ortho	1.35	2.20
Cesium	Cs	55	132.9	1.9	bcc	2.62	1.65
Barium	Ba	56	137.3	3.5	bcc	2.17	1.43
Tantalum	Ta	73	180.95	16.6	bcc	1.429	0.68
Tungsten	W	74	183.9	19.3	bcc	1.369	0.68
Osmium	Os	76	190.2	22.5	hcp	1.367	0.67
Iridium	Ir	77	192.2	22.5	fcc	1.357	0.66
Platinum	Pt	78	195.1	21.45	fcc	1.387	
Gold	Au	79	197.0	19.32	fcc	1.441	1.37
Thallium	Tl	81	204.4	11.85	hcp	1.704	1.05
Lead	Pb	82	207.02	11.34	fcc	1.750	1.32
Thorium	Th	90	232.0	11.5	fcc	1.800	1.10

crystal. The ionic radius of sodium (0.98 Å) was obtained by averaging sodium radii measured from a series of sodium salts such as sodium chloride, sodium iodide, sodium carbonate, and so on.

Determining the number of atoms contained in a particular unit cell at first seems like a trivial task. A quick look at Figure 6.3a and the reader might jump to the conclusion that simple cubic metal unit cells contain 8 atoms. The correct answer, however, is *one* atom. Atoms are not considered to be completely inside the cell (and therefore deserving of a full count of one) unless the atom centers are completely inside the cell. Any atom centered on the outer surface of the cell (for example, a vertex atom) gets only a fractional count, since part of the spherical atom volume is outside the cell boundary. In the sc metal structure, only one-eighth of each vertex atom volume falls within the confines of the cell. (Note that each vertex atom is shared equally among eight unit cells.) Since there are eight vertices in a cube, the total number must be one atom per unit cell.

Referring to the problem statement and Figure 3b, there is one atom with a center completely inside the cell and eight with centers at the cube vertices. Since each vertex atom gets a 1/8 count, the total number of atoms is *two*. *Partial answer to (1)*.

Referring to Figure 6.3c, determine the number of vertex atoms and the number of face atoms. Calculate the number of atoms inside the fcc metal unit cell.

There are eight vertex atoms, each of which gets a 1/8 count. There are six face atoms, each of which gets a 1/2 count. There are no atoms with centers completely inside the cell. The total number of atoms is therefore four. Partial answer to (1).

Referring now to Figure 6.4, determine the number of atoms with centers completely inside the cell and the number of vertex atoms. Determine what fraction of each vertex atom is inside the cell. Calculate the number of atoms inside the hcp unit cell.

There are three atoms with centers entirely inside the cell. There is one atom in the center of each of the two hexagons; these get a $1/2$ count. There are 12 vertex atoms, each one of which counts $1/6$. The total count is therefore *six* atoms. *Partial answer to (1)*.

For the bcc unit cell, the body diagonal, b, can be expressed in terms of the atomic radius, r, where $b = 4r$. By relating the edge length a to the body diagonal, b, and using the above result one can generate an expression for the edge of a bcc metal unit cell in terms of the atomic radius.

From Problem 6.3, $b = \sqrt{3}a$. Combining this equation with that above leads to

$$a = 4r/\sqrt{3} \qquad \text{Answer to (2)}$$

For the fcc unit cell, represent the face diagonal, f, in terms of the atomic radius, r:

$$f = 4r$$

By relating the edge length, a, to the face diagonal f, and using the result above, find an expression for the edge of a fcc metal unit cell in terms of the atomic radius. From Problem 6.3

$$f = \sqrt{2}\,a$$

Combining this equation with that above gives

$$a = 4r/\sqrt{2} \qquad \text{Answer to (3)}$$

Express the edge, a, of the hexagon in terms of the atomic radius:

$$a = 2r$$

Also, express the height, c, of the unit cell in terms of the hexagon edge. The height of the cell, c, is twice the altitude of tetrahedron ABCD. Therefore,

$$c = 2\sqrt{2/3}a$$

By combining the above two results the cell height in terms of the radius can be determined.

$$c = 4\sqrt{2/3}\,r \qquad \text{Answer to (4)}$$

The answers to (2) and (3) should prove useful in solving some of the remaining problems in this Chapter. A word of warning is appropriate, however. It must be remembered that the two relationships were derived for metallic crystals and are therefore valid only for bcc and fcc metals. They cannot be applied to any other type of bcc or fcc crystal (ionic or covalent, for example).

It should also be pointed out to the reader that, even though a simple cubic metal structure was used in an example in Problem 6.3, simple cubic *metals* do not exist in nature. The packing factor for simple cubic metal is 0.52, which means that almost half of the crystal's volume would be empty space. This poor efficiency of packing would cause such a crystal to be unstable. Note, however, that simple cubic *ionic* crystals do exist (CsCl, for example) and have high packing factors.

The symbol a, which represents the edge length of the cubic unit cell, is one of six *cell constants* used to describe cell shapes. It takes six-constants to define the shapes of six sided cells; three of these are length dimensions, a, b, and c, and three are angles, α, β, and γ. In this Chapter only cubic and hexagonal systems are dealt with. In cubic unit cells when all six sides are squares, $a = b = c$ and $\alpha = \beta = \gamma = 90°$. Only one cell constant need therefore be specified to completely define the cubic unit cell – the edge length, a. Hexagonal unit cells are eight-sided; the top and bottom sides are hexagons and the other six sided are rectangles. The edge of the hexagon is assigned the symbol, a, and the height of the cell is c. The rectangular sides therefore have the dimensions, $a \times c$. The only surface angles to be found in the hexagonal cell are the hexagon angle, $120°$, and the rectangle angle, $90°$.

6.5 GEOMETRY OF IONIC UNIT CELLS

Answer the following questions

1. Determine the numbers of positive and negative ions in a fcc ionic unit cell such as sodium chloride.

2. Determine the numbers of positive and negative ions in a sc ionic unit cell such as cesium chloride.

3. Find an expression for the edge of the fcc ionic unit cell in terms of the ionic radii, r_+ and r_-.

4. Find and expression for the edge of the sc ionic unit cell in terms of the ionic radii, r_+ and r_-.

SOLUTION

In the previous Problem, it was seen that, for bcc and fcc metallic crystals, the transformation from the Bravais lattice to the unit cell is a matter of replacing a point by a single spherical atom. For the hcp metallic crystal, each Bravais lattice point is replaced by two spherical atoms. In ionic materials, a *single* ion can never replace a Bravais lattice point; all points of the Bravais lattice must be identical and must have identical neighborhoods. The sodium chloride structure shown in Figure 6.5, e.g., could have been arrived at by replacing a point of the sc Bravais lattice by a

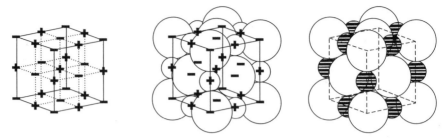

Figure 6.5   A fcc ionic unit cell (NaCl).

sodium ion and an adjacent point by a chloride ion, and so on. The sodium chloride structure, however, is not simple cubic; it is face centered cubic. To demonstrate this, if each point of the fcc Bravais lattice pictured in Figure 6.3c is replaced by an ion pair, e.g., the structure shown in Figure 6.5 results. (Note that the sodium ions paired with the five chlorides on the right face of the unit cell are not shown and do not belong in Figure 6.5; the centers of these sodium ions are outside of the unit cell shown, and belong in a neighboring cell.) In this structure, the spherical ions along each edge are in contact, but not those along the body and face diagonal. In an ionic lattice, the only spheres that should be in contact are those representing ions of opposite charge.

Although simple cubic *metals* do not exist in nature, simple cubic *salts* (ionic crystals) do. Cesium chloride is an example. The CsCl structure can be demonstrated by replacing each point of the sc lattice shown in Figure 6.3a by a cesium ion–chloride ion pair to achieve the structure shown in Figure 6.6. In this case the chloride ions were placed at each vertex and only one cesium ion, the one in the center of the cube, is shown. The centers of the other seven cesium ions fall outside of the cube and are members of adjacent unit cells. In this structure, the spherical ions are in contact along the body diagonal only. Again, only ions of opposite charge should be touching.

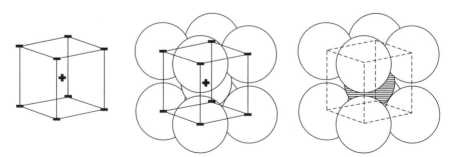

Figure 6.6   A simple cubic ionic unit cell (CsCl).

With respect to the Problem statement, refer to Figure 6.5. There are eight negative ions with centers at a vertex; one eighth of each is inside the cell. There are six negative ions with centers on a face; one-half of each is inside the cell. The total is therefore

$$8(1/8) + 6(1/2) = 4 \text{ negative ions per fcc ionic cell} \qquad \text{Partial answer to (1)}$$

Using the same diagram, one notes that there are 12 positive ions with centers in the middle of an edge; one fourth of these atoms are inside the cell. There is only one positive ion with a center completely inside the cell. The total is

$$12(1/4) + 1 = 4 \text{ positive ions per fcc ionic cell.} \qquad \text{Partial answer to (1)}$$

Referring to Figure 6.6, there are eight negative ions with centers at a vertex; one eighth of each is inside the cell. Therefore there is

$$8(1/8) = 1 \text{ negative ion per sc ionic cell.} \qquad \text{Partial answer to (2)}$$

Using the same diagram, there is one positive ion per sc ionic cell. *Partial answer to (2)*.

Referring to Figure 6.5, determine the edge length, a, in terms of r_+ and r_-.

$$a = 2r_+ + 2r_- \qquad \textit{Answer to (3)}$$

Referring to Figure 6.6, one can determine a distance (e.g., one of the cube diagonals) that can be represented directly in terms of the radii. By relating this distance to the edge length, the edge length can be expressed in terms of r_+ and r_-. The body diagonal can be represented in terms of radii by:

$$b = 2r_+ + 2r_-$$

Since $b = \sqrt{3}\, a$,

$$a = (2r_+ + 2r_-)/\sqrt{3} \qquad \textit{Answer to (4)}$$

It should once again be emphasized to the reader that a very simple model of the atom is being used here to describe crystal structure. No one really knows what an atom actually looks like. The best the scientist can do is to describe the atom in terms of mathematical and physical "models." Modern chemistry and physics employ models that are far more sophisticated than the incompressible sphere model used here. For the purposes of this text, however, this simple model is adequate.

6.6 PACKING FACTOR

Calculate the packing factors for:

1. an fcc metal;
2. the fcc ionic crystal, NaCl; and
3. the sc ionic crystal, CsCl.

The ionic radii of Na^+, Cs^+, and Cl^- can be found in Table 6.2 in Section 6.4.

SOLUTION

As explained earlier, the *packing factor* is defined as the fraction of space in the crystal occupied by atoms, where once again, it is assumed that the atom is an incompressible sphere. The packing factor is conveniently calculated by first determing the number of atoms in the unit cell of the crystal, calculating the spherical volumes of those atoms and dividing by the total volume of the unit cell. For example, a simple cubic metal, if it did exist, would have a packing factor of 0.52. This number is obtained by first finding the volume of *one* sphere of radius, r. (Figure 6.2a shows that there is *one* single atom per unit cell.) This volume is then divided by the unit cell volume, which is determined by cubing the edge length, $2r$, as is shown by the following equation:

$$(4/3)\pi r^3/(2r)^3 = 0.52$$

For metals, the calculation of the packing factor depends only on the metal structure and not on the atom size. This is not the case for ionic solids, when at least two different atom sizes must be taken into account.

With reference to the Problem statement, there are four atoms in a fcc metal cell. The volume of the atoms, V_{atoms} is

$$V_{atoms} = 4(4/3)\pi r^3$$

The volume of the fcc metal unit cell in terms of the atomic radius is

$$a = 4r/\sqrt{2}$$

The unit cell volume, v_{cell}, is given by

$$v_{cell} = a^3 = (4r/\sqrt{2})^3$$

The packing factor, pf, is then given by

$$pf = 4(4/3)\pi r^3/(4r/\sqrt{2})^3$$
$$= 4(4/3)\pi/(4/\sqrt{2})^3$$
$$= 0.74 \quad Answer\ to\ (1)$$

From Table 6.2, $r_+ = 0.98$Å for sodium and $r_- = 1.81$ Å for chlorine.
Determine the number of Na^+ anc Cl^- ions in the NaCl unit cell and calculate the total volume occupied by those ions in Å^3. There are four sodium and four chloride ions per unit cell. The volume of the ions is

$$V_{ions} = 4(4/3)\pi r_+^3 + 4(4/3)\pi r_-^3$$
$$= 4(4/3)\pi(0.98)^3 + 4(4/3)\pi(1.81)^3$$
$$= 115.1\ \text{Å}^3$$

For the fcc ionic cell, calculate the volume of the NaCl unit cell in Å^3.

$$a = 2r_+ + 2r_-$$
$$= 2(0.98) + 2(1.81)$$
$$= 5.58\ \text{Å}^3$$
$$V_{cell} = a^3$$
$$= (5.58)^3$$
$$= 173.7\ \text{Å}^3$$

Also determine the packing factor for NaCl.

$$pf = V_{ions}/V_{cell}$$
$$= 115.1/173.7$$
$$= 0.66 \quad Answer\ to\ (2)$$

Find the ionic radius of the Cs^+ ion. From Table 6.2, $r_+ = 1.65$ Å for cesium. There is one cesium and one chloride ion per unit cell. The volume of the ions is

$$V_{ions} = (4/3)\pi r_+^3 + (4/3)\pi r_-^3$$
$$= (4/3)\pi(1.65)^3 + (4/3)\pi(1.81)^3$$
$$= 43.66\ \text{Å}^3$$

For the sc ionic cell, also calculate the volume of the CsCl unit cell in Å^3.

$$a = (2r_+ + 2r_-)/\sqrt{3}$$
$$= [2(1.65) + 2(1.81)]/\sqrt{3}$$
$$= 3.995 \ \text{Å}^3$$
$$V_{cell} = a^3$$
$$= (3.995)^3$$
$$= 63.77 \ \text{Å}^3$$

Finally, determine the packing factor for CsCl.

$$\text{pf} = V_{ions}/V_{cell}$$
$$= 43.66/63.77$$
$$= 0.685 \qquad \text{Answer to (3)}$$

In Problem 6.3, the point was made that, in the construction of ionic and metallic crystals, the most important factor in determining crystal structure is efficiency of packing. As an exercise, the reader is encouraged to calculate (a) what the packing factor of sodium chloride would be if it had the same structure as cesium chloride (i.e., simple cubic) and (b) what the packing factor of cesium chloride would be if it had the same structure as sodium chloride (i.e., fcc). The results will show why these two ionic solids are packed the way they are and will demonstrate the significance of the packing factor in determining crystal structure.

6.7   DENSITY CALCULATION

The crystal structures, atomic weights and atomic radii of many elements can be found in Table 6.2. From this information, calculate (a) the density of gold, and (b) the density of titanium, in g/cc.

SOLUTION

Properties of materials fall into one of two categories: *extensive* and *intensive*. An *extensive* property is one that depends on the size of the material; volume and mass are examples of such properties. An *intensive* property is independent of size and the measurement of that property always yields the same value whether a small or large amount of the material is being used to make the measurement. *Density* is one such property.

The *unit cell* was described in Problem 6.4 as a *representative* portion of the crystal lattice. If the cell is truly representative, it must possess the same qualities as the crystal that it represents. It is therefore theoretically possible to use a single unit cell as a basis for calculating intensive properties. (Note: As explained earlier the term *basis* is used to indicate that amount of material on which a calculation is being performed. In many problems, amounts are not specified and as long as the problem involves *intensive* properties, the choice of a *basis* is arbitrary and left to the person performing the calculation.)

In this Problem, a single unit cell should be chosen as the *basis*. To solve the problem, information from Table 6.2, must be used. In this table, atomic weights are listed in *atomic mass units* (amu). As indicated earlier, the following equation can be used to convert from amu to grams:

$$6.023 \times 10^{23} \text{ amu} = 1 \text{ g}$$

The reader will quickly recognize the number, 6.023×10^{23}, as the *Avogadro number*, which was defined earlier as the number of atoms or molecules in a *mole*. It can be seen from the equation above that an equivalent definition of the Avogadro number is *the number of atomic mass units, amu, in one gram*.

Return to the Problem statement. First find the structure, atomic weight, and atomic radius of gold. From Table 6.2, gold is fcc, has an atomic weight of 197.0, and an atomic radius, r, of 1.441 Å. There are four atoms per unit cell for gold. Calculate the mass of the unit cell in grams.

$$\left(\frac{4 \text{ atoms}}{\text{cell}}\right)\left(\frac{197.0 \text{ amu}}{\text{atom}}\right)\left(\frac{1 \text{ g}}{6.023 \times 10^{23} \text{ amu}}\right) = 1.308 \times 10^{-21} \text{ g/cell}$$

The cell constant, a, of the unit cell in cm is

$$a = 4r/\sqrt{2}$$
$$= 4(1.441/\sqrt{2})$$
$$= 4.076$$
$$= 4.076 \times 10^{-8} \text{ cm} \qquad (1 \text{ cm} = 10^8 \text{ Å})$$

The volume of the unit cell in cc is then

$$V_{\text{cell}} = a^3$$
$$= (4.076 \times 10^{-8})^3$$
$$= 6.772 \times 10^{-23} \text{ cc/cell}$$

The density of gold may now be calculated:

$$\text{Density} = (1.308 \times 10^{-21} \text{ g/cell})/(6.772 \times 10^{-23} \text{ cc/cell})$$
$$= 19.3 \text{ g/cc} \quad \textit{Answer to (a)}$$

From Table 6.2, titanium is hcp, has an atomic weight of 47.9 and an atomic radius, r, of 1.458 Å. There are six atoms per unit cell. The mass of the unit cell in grams is then

$$\left(\frac{6 \text{ atoms}}{\text{cell}}\right)\left(\frac{47.90 \text{ amu}}{\text{atom}}\right)\left(\frac{1 \text{ g}}{6.023 \times 10^{23} \text{ amu}}\right) = 4.772 \times 10^{-22} \text{ g/cell}$$

Determine the unit cell constant, a (the edge of the hexagon), from the atomic radius.

$$a = 2r$$
$$= 2(1.458)$$
$$= 2.916 \text{ Å}$$

Determine the unit cell constant, c (the height of the cell).

$$c = \sqrt{2/3}\,(2)\,(2r)$$
$$= \sqrt{2/3}\,(2)\,(2)\,(1.458)$$
$$= 4.762 \text{ Å}$$

From the cell constants, calculate the volume of the unit cell in cc. Dividing the hexagon into six equilateral triangles (as shown below), the area of one triangle is

$$1/2\,a\,(a \sin 60°)$$

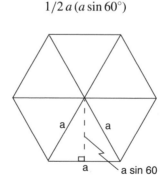

The hexagon area is

$$(6)(1/2)(a)(a \sin 60°)$$

Therefore,

$$
\begin{aligned}
V_{cell} &= (6)(1/2)(a)(a \sin 60°)(c) \\
&= (6)(1/2)(2.916)(2.916)(0.8660)(4.762) \\
&= 105.2 \\
&= 1.052 \times 10^{-22} \, cc/cell
\end{aligned}
$$

Finally, calculate the density of titanium.

$$
\begin{aligned}
density &= (4.772 \times 10^{-22} \, g/cell)/(1.052 \times 10^{-22} \, cc/cell) \\
&= 4.54 \, g/cc \qquad \textit{Answer to (b)}
\end{aligned}
$$

The densities of gold and titanium that were arrived at in this Problem are calculated values based on the atomic radius of those atoms, as measured by x-ray diffraction. The density of pure gold and pure titanium can also be readily measured in the laboratory; values for these densities are also given in Table 6.2. Although the calculated and measured values of densities are usually quite close, they do not agree perfectly. Generally, the value of the density calculated in this manner is slightly higher than the measured value. The reason for this discrepancy is that the method demonstrated in this problem assumes that the crystal is "perfect". This is not the case. Real crystals are riddled with all types of imperfections (except perhaps at a temperature of absolute zero). As the temperature increases toward the crystal's melting point and the atomic activity inside the crystal becomes more frenetic, the number of imperfections in the crystal lattice also increases. One common type of imperfection is the *vacancy*, where an atom or ion is missing from a lattice site. The presence of such vacancies could also help explain why the calculated densities of gold and titanium are higher than their measured densities.

6.8 DIRECTIONS AND PLANES

Answer the following questions:

1. Find the indices for the three directions shown in Figure 9 below.
2. In the cube below (see Figure 11b), draw an arrow representing the ($\bar{1}10$) direction.
3. Find the Miller indices for the three planes shown in Figure 10 below.
4. In the cube below (see Figure 11d), sketch the (201) plane.

SOLUTION

Many crystal phenomena are directional. For example, when a metal crystal is plastically deformed (i.e., permanently distorted), parts of the crystal move relative to other parts. This slippage occurs in certain predictable directions and along certain predictable atom planes. Many crystal properties are also directional, for example, the elastic modulus, ductility, and conductivity. It is important, therefore, that directions and planes of a crystal be identifiable.

On the left side of Figure 6.7, the right-handed three-dimensional coordinate system that will be used throughout this Problem set is given. In the two cubes shown in Figure 6.7, five directions (indicated by vectors) are shown. Directions are represented symbolically by [hkl], where h, k and l (lowercase el) are three small whole numbers called *indices*. Note that there are no commas between the numbers. The indices for a given direction are found as follows. The base of the vector is chosen as the origin of the coordinate system. The coordinates of the tip of the arrowhead are then determined. Using direction (a) as an example, the coordinates of the arrowhead are (0,0,1). (Note: the *physical* coordinates of the arrowhead would have length dimensions and be given in terms of a, the cell constant. In this example, those coordinates are (0,0,a). For purposes of determining the indices, however, the cell constant is considered to be one mathematical unit and the coordinates are given as pure numbers.) The coordinates of the arrowhead tip yield the indices. The directions (a), (b), and (c) are represented by [001], [111], and [110], respectively. The tip of the arrowhead for direction (d) has the coordinates $1,0,-1$) and is represented by [10$\bar{1}$]. Note that a negative direction is indicated by a bar over the index. For (e), the arrowhead has the coordinates $(1,0,1/2)$, which, following the rules for finding the indices given so far, would yield [1,0,1/2]. However, the indices should be small whole numbers. Since the direction indicated by vector (e) also passes through the points (2,0,1) and (4,0,2), that direction could also be feasibly represented by [201] and [402]. In other words, multiplying the coordinates and indices by a positive number does not alter the direction. The convention used here is to multiply the indices by the smallest positive integer that will convert all three to small whole numbers. Direction (e) should be therefore represented by [201].

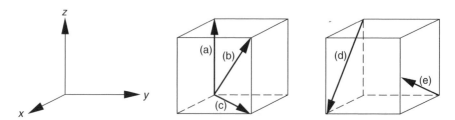

Figure 6.7 Crystal directions.

(a)   (b)   (c)

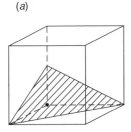 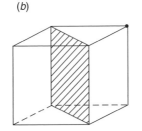 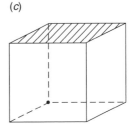

Figure 6.8   Crystal planes.

The atom planes of a crystal are represented by the symbol (hkl) where once again h, k, and l are small whole numbers. (It is important that the reader distinguish between the symbols used for coordinates and planes. Both employ parentheses, but the coordinates of a point are separated by commas and need not be small whole numbers. No commas appear in the symbol for a plane.) It takes three points to fix the position of a plane in space, unless the points are along the same straight line. The indices of a plane (sometimes referred to as *Miller* indices) are found as follows. (1) Choose an origin at some vertex close to *but not in the plane in question.* For example (a) in Figure 6.8, the lower left back vertex has been chosen as the origin. (2) Find the three *intercepts.* An *intercept* of an axis is the point of intersection of the axis and the plane. In Figure 6.8a the x-, y- and z- intercepts are 1, 1, and 1/2, respectively. (3) Take the reciprocals of the three intercepts; these reciprocals are the Miller indices. Thus, the plane in Figure 6.8a is represented as (112). In Figure 6.8b, the origin has been chosen as the upper back right vertex of the cell. The x- and y-intercepts are 1 and -1, respectively, but there is no z-intercept because the plane is parallel to the z-axis. In this case, the intercept is considered parallel to the z-axis and the intercept is considered to be infinity (∞). Since the value of ($1/\infty$) is zero, this plane is represented by (110). In other words, whenever one of the Miller indices is zero, the plane must be parallel to the axis associated with that index. Note that if the origin was chosen as the front left upper vertex, the final result would have been (110). Multiplying the three Miller indices by -1 does not alter the position of the plane.

In Figure 6.8c, the indicated plane, which is the top side of the unit cell, is simultaneously parallel to both the x- and y-axes. Its designation is therefore (001) or (00$\bar{1}$). The bottom of the unit cell has the same designation, since obviously it is an equivalent plane.

On to the solution to this Problem. In Figure 6.9, find the coordinates of the point at the tip of the arrowhead and determine the indices for direction (a).

The coordinates are $(0,0,-1)$; the direction is [00$\bar{1}$].

In Figure 6.9, find the coordinates of the point at the tip of the arrowhead and determine the indices for direction (b).

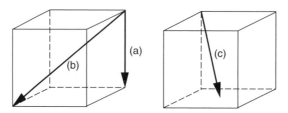

Figure 6.9   Unknown crystal directions for problem 6.8(*a*).

The coordinates are $(1, -1, -1)$; the direction is $[11\bar{1}]$.

In Figure 6.9, find the coordinates of the point at the tip of the arrowhead and determine the indices for direction (c).

The coordinates are $(1/2, 1/2, -1)$; the direction is $[11\bar{2}]$.

Draw a cube and choose one of the vertices as the origin. Locate the point with the coordinates corresponding to the indices given in part (2) of the problem and draw an arrow from the origin to the point.

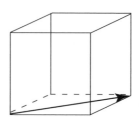

Using Figure 6.10*a*, choose an origin and find the three intercepts. Determine the Miller indices of the plane.

The chosen origin is the lower front right vertex. The x-, y-, and z-intercepts are $-1, -1/2$, and $1/2$, respectively, and their reciprocals are, -1, -2, and 2. The plane is $(\bar{1}\bar{2}2)$.

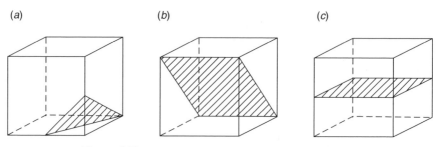

Figure 6.10   Unknown crystal planes for Problem 6.8(*c*).

Using Figure 6.10*b*, choose an origin and find the three intercepts. From the intercepts, determine the Miller indices of the plane.

The chosen origin is the lower front right vertex. The x-, y-, and z-intercepts are -1, ∞, and 1, respectively, and their reciprocals are, -1, 0, and 1. The plane is ($\bar{1}$01).

Using Figure 6.10*c*, choose an origin and find the three intercepts. From the intercepts, determine the Miller indices of the plane.

The chosen origin is the lower front right vertex. The x-, y-, and z-intercepts are ∞, ∞, and $1/2$, respectively, and their reciprocals are, 0, 0, and 2. The plane is (002).

Finally, draw a cube and choose one of the vertices as the origin. Using the Miller indices given in part (4) of the Problem, determine the intercepts and draw the plane.

The chosen origin is the lower back right vertex. The x-, y-, and z-intercepts of the (201) plane are $1/2$, ∞, and 1, respectively. Draw small x-marks at $1/2$ on the x-axis and 1 on the z-axis. The plane that passes through the two x-marks provided below is parallel to the y-axis is

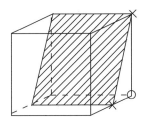

The choice of an appropriate origin to identify a particular plane is not always obvious. As a rule, the origin should be the vertex closest to the plane without actually being in the plane. In some cases, the origin is best chosen by trial and error. If it is difficult to locate the intercepts, a change of origin can often make this task a lot easier. The plane of Figure 6.10*a*, for example, is much more difficult to identify when the front upper right vertex is chosen as the origin instead of the front lower right vertex. Using the upper vertex necessitates the drawing of several more cells to determine the intercepts.

The reader may note that the direction shown in Figure 6.7, direction (d), is perpendicular to the plane shown in Figure 6.10*b*. This direction, [$\bar{1}$01], and this plane, ($\bar{1}$01), also have the same three indices. This relationship is always true as long as the system is cubic. Directions and planes with the same set of indices must therefore be mutually perpendicular.

The reader may also note that the plane of Figure 6.10c is the (002) plane, while the parallel plane of Figure 6.8c is the (001) plane. By convention, all (001) planes are also considered to be (002) planes. The opposite is not true, however, and in Figure 6.10c, the top, intermediate, and bottom planes are all (002) planes, but only the top and bottom are also (001) planes.

6.9 LINEAR DENSITY

Calculate the linear density in atom/cm along the [111] direction of gold.

SOLUTION

In Problem 6.8 it was stated that crystalline materials often have directional characteristics. The slipping of planes during plastic deformation of metals was given as an example. The direction of slip is generally that direction along which the atoms have the shortest internuclear spacing or, equivalently, the direction of the highest *linear density*. *Linear density* is defined as *the number of atom centers per unit length along a given direction*. In Figure 6.11, the intersection of the [111] direction with the unit cell of a bcc metal is shown. Theoretically, the [111] direction maybe considered to be a line of infinite length. To calculate the linear density, a line segment extending from one atom center to some other atom center should be chosen as a basis. Once again, since linear density is an intensive property, the choice of a basis is arbitrary. In the example of Figure 6.12, that part of the line that intersects the cubic cell is chosen as the basis. The number of atom centers that are on this line segment looks like three, at first glance. However, since the atom centers at the ends of the line segment are shared equally with adjacent line segments, these *border* atoms contribute only $1/2$ atom to the basis. The number of atom centers on the line is therefore *two*. The next step in calculating the linear density is to determine the length of the line segment. Since this segment is a body diagonal, it can be readily calculated from the edge length, a, which, in turn, can be calculated from

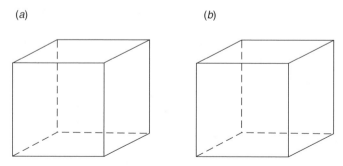

(a)   *(b)*

Figure 6.11 For answers to Problem 6.8.

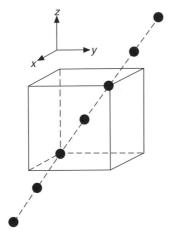

Figure 6.12   Linear density of atoms along the [111] direction of bcc lithium.

the atom radius. The following equation provides an alternate, more general, and perhaps quicker, method. The distance, L, between any two points in the cubic system may be determined from

$$L = ([h_2 - h_1]^2 + [k_2 - k_1]^2 + [l_2 - l_1]^2)^{1/2}\, a$$

where the coordinates of the two points are (h_1, k_1, l_1) and (h_2, k_2, l_2). If one of the points is chosen as the origin, the equation reduces to

$$L = (h^2 + k^2 + l^2)^{1/2}\, a$$

where (h, k, l) are the coordinates of the second point. Obviously, L has the same length units as a.

The final step in the linear density calculation is to divide the number of atoms on the line segment by the line segment length. Typical units for linear density are atom/cm.

From Table 6.2, the structure of gold is fcc and its atomic radius is 1.441 Å. The given direction is sketched in Figure 6.13.

From the crystal structure of gold, one can mark the atom centers in the cube drawn in Figure 6.13. The number of atom centers that fall on the vector representing the given direction can now be counted. Each of the two atoms pictured Figure 6.13b gets a count of 1/2. The total is 1 atom.

(a)   (b)

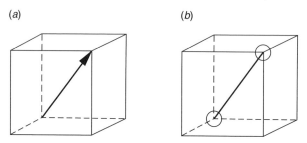

Figure 6.13   Figure for Problem 6.9.

From the atom radius, determine the length of the vector in Å and convert to cm.

$$a = 4r/\sqrt{2}$$
$$= 4(1.441)/\sqrt{2}$$
$$= 4.076\,\text{Å}$$
$$L = (h^2 + k^2 + l^2)^{1/2}a$$
$$= (1^2 + 1^2 + 1^2)^{1/2}(4.076)$$
$$= 7.059\,\text{Å}$$
$$= 7.059 \times 10^{-8}\,\text{cm}$$

Calculate the linear density

$$\text{Linear density} = \text{number of atoms/cm}$$
$$= 1/7.059 \times 10^{-8}$$
$$= 1.417 \times 10^7\,\text{atom/cm}$$

When counting atoms on a line segment, the reader has to be careful not to be confused with what was done earlier. Referring to Problem 6.6, the vertex atom of the unit cell is given a count of 1/8 atom because that atom is shared equally among 8 unit cells. In the linear density calculation, the same atom is given a count of 1/2. Note that, in the latter case, the *sharing* of the vertex atom is only between *two* adjacent line segments.

6.10   PLANAR DENSITY

Calculate the planar density in atoms/cm^2 on the (101) plane of sodium.

SOLUTION

Planar density is similar to *linear density*, except instead of measuring how well a *direction* is packed with atoms, the *planar density* measures how well a *plane* is packed with atoms, In Problem 6.9, it was pointed out that, when the planes of a metallic crystal slip during plastic distortion, the most likely *direction* of slip is the one with the highest linear density. By the same token, the most likely *plane* of slip is the one with the highest planar density.

Planar density is defined as the number of atom centers per unit area on a given plane. In Figure 6.14, the intersection of the (100) plane with the unit cell of a fcc metal is shown. Theoretically, the (10$\bar{0}$) plane may be considered to be of infinite area, but only a small segment of that area (i.e., the basis for the calculation) need be used. Taking that segment of the (100) plane that intersects the unit cell as the basis, the number of atoms with centers in that plane segment are first counted. Once again, it should be noted that only atoms with centers completely within the boundaries of the plane segment get a full count. Atoms centered any-where on the boundary receive a fractional count. The vertex atoms shown in Figure 6.14 each receive a count of $1/4$, since each of those atoms is equally shared among four such area
segments. The total atom count in this example is therefore *two*. Next, the area of the plane segment must be calculated. In this case, the area is simply the square of the cell constant, a. The planar density can now be determined by dividing the number of atoms on the plane segment by the plane segment area. Typical units are atoms/m^2 or atoms/cm^2.

In Table 6.2, the structure of sodium is bcc and its atomic radius is 1.857 Å. The given plane is sketched in the cube in Figure 6.15a.

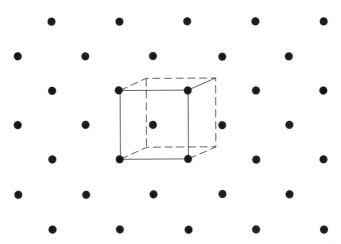

Figure 6.14 Planar density of atoms on the (100) plane of fcc gold.

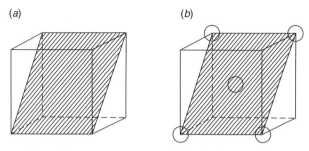

Figure 6.15   Figure for Problem 6.10.

From the crystal structure of sodium, one can mark the atom centers in the cube of Figure 6.15*b*.

The number of atom centers that fall on the plane segment drawn in Figure 6.15*b* can be counted. Each of the four vertex atoms get a count of 1/4. The central atom gets a count of 1. The total is 2 atoms.

The plane segment of Figure 6.15 is a rectangle bonded by two cube edges, *a*, and two face diagonals, *f*. The rectangle area is $a \times f$. Therefore,

$$a = 4r/\sqrt{3}$$
$$= 4(1.857)/\sqrt{3}$$
$$= 4.289$$
$$f = \sqrt{2}\,a$$
$$= \sqrt{2}\,(4.289)$$
$$= 6.065$$
$$\text{Rectangle area} = af$$
$$= (4.289)\,(6.065)$$
$$= 26.01$$
$$= 2.601 \times 10^{-15}\,\text{cm}^2; \quad (1\,\text{Å} = 10^{-8}\,\text{cm})$$

The planar density can now be calculated.

$$\text{Planar density} = \text{atoms/area}$$
$$= 2/2.601 \times 10^{-15}$$
$$= 7.69 \times 10^{14}\,\text{atom/cm}^2$$

As noted earlier, the reader is once again warned about confusing these atom counts. The same vertex atom that is given a 1/8 count in a *density* (g/cc) calculation and a $1/2$ count in a *linear density* calculation (atom/cm), may be given a $1/4$ count in a *planar density* calculation (atom/cm^2). The first and key consideration in counting border atoms in these *density* calculations must be the determination of how the atom is shared.

7 Physical and Chemical Property Estimation

This is probably the most important Chapter in Part 1. Procedures need to be put in place that will allow the practitioner to estimate key physical and chemical properties of new materials that are developed at the nanoscale level. Although the scientific community has traditionally resorted to experimental methods to accurately determine the aforementioned properties, that option may very well not be available when dealing with nanomaterials that exist as nanoparticles.

Predictive methods, albeit traditional ones, may be the only option available to obtain a first estimate of these properties. It should be noted that significant errors may be involved since extrapolating (or extending) satisfactory estimation procedures at the macroscale level may not be applicable at the nanoscale. Notwithstanding these concerns, examples to estimate some of the key physical and chemical properties are given below. Properties receiving attention include:

- Vapor pressure
- Latent enthalpy
- Critical properties
- Viscosity
- Thermal conductivity
- Heat capacity

Is property estimation important? Absolutely. As indicated above, there are times and situations when experimental procedures cannot be implemented. For this scenario, one can turn to theoretical and semitheoretical methods and equations to obtain first estimates of important property information.

One could argue that the present procedures available to estimate the properties of materials are based on macroscopic quantities and would not apply to nanoparticles and nanosubstances. The author believes this argument is valid since the properties of materials in the nanosize range are a strong function of size, rendering it impossible to obtain a unique property to describe the material. Nonetheless, the traditional methods available for property estimation either may be applicable or may suggest alternative theoretical approaches.

Nanotechnology: Basic Calculations for Engineers and Scientists, by Louis Theodore
Copyright © 2006 John Wiley & Sons, Inc.

The above discussion is further complicated since there appears to be some question whether the traditional macroscopic laws of physics and chemistry apply to particles in the nanorange. The author believes the conservation laws (momentum, mass, and so on) and the laws of thermodynamics are still applicable. But recent research studies and comments in the literature appear to shed doubt on these macroscopic laws.

Four key references [6, 8, 13, 14] were employed in developing some of the examples that follow. Although the references can be found in the reference section at the end of this Part, they are now listed because of their importance.

R. Perry and D. Green (Eds), *Perry's Chemical Engineers' Handbook*, 7th edition, McGraw-Hill, New York, 1997.

J. Smith, H. Van Ness and M. Abbott, *Introduction to Chemical Engineering Thermodynamics*, 7th edition, McGraw-Hill, New York, 2005.

N. Chopey, *Handbook of Chemical Engineering Calculations*, 2nd edition, McGraw-Hill, New York, 1994.

R. Bird, W. Stewart and E. Lightfoot, *Transport Phemenon*, 2nd edition, John Wiley and Sons, Hoboken, NJ, 2002.

The first two references are somewhat complementary, but each provides extensive information on this topic. This includes equations and procedures on several other properties not addressed in the example problems. The interested reader should check these references for more details.

7.1 PROPERTY DIFFERENCES

Describe the difference between chemical and physical properties.

SOLUTION

Every compound has a unique set of *properties* that allows one to recognize and distinguish it from other compounds. These properties can be grouped into two main categories: physical and chemical. *Physical properties* are defined as those that can be measured without changing the identity and composition of the substance. Key properties include viscosity, density, surface tension, melting point, boiling point, and so on. *Chemical properties* are defined as those that may be altered via reaction to form other compounds or substances. Key chemical properties include upper and lower flammability limits, enthalpy of reaction, autoignition temperature, and so on.

These properties may be further divided into two categories – intensive and extensive. As indicated earlier, *intensive properties* are not a function of the quantity of the substance, while *extensive properties* depend on the quantity of the substance.

7.2 MATERIAL SELECTION

Select the compound/substance for each of the three tasks developed below. Justify the answer.

1. A fluid to cool a hot hydrocarbon stream.
2. Materials to reduce the heat loss from a hazardous waste incinerator.
3. Materials for an armored tank to be employed in Iraq.

SOLUTION

Each case is treated separately.

1. Water would be an excellent choice since it has a high heat capacity and low viscosity. In addition, it is inexpensive, nontoxic, and readily available.
2. A material with excellent insulating properties, i.e., low thermal conductivity, should be selected. Fiberglass, asbestos and certain polymers are options, but there are environmental concerns with asbestos.
3. A metal that possesses high strength, can withstand impact and shock, and survive a high temperature excursion should be selected. The metal should also be as light as possible, suggesting a requirement of low density or specific gravity.

7.3 VAPOR PRESSURE

Define vapor pressure.

SOLUTION

Vapor pressure is an important property of liquids, and to a somewhat lesser extent, of solids. If a liquid (or solid) is allowed to evaporate in a confined space, the pressure of the vapor phase increases as the amount of vapor increases. If there is sufficient liquid present, the pressure in the vapor space eventually comes to equal exactly the pressure exerted by the liquid at its own surface. At this point, a dynamic equilibrium exists in which vaporization and condensation take place at equal rates and the pressure in the vapor space remains constant. The pressure exerted at equilibrium is called the vapor pressure of the liquid. Solids, like liquids, also exert a vapor pressure. Evaporation of solids (sublimation) is noticeable for solids characterized by appreciable vapor pressures.

The reader should note that vapor pressure is a function only of temperature. Unfortunately it is not a particularly simple function. Both experimental data and correlations are available to the engineer and scientist. The two most often used empirical equations are reviewed in the next Problem.

7.4   VAPOR PRESSURE CALCULATION

Two popular equations that are used to estimate the vapor pressure of compounds are the Clapeyron and Antoine equations. The Clapeyron equation is given by

$$\ln p' = A - (B/T)$$

where p' and T are the vapor pressure and temperature, respectively, and A and B are the experimentally determined Clapeyron coefficients. The Antoine equation is given by

$$\ln p' = A - [(B/(T + C)]$$

where A, B and C are the experimentally determined Antoine coefficients.

Use the Clapeyron and Antoine equations to estimate the vapor pressure of acetone at 0°C

The Clapeyron coefficients have been experimentally determined to be

A = 15.03
B = 2817

for p and T in mm Hg and K, respectively. The Antoine coefficients are

A = 16.65
B = 2940
C = − 35.93

with p and T in the same units.

SOLUTIONS

First, calculate the vapor pressure, p', of acetone at 0°C using the Clapeyron equation.

$$\ln p' = 15.03 - [2817/(0 + 273)]$$
$$= 4.7113$$
$$p' = 111.2 \, \text{mm Hg}$$

Calculate the vapor pressure of acetone at $0°C$ using the Antoine equation.

$$ln\, p' = 16.65 - [2940/(273 - 35.93)]$$
$$= 4.2486$$
$$p' = 70.01\, mm\, Hg$$

The Clapeyron equation generally overpredicts the vapor pressure at or near ambient conditions. The Antoine equation is widely used in industry and usually provides excellent results. Also note that, contrary to statements appearing in the Federal Register and some US EPA publications, vapor pressure is not a funtion of pressure.

7.5 HEAT OF VAPORIZATION FROM VAPOR PRESSURE DATA

Calculate the enthalpy of vaporization ΔH of n-butane at $270°F$ using the data provided below.

T (°F)	P' (atm)	V_1 (ft^3/lb)	V_s (ft^3, lb)
260	24.66	0.0393	0.222
270	27.13	0.0408	0.192
280	29.79	0.0429	0.165

The experimental value for ΔH is approximately $67.0\, Btu/lb$.

SOLUTION

When a pure substance is vaporized from the liquid phase at constant pressure, there is also no change in temperature. However, there is a finite quantity of heat transferred to the substance. A characteristic feature of this process is the coexistence of two phases. There is a basic relationship between the latent enthalpy (heat) accompanying the change in phase and PVT (pressure, volume, temperature) data. This relationship is defined as the Clausius–Clapeyron (C–C) equation.

$$\Delta H = T \Delta V (dP'/dT)$$

where ΔV = specific volume change accompanying the phase change at temperature T, volume/mass; ΔH = latent enthalpy of vaporization or condensation, energy/ mass, and (dP'/dT) = rate of change of vapor pressure with temperature, pressure/degrees absolute. The term (dP'/dT) is the slope of the vapor pressure vs. temperature curve at the temperature in question.

Estimate the slope dP'/dT at $270°F$ in $atm/°R$.

$$dP'/dT \cong \Delta P'/\Delta T$$

"Straddling" temperature 260° and 280°F gives

$$
\begin{aligned}
\Delta P'/dT &= (29.79 - 24.66)/(280 - 260) \\
&= (29.79 - 24.66)/(740 - 720) \\
&= 0.257 \text{ atm}/°R
\end{aligned}
$$

Calculate V in ft^3/lb. At $T = 270°F$,

$$
\begin{aligned}
\Delta V &= 0.192 - 0.0408 \\
&= 0.151 \text{ ft}^3/\text{lb}
\end{aligned}
$$

Calculate the enthalpy of vaporization in Btu/lb from the Clausius–Clapeyron equation.

$$
\begin{aligned}
T &= 270 + 460 = 730°R \\
\Delta H &= (730)(0.151)(0.257)(14.7)(144)/778 \\
&= 77.1 \text{ Btu/lb}
\end{aligned}
$$

The agreement is good considered the crude approach employed in estimating the derivative.

7.6 CRITICAL AND REDUCED PROPERTIES

There are some engineering calculations that require that deviations from ideality be included in the analysis. Many of the non-ideal correlations involve the critical temperature T_c, the critical pressure P_c, and a term defined as the acentric factor ω. An abbreviated list of these properties is presented in the Table 7.1. The reduced temperature and reduced pressure, T_r and P_r, respectively, are defined as

$$
T_r = T/T_c
$$
$$
P_r = P/P_c
$$

Calculate the reduced properties for chlorine at 230°C and 2500 psia.

SOLUTION

Obtain T_c, P_c, and ω from Table 7.1 for chlorine.

$$T_c = 417\,K$$
$$P_c = 76\,atm$$
$$\omega = 0.074$$

Calculate T_r.

$$T = 230 + 273$$
$$= 503\,K$$
$$T_r = 503/417$$
$$= 1.21$$

Calculate P_r.

$$P_r = 2500\,psia\,(14.7)(76)$$
$$= 2.24$$

7.7   ESTIMATING ENTHALPY OF VAPORIZATION

Estimate the enthalpy of vaporization of water at its normal boiling point. Compare the calculated value with the experimental value of 970 Btu/lb.

SOLUTION

The key latent enthalpy change is that associated with vaporization/condensation processes. One normally obtains numerical values for the enthalpy of vaporization (or condensation) by looking them up in a reference or physical property text. However, if these data are not available, the Reidel equation [19] may be used to estimate the enthalpy of vaporization at the normal (1 atm) boiling point of the substance in question. This semi-empirical equation is given by

$$\Delta H_n/T_n = 2.17(\ln P_c - 1)/(0.930 - T_{rn})$$

where T_n = normal boiling point, K; ΔH = enthalpy of vaporization at the normal boiling point, cal/gmol; P_c = critical pressure, atm; and T_{rn} = reduced temperature at the normal boiling point, dimensionless.

Obtain the key critical properties of water from Table 7.1.

$$T_c = 647\,K$$
$$P_c = 217.6\,atm$$

TABLE 7.1   Critical Properties

Compound/Element	Critical Temperature (K)	Critical Pressure (atm)	Acentric Factor (ω)
Acetylene	308.3	60.6	0.814
Ammonia	405.6	111.3	0.250
Bensene	562.1	48.3	0.210
Bromine	584.0	102.0	0.132
i-Butane	408.1	36.0	0.176
n-Butane	425.2	37.5	0.193
1-Butene	419.6	39.7	0.187
Carbon dioxide	304.2	72.8	0.225
Carbon monoxide	132.9	34.5	0.041
Carbon tetrachloride	556.4	45.0	0.193
Chlorine	417.0	76.0	0.074
Chloroform	536.4	54.0	0.214
Chlorobenzene	632.4	44.6	0.255
Ethane	305.4	48.2	0.091
Ethanol	516.2	63.0	0.635
Ethylene	282.4	49.7	0.086
Ethylene oxide	469.0	71.0	0.157
n-Hexane	407.4	29.3	0.296
n-Heptane	540.2	27.0	0.351
Hydrogen	33.2	12.8	0.000
Hydrogen chloride	324.6	82.0	0.266
Hydrogen cyanide	456.8	53.2	0.399
Hydrogen sulfide	373.2	88.2	0.100
Methane	190.6	45.4	0.007
Methanol	512.6	79.9	0.556
Methyl chloride	416.2	65.9	0.158
Nitric oxide (NO)	180.0	64.0	0.600
Nitrogen	126.2	33.5	0.040
Nitrous oxide (N_2O)	309.6	71.5	0.160
n-Octane	568.8	24.5	0.394
Oxygen	154.6	49.8	0.021
i-Pentane	460.4	33.4	0.227
n-Pentane	469.6	33.3	0.251
1-Pentane	464.7	40.0	0.245
Propane	369.8	41.9	0.145
Propylene	365.0	45.6	0.148
Sulfur	1314.0	116.0	0.070
Sulfur dioxide	430.8	77.8	0.273
Sulfur trioxide	491.0	81.0	0.510
Toluene	591.7	40.6	0.257
Water	647.1	217.6	0.348

Sources: Kudchadker et al. [15]; Mathews [16]; Reed and Sherwood [17]; Passut and Damer [18].

Calculate the reduced temperature at the normal boiling point.

$$T_n = 100 + 273$$
$$= 373\,K$$
$$T_{rn} = 373/647$$
$$T_{rm} = 0.577$$

Calculate the enthalpy of vaporization at the normal boiling point in cal/gmol employing the Reidel equation:

$$\Delta H_n/T_n = (2.17)(\ln P_c - 1.)/(0.930 - T_{rn})$$
$$\Delta H_n/T_n = (2.17)(\ln 217.6 - 1.)/(0.930 - 0.577)$$
$$= 26.94$$
$$\Delta H_n = (26.94)(373)$$
$$= 10,049\,cal/gmol$$

Convert the enthalpy of vaporization to Btu/lb.

$$\Delta H_n = (10\,049)(454)/(252)(18)$$
$$= 1006\,Btu/lb$$

Compare the calculated value with the experimental value.

$$\%difference = (1006 - 970)/970$$
$$= 0.037$$
$$= 3.7\%$$

The correlation is excellent.

Enthalpy of vaporization data are usually required at conditions other than the normal boiling point. Watson's correlation [20] may be used to estimate the enthalpy of vaporization at a second condition, if a value is available at another condition, including the normal boiling point. This semi-empirical equation takes the form

$$\Delta H_2/\Delta H_1 = [(1 - T_{r2})/(1 - T_r)]^{0.38}$$

Rough estimates of latent heats of vaporization for pure liquids at their normal boiling points are given by *Trouton's rule:*

$$\frac{\Delta H_n}{RT_n} \sim 10$$

where T_n is the absolute temperature of the normal boiling point. The units of ΔH_n, R, and T_n must be chosen so that $\Delta H_n / R T_n$ is dimensionless. This rule provides a simple check on whether values calculated by other methods are reasonable. Representative experimental values for this ratio are Ar, 8.0 N_2 8.7; O_2, 9.1; HCL 10.4; C_6H_6, 10.5; H_2S, 10.6 and H_2O, 13.1 [8].

7.8 VISCOSITY

Calculate the viscosity of air at 50°C and 854 atm.

SOLUTION

Viscosity is a property associated with a fluid's resistance to flow. More precisely, this property accounts for energy losses that result from shear stresses that occur between different portions of the fluid which are moving at different velocities. As presented earlier, *absolute viscosity* (μ) has units of mass per length-time; the fundamental unit is the *poise*, which is defined as 1 g/cm-s. This unit is inconveniently large for many practical purposes and viscosities are frequently given in *centipoises* (0.01 poise) which is abbreviated cP. The viscosity of pure water at 68.6 °F is 1.00 cP. In English units, absolute viscosity is expressed either as pounds (mass) per feet second (lb/ft-s) or pounds per foot hour (lb/ft-h). The absolute viscosity depends primarily on temperature and to a lesser degree on pressure. As noted, *kinematic visocosity* (ν) is the absolute viscosity divided by the density of the fluid and is useful in certain fluid flow problems; the units for this quantity are length squared per time, e.g., square foot per second (ft^2/s) or square meters per hour (m^2/h). A kinematic viscosity of 1 cm^2/s is called a *stoke*. For pure water at 70°F, $\nu = 0.983$ cs (centistokes). Because fluid viscosity changes rapidly with temperature, a numerical value of viscosity has no significance unless the temperature is specified.

Liquid viscosity is usually measured experimentally by the amount of time it takes for a given volume of liquid to flow through an orifice. The *Saybolt Universal viscometer* is the most widely used device in the United States for the determination of the viscosity of fuel oils and liquid wastes. It should be stressed that Saybolt viscosities, which are expressed in *Saybolt seconds* (SSU), are not even approximately proportional to absolute viscosities except in the range above 200 SSU; hence, converting units from Saybolt seconds to other units requires the use of special conversion tables. As the time of flow decreases, the deviation becomes more marked. In any event, viscosity is an important property because of potential flow problems.

When experimental data are lacking and there is not time to obtain them, the viscosity can be estimated by semi-empirical methods, making use of other data on the given substance. The *corresponding-states correlation*, which facilitates such estimates and illustrates general trends of viscosity with temperature and pressure for ordinary fluids may be applied. The principle of corresponding states, as described in the previous Problem, is widely used for correlating equation-of-state and thermo-dynamic data.

Bird et al [14] provide the pressure and temperature dependence of viscosity where the reduced viscosity $\mu_r = \mu/\mu_c$ is plotted versus the reduced temperature $T_r = T/T_c$ for various values of the reduced pressure, $P_r = P/P_c$. The chart shows that the viscosity of a gas approaches a limit (the low-density limit) as the pressure becomes smaller for most gases; this limit is usually attained at 1 atm pressure. The viscosity of a gas at low density *increases* with increasing temperature, whereas the viscosity of a liquid *decreases* with increasing temperature [14].

If critical data are available, then μ_c may be estimated from either of these empirical relations:

$$\mu_c = 61.6[(MW)(T_c)]^{1/2}(V_c)^{-2/3}; \qquad \mu_c = 7.70(MW)^{1/2}P_c^{2/3}T_c^{-1/6}$$

Here μ_c is in micropoises, P_c in atm, T_c in K, and V_c in cm^3/gmole.

For a multicomponent mixture with n components with mole fraction x, pseudo-critical properties (bars) may be calculated using a procedure referred to as Kay's rule.

$$\bar{P}_c = \sum_{i=1}^{n} x_i P_{ci} \qquad \bar{T}_c = \sum_{i=1}^{n} x_i T_{ci} \qquad \bar{\mu}_c = \sum_{i=1}^{n} x_i \mu_{ci}$$

Thus, one uses the chart exactly as for pure fluids, but employing pseudocritical properties instead of the critical properties. This empirical procedure works reason-ably well for most engineering applications.

Since the problem requests the viscosity of air, and air consists of 79% (by mole) nitrogen and 21% oxygen, one may obtain a first estimate by assuming the air to be nitrogen. For this condition,

$$MW = 28$$
$$P_c = 33.5 \text{ atm}$$
$$T_c = 126.2 \text{ K}$$

Using the latter of the two critical viscosity equations leads to:

$$\mu_c = 7.70(MW)^{1/2}P_c^{2/3}T_c^{-1/6}$$
$$= (7.70)(28)^{1/2}(33.5)^{2/3}(126.2)^{-1/6}$$
$$= 189.1 \text{ micropoise} = 189 \times 10^{-6} \text{ poise (P)}$$

The reduced temperatures and pressures are

$$T_r = \frac{273 + 50}{126.2} = 2.559$$

$$P_r = \frac{854}{33.5} = 25.49$$

From Figure 1.3-1 in Bird et al. [14]

$$\mu_r = \mu/\mu_c$$
$$= 2.5$$

Therefore

$$\mu = (189.1 \times 10^{-6})(2.5)$$
$$= 472.8 \ P$$

This is in reasonable agreement with both the literature value for air.

7.9 THERMAL CONDUCTIVITY

Define thermal conductivity.

SOLUTION

Experience has shown that when a temperature difference exists across a solid body, energy in the form of heat will transfer from the high temperature region to the low temperature region until thermal equilibrium (same temperature) is reached. This mode of heat transfer in which vibrating molecules pass along kinetic energy through the solid is called *conduction*. Liquids and gases may also transport heat in this fashion. The property of *thermal conductivity* provides a measure of how fast (or how easily) heat flows through a substance. It is defined as the amount of heat that flows in unit time (q) through a unit surface area (A) of unit thickness (χ) as a result of a unit difference in temperature (ΔT). The describing equation is

$$k = \frac{(q/A)\chi}{\Delta T}$$

in consistent units. Typical units for conductivity are Btu-ft/h-ft²-°F or Btu/h-ft-°F. Thermal conductivities for some liquids, gases, and solids are given in Table 7.2.

TABLE 7.2 Thermal Conductivities of Gases, Liquids, and Solids

Gases and Vapors (1 atm)	Thermal Conductivity $(k)^a$	
	32°F	212°F
Acetone	0.0057	0.0099
Acetylene	0.0108	0.0172
Air	0.0140	0.0184
Ammonia	0.0126	0.0192
Benzene	0.0052	0.0103
Carbon dioxide	0.0084	0.0128
Carbon monoxide	0.0134	0.0176
Carbon tetrachloride		0.0052
Chlorine	0.0043	
Ethane	0.0106	0.0175
Ethyl alcohol		0.0124
Ethyl ether	0.0077	0.0131
Ethylene	0.0101	0.0161
Helium	0.0818	0.0988
Hydrogen	0.0966	0.1240
Methane	0.0176	0.0255
Methyl alcohol	0.0083	0.0128
Nitrogen	0.0139	0.0181
Nitrous oxide	0.0088	0.0138
Oxygen	0.0142	0.0188
Propane	0.0087	0.0151
Sulfur dioxide	0.0050	0.0069
Water vapor		0.0136
Liquids	Temperature (°F)	k^a
Acetic acid	68	0.909
Acetone	86	0.102
Ammonia	5–86	0.29
Aniline	32–68	0.100
Benzene	86	0.092
n-Butyl alcohol	86	0.097
Carbon bisulfide	86	0.093
Carbon tetrachloride	32	0.107
Chlorobenzene	50	0.083
Ethyl acetate	68	0.101
Ethyl alcohol	68	0.105
Ethyl ether	86	0.080
Ethyl glycol	32	0.153
Gasoline	86	0.078
Glycerine	68	0.164
n-Heptane	86	0.081
Kerosene	68	0.086

(*Continued*)

TABLE 7.2   *Continued*

Gases and Vapors (1 atm)	Thermal Conductivity $(k)^a$	
	32°F	212°F
Methyl alcohol	68	0.124
Nitrobenzene	86	0.095
n-Octane	86	0.083
Sulfur dioxide	5	0.128
Sulfuric oxide	86	0.21
Toluene	86	0.086
Trichloroethylene	122	0.080
o-Xylene	68	0.090
Solids	k^a	
Asbestos-cement boards	0.43	
Sheets	0.096	
Asbestos cement	0.202	
Celotex	0.028	
Corkboard, 10 lb/ft³	0.025	
Diatomaceous earth (Sil-o-cel)	0.035	
Fiber, insulting board	0.028	
Glass wool 1.5 lb/ft³	0.022	
Magnesia, 85%	0.039	
Rock wool 10 lb/ft³	0.023	

$^a k = $ Btu/-ft-°F.

7.10   THERMAL CONDUCTIVITY APPLICATION

What is the heat flux (Btu/h-ft^2) through a copper plate 0.15 in. thick if one side is at 100°C and the other side is at 250°C? The thermal conductivity of copper is 214 Btu/h-ft-°F.

SOLUTION

From the definition of thermal conductivity (see Problem 7.9 for additional details),

$$\frac{q}{A} = \frac{k\Delta T}{x} = (214\,\text{Btu/h-ft-}°\text{F})[(250 - 100)(1.8)]°\text{F}/(0.15/12)\text{ft}$$
$$= 4.62 \times 10^6 \text{Btu/h-ft}^2$$

where q/A = heat flux and, x = thickness.

Note: The direction of heat transfer is from the high temperature region to the low temperature region.

When thermal conductivity data for a particular compound are not available one can obtain an estimate by using the corresponding-states chart provided in Bird et al [14], which is based on thermal conductivity data for several nonatomic substances. This chart, which is similar to that for viscosity provided in Problem 7.8, is a plot of the reduced thermal conductivity k_r, or k/k_c, which is the thermal conductivity at pressure P and temperature T divided by the thermal conductivity at the critical point. This quantity is plotted as a function of the reduced temperature $T_r = T/T_c$ and the reduced pressure $P_r = P/P_c$.

7.11 NOKAY EQUATION AND LYDERSEN'S METHOD

A new compound with a normal boiling point and specific gravity of approximately 500 K and 0.9, respectively, has been developed by Nanotheodore International using nanotechnology procedures. The structure of this compound is:

1. Calculate the critical temperature of the new compound using the Nokay equation.
2. Compute the compound's key critical properties(temperature, pressure, volume, and compressibility factor) using Lydersen's method of group contributions.

SOLUTION [21]

The Nokay equation is given by

$$\log T_c = A + B \log(SG) + C \log T_b$$

where T_c is the critical temperature in K, T_b in the normal boiling point in K, and SG is the specific gravity. The correlation coefficients A, B, and C are given in the Table 7.3 below [13].

Employ A, B, and C for an aromatic compound,

$$A = 1.057019$$
$$B = 0.227320$$
$$C = 0.669286$$

TABLE 7.3   Constants for Nokay's Equation

Family of Compounds	A	B	C
Alkanes (paraffins)	1.359397	0.436843	0.562244
Cycloalkanes (naphthenes)	0.658122	−0.071646	0.811961
Alkenes (olefins)	1.095340	0.277495	0.655628
Alkynes (acetylenes)	0.746733	0.303809	0.799872
Alkadienes (diolefins)	0.147578	−0.396178	0.994809
Aromatics	1.057019	0.227320	0.669286

Substituting in Nokay's equation gives

$$\log T_c = 1.057019 + 0.227320 \log(0.9) + 0.669286 \log(500)$$
$$= 1.057019 - 0.0104016 + 1.806383$$
$$= 2.853$$

and

$$T_c = 713\,\text{K} = 1283\,\text{R}$$

The Lydersen formulas are [13, 22]:

$$T_c = T_b\{(0.567) + \sum(N)(\Delta T) - [\sum(N)(\Delta T)]^2\}^{-1}$$
$$P_c = MW[0.34 + (N)(\Delta P)]^{-2}$$
$$V_c = [40 + (N)(\Delta V)]$$
$$Z_c = P_c V_c / RT_c$$

where T_c, P_c, V_c, and Z_c are the critical temperature, critical pressure, critical volume, and critical compressibility factor, respectively. The group contributions are provided in Table 7.4.

Employing the Lydersen approach, first calculate the molecular weight (MW)

$$MW = 3(N) + 6(C) + 3(O) + 7(H)$$
$$= 3(14) + 6(12) + 3(16) + 7(1)$$
$$= 169$$

TABLE 7.4 **Lydersen's Structural Group Contributions**

Symbols	ΔT	ΔP	ΔV
Nonring increments			
—CH^3	0.020	0.227	55
—CH$_2$	0.020	0.227	55
—CH	0.012	0.210	51
—C—	0.00	0.210	41
=CH$_2$	0.018	0.198	45
=CH	0.018	0.198	45
=C—	0.0	0.198	36
=C=	0.0	0.198	36
≡CH	0.005	0.153	(36)
≡C—	0.005	0.153	(36)
Ring increments			
—CH$_2$—	0.013	0.184	44.5
—CH	0.012	0.192	46
—C—	(−0.007)	(0.154)	(31)
=CH	0.011	0.154	37
=C—	0.011	0.154	36
=C=	0.011	0.154	36
Halogen increments			
—F	0.018	0.221	18
—Cl	0.017	0.320	49
—Br	0.010	(0.50)	(70)
—I	0.012	(0.83)	(95)
Oxygen increments			
—OH (alcohols)	0.082	0.06	(18)
—OH (phenols)	0.031	(−0.02)	(3)
—O— (nonring)	0.021	0.16	20
—O— (ring)	(0.014)	(0.12)	(8)
—C=O (nonring)	0.040	0.29	60

(Continued)

TABLE 7.4 *Continued*

Symbols	ΔT	ΔP	ΔV
$\overset{\mid}{\underset{\mid}{-}}C=$O (ring)	(0.033)	(0.2)	(50)
HC $=$O (aldehyde)	0.048	0.33	73
—COOH (acid)	0.085	(0.4)	80
—COO— (ester)	0.047	0.47	80
$=$(except for combinations above)	(0.02)	(0.12)	(11)
Nitrogen increments			
—NH$_2$	0.031	0.095	28
$\overset{\mid}{-}$NH (nonring)	0.031	0.135	(37)
$\overset{\mid}{-}$NH (ring)	(0.024)	(0.09)	(27)
$\overset{\mid}{-}$N— (nonring)	0.014	0.17	(42)
$\overset{\mid}{-}$N— (ring)	(0.007)	(0.13)	(32)
—CN	(0.060)	(0.36)	(80)
—NO^2	(0.055)	(0.42)	(78)
Sulfur increments			
—SH	0.015	0.27	55
—S— (nonring)	0.015	0.27	55
—S— (ring)	(0.008)	(0.24)	(45)
$=$S	(0.003)	(0.24)	(47)
Miscellaneous			
$\overset{\mid}{\underset{\mid}{-}}$Si—	0.03	(0.54)	
$\overset{\mid}{\underset{\mid}{-}}$B—	(0.03)		

Source: A. L. Lydersen, U. of Wisconsin Eng. Exp. Station, 1995 [23].
Note: There are no increments for hydrogen.

The contributions from each group are shown in Table 7.5. The critical properties may now be calculated:

$$T_c = T_b\{(0.567) + \sum (N)(\Delta T) - [\sum (N)(\Delta T)]^2\}^{-1}$$
$$= 500\{(0.567) + (0.214) - (0.214)^2\}^{-1}$$
$$= 500/0.7352$$
$$= 680.1\,\text{K}$$

TABLE 7.5 Contributions From Each Group

Group Type	N	ΔT	ΔP	ΔV	$(N)(\Delta T)$	$(N)(\Delta P)$	$(N)(\Delta V)$
NH^2 (nonring)	2	0.031	0.095	28	0.062	0.19	56
$-\overset{\displaystyle |}{\underset{\displaystyle |}{C}}=$ (ring)	4	0.011	0.154	36	0.044	0.616	144
$H-\overset{\displaystyle |}{C}=$ (ring)	2	0.011	0.154	37	0.022	0.308	74
OH (nonring)	1	0.031	(−0.02)	3	0.031	−0.02	3
NO_2 (nonring)	1	0.055	0.42	78	0.055	0.42	78
Total					0.214	1.514	355

$$P_c = MW[0.34 + (N)(\Delta P)]^{-2}$$
$$= 169.15[0.34 + (1.514)]^{-2}$$
$$= 49.21 \text{ atm}$$
$$V_c = [40 + (N)(\Delta V)]$$
$$= 40 + 355$$
$$= 395 \text{ cm}^3/(\text{gmol})$$
$$Z_c = P_c V_c / RT_c$$
$$= (49.21)(395)/(82.06)(680.1)$$
$$= 0.348$$

7.12 THE RIHANI AND DORAISWAMY PROCEDURE, AND THE LEE–KESLER EQUATION

Refer to Problem 7.11 and answer the following two questions.

1. Outline a procedure to estimate the heat capacity of the amine using the procedure recommended by Rihani and Doraiswamy.
2. If the accentric factor is approximated as the critical compressibility factor, estimate the vapor pressure at 550 K using the Lee–Kesler equation.

SOLUTION

1. As discussed earlier, the heat capacity of a substance is defined as the quality of heat required to raise the temperature of the substance by 1°; the specific heat

capacity is the heat capacity on a unit mass basis [21]. The term specific heat is frequently used in place of specific heat capacity. This is not strictly correct, because specific heat has traditionally been defined as the ratio of the heat capacity of a substance to the heat capacity of water. However, since the specific heat of water is approximately 1 cal/g-°C or 1 Btu/lb-°F, the term *specific heat* has come to imply heat capacity per unit mass. For gases, the addition of heat to cause the 1° temperature rise may be accomplished either at constant pressure or at constant volume. Since the amounts of heat necessary are different for the two cases, subscripts are used to identify which heat capacity is being used – C_P, for constant pressure, and C_V, for constant volume. For liquids and solids, this distinction does not have to be made since there is little difference between the two. Values of specific heats are available in the literature [6, 8].

Heat capacities are functions of both the temperature and pressure, although the effect of pressure is generally small and is neglected in almost all engineering calculations. The effect of temperature on C_p can be described by a host of equations, including the following four.

$$C_p = \alpha + \beta T + \gamma T^2$$

$$C_p = \alpha + \beta T + \gamma T^2 + \varepsilon T^3$$

$$C_p = a + bT + cT^{-2}$$

$$C_p = a + bT + cT^2 + dT^{-2}$$

The Rihani–Doraiswamy [23] method is based on the equation

$$C_p^0 = \sum_i n_i a_i + \sum_i n_i b_i T + \sum_i n_i c_i T^2 + \sum_i n_i d_i T^3$$

where n_i is the number of groups of type i, T is the temperature in K. The terms a_i, b_i c_i, and d_i are the additive group parameters which are available in the literature. [13, 23]. The procedure to follow is to sum up the group contributions for each compound in a manner similar to that provided by Lydersen in Problem 7.11. The $\sum_i n_i a_i$, $\sum_i n_i b_i$, and so on, are then substituted into the above equation.

2. The Lee–Kesler equation [24] is

$$(\ln P_r^*) = (\ln P_r^*)^{(0)} + \omega (\ln P_r^*)^{(1)}$$

where $P_r^* = P^*/P_c$, the reduced vapor pressure, $P^* =$ vapor pressure at T_r, $P_c =$ critical pressure, $\omega =$ accentric factor, $T_r = T/T_c$, the reduced temperature. The terms $\ln P_r^{*0}$ and $\ln P_r^{*1}$ are correlation functions given in Table 7.6.

TABLE 7.6   Correlation Terms for the Lee–Kesler Equation [25]

T_r	$-\ln(P_r^*)^{(0)}$	$-\ln(P_r^*)^{(1)}$	T_r	$-\ln P_r^{*0}$	$-\ln P_r^{*1}$
1.00	0.000	0.000	0.64	3.012	3.218
0.98	0.118	0.098	0.62	3.280	3.586
0.96	0.238	0.198	0.60	3.568	3.992
0.94	0.362.	0.303	0.58	3.876	4.440
0.92	0.489	0.412	0.56	4.207	4.937
0.90	0.621	0.528	0.54	4.564	5.487
0.88	0.757	0.650	0.52	4.951	6.098
0.86	0.899	0.781	0.50	5.370	6.778
0.84	1.046	0.922	0.48	5.826	7.537
0.82	1.200	1.073	0.46	6.324	8.386
0.80	1.362	1.237	0.44	6.869	9.338
0.78	1.531	1.415	0.42	7.470	10.410
0.76	1.708	1.608	0.40	8.133	11.621
0.74	1.896	1.819	0.38	8.869	12.995
0.72	2.093	2.050	0.36	9.691	14.560
0.70	2.303	2.303	0.34	10.613	16.354
0.68	2.525	2.579	0.32	11.656	18.421
0.66	2.761	2.883	0.30	12.843	20.820

Employ an average of the two previously determined critical temperatures.

$$T_c = (680.1 + 713)/2 = 696.5 \text{ K}$$

Thus,

$$T_c = 696.5 \text{ K}$$
$$P_c = 49.21 \text{ atm}$$
$$Z_c = 0.348$$
$$\omega = Z_c = 0.348$$

The reduced temperature can now be calculated

$$T_r = T/T_c = 550/696.5$$
$$= 0.7897$$

Linearly interpolating from Table 7.6,

$$-\ln P_r^{*0} = 1.445$$
$$-\ln P_r^{*1} = 1.329$$

Using the Lee–Kesler equation,

$$\ln P_r^* = \ln P_r^{*0} + \omega \ln P_r^{*1}$$
$$= -1.449 + 0.348(-1.329)$$
$$= -1.9115$$

so that

$$P_r^* = 0.1479$$

Finally,

$$P^* = (0.1479)(49.21)$$
$$= 7.278 \text{ atm @ } 550 \text{ K}$$

References: Part 1

1. L. Theodore and R. Kunz, *Nanotechnology: Environmental Implication and Solution*, Hoboken NJ: John Wiley and Sons, 2005.

2. J. Reynolds, J. Jeris and L. Theodore, *Chemical and Environmental Engineering Calculation Handbook*, Hoboken NJ: John Wiley & Sons, 2002.

3. J. Spero, B. Devito and L. Theodore, *Regulatory Chemicals Handbook*, New York City: Marcel Dekker, 2000

4. J. Santoleri, J. Reynolds, and L. Theodore, *Introduction to Hazardous Waste Incineration*, Hoboken NJ: John Wiley & Sons, 2000.

5. Personal Communication, R. D'Aquino, 2005.

6. R. Perry and D. Green (editors), *Perry's Chemical Engineers Handbook*, 7th Edition, New York City: McGraw-Hill, 1997.

7. R. Dupont, L. Theodore, and K. Ganesan, *Pollution Prevention*, Boca Raton FL: CRC–Lewis Publishers, 2000.

8. J. Smith, H. Van Ness, and M. Abbott, *Introduction to Chemical Engineering Thermodynamic*, 7th Edition, New York City: McGraw-Hill, 2005.

9. S. Shelley and G. Ondrey, Nanotechnology – The Sky's the Limit, *Chemical Engineering*, pp. 23–27, New York City, (December 2002).

10. J. Reynolds, *Materials Science and Engineering*, A Theodore Tutorial, East Williston NY: Theodore Tutorials, 1994.

11. W. Callister, *Materials Science and Engineering An Introduction*, 3rd Edition, Hoboken NJ: John Wiley & Sons, 1994.

12. R. Eastman, *General Chemistry: Experiment and Theory*, New York City: Holt, Rinehart and Winston, 1970.

13. N. Chopey, *Handbook of Chemical Engineering Calculations*, 2nd Edition, New York City: McGraw-Hill, 1994.

14. R. Bird, W. Stewart, and E. Lightfoot, *Transport Phenomena*, 2nd Edition, Hoboken NJ: John Wiley & Sons, 2002.

15. R. Ried, J. Prausnitz, and T. Sherwood, *Properties of Gasses and Liquids*, New York City: McGraw-Hill, 1997.

16. J. F. Mathews, *Chem. Rev.*, 72, 71 (1972).

17. R. C. Reed and T. K. Sherwood, *The Properties of Liquids*, 2nd Edition, New York: McGraw-Hill, 1986.

18. C. A. Passut and R. P. Danner, *Ind. Eng. Chem. Process. Res. Develop.*, 12, 365 (1974).

19. L. Reidel, *Chem. Eng. Tech.*, 26, 679 (1954).

20. K. Watson, *Ind. Eng. Chem.*, 35, 3985 (1943).

21. P. Bogdan; takehome exam solution submitted to L. Theodore, 2004.

22. A. Lyderson, University of Wisconsin Engineering Experimental System, 1955.

23. D. Rihani and L. Doraisawamy, *Ind. Eng. Chem. Fund.*, 4, 17 (1965).

24. R. Lee and R. Kessler, *AIChE Journal*, 21, 510 (1975).

PART 2
Particle Technology

It is relatively safe to say that there are/will be two classifications of nanoemissions from nanoprocesses: particulates and gases. Additional details cannot be provided at this time since many of these new processes, and their generation and corresponding emissions, have yet to be formulated. However, the area of major concern will be particulates – thus the inclusion of this Part on Particle Technology. Nonetheless, it seems reasonable to conclude that many of these emissions will be similar in classification to what presently exists. The classification and sources of particulates of interest have to be related to "traditional" materials. At the time of this writing, little could be said about those particulates generated from nanoapplications [1].

Particulates in the traditional sense include dust, smoke, metals, and aerosols. Major sources include steel mills, power plants, cotton gins, cement plants, smelters, and diesel engines. Other sources are grain storage elevators, industrial haul roads, construction work, and demolition. Wood-burning stoves and fireplaces can also be significant sources of particulates. Urban areas are likely to have wind-blown dust from roads, parking lots, and construction work [1].

Of primary concern on evaluating "unknown" situations are atmospheric and ambient health and safety hazards, a topic that is treated in Part 4. Concentrations (or potential concentrations) of particulates, and any potential time variation of these variables, all present immediate concerns. Initial resolutions will unquestionably be based on experience, judgment, and professional knowledge.

Extending standard procedures and the design of equipment to recover and control emissions from nanotechnology processes are also subject to question at this time. The chemical and physical properties of these nanoemissions are not fully known, thus rendering judgments regarding equipment somewhat mute. However, information is critical to the advancement and development of new recovery control technologies. This is especially true of ultrafine particles, that is, at the nanosize, since their mass, volume, and size is extremely small.

Particulates from both manmade and natural sources do not consist in any one size. Particulates discharged from an operation consist of a size distribution ranging anywhere from extremely small particles (less than 1 μm) to very large particles (greater than 100 μm). Data on particle size and particle concentration

Nanotechnology: Basic Calculations for Engineers and Scientists, by Louis Theodore
Copyright © 2006 John Wiley & Sons, Inc.

for some typical industrial aerosols are available in the literature [1]. This subject receives significant treatment later in this Part.

The five major types of particulate recovery/control equipment are:

1. Gravity settlers
2. Cyclones
3. Electrostatic precipitators
4. Venturi scrubbers
5. Baghouses

The performance of these devices receives attention in the last Chapter in this Part – particularly with respect to collection efficiency. The analysis in this last Chapter clearly and interestingly demonstrates that nanometer particles are easier to control than micrometer-plus larger counterparts.

The second Part of the text addresses the following 6 subject areas in Particle Technology:

1. Nature of particulates
2. Particle size distribution
3. Particle sizing and measurement methods
4. Fluid–particle dynamics
5. Particle collection mechanisms
6. Particle collection efficiency

8 Nature of Particulates

Theodore et al [2] have defined particulates as a small, discrete mass of solid or liquid matter; a fine liquid or, solid particle that is found in the air or emissions; any solid or liquid matter that is dispersed in a gas; or, insoluble solid matter dispersed in a liquid so as to produce a heterogeneous mixture. Further, particulate matter 10 (PM-10) is defined as particulate matter with a diameter less than or equal to 10 micrometers (μm) while particulate matter 2.5 (PM-2.5) is particulate matter with a diameter less than or equal to 2.5 micrometers.

Particulates may be classified into two broad categories: (1) natural and (2) human-made. Natural sources include (but are not limited to):

1. Windblown dust;
2. Volcanic ash and gases;
3. Smoke and fly ash from forest fires;
4. Pollens and other aeroallergens.

These sources contribute to background values over which control activities can have little, if any, effect.

Human-made sources cover a wide spectrum of chemical and physical activities and are contributors to urban air pollution. Particulates in the United States pour out from over 200 million vehicles, from the refuse of nearly 300 million people, the generation of billions of kilowatts of electricity, and the production of innumerable products demanded by everyday living.

Particulates may also be classified by their origin.

1. Primary – emitted directly from a process;
2. Secondary – subsequently formed as a result of a chemical reaction

8.1 DEFINITION OF PARTICULATES

Qualitatively define particulates.

Nanotechnology: Basic Calculations for Engineers and Scientists, by Louis Theodore
Copyright © 2006 John Wiley & Sons, Inc.

SOLUTION

Particulates may be defined as solid or liquid matter whose effective diameter is larger than a molecule but smaller than approximately 1000 μm. Particulates dispersed in a gaseous medium are collectively termed an aerosol. The terms smoke, fog, haze, and dust are commonly used to describe particular types of aerosols, depending on the size, shape, and characteristic behavior of the dispersed particles. Aerosols are rather difficult to classify on a scientific basis in terms of their fundamental properties such as settling rate under the influence of external forces, optical activity, ability to absorb electric charge, particle size and structure, surface-to-volume ratio, reaction activity, physiological action, and so on. In general, particle size and settling rate have been the most characteristic properties emphasized by engineers and scientists. For example, particles larger than 100 μm may be excluded from the category of dispersions because they settle too rapidly. On the other hand, particles on the order of 1 μm or less settle so slowly that, for all practical purposes, they are regarded as permanent suspensions. Despite possible advantages of scientific classifications schemes, the use of popular descriptive terms such as smoke, dust, and mist, which are traditionally and essentially based on the mode of formation, appears to be a satisfactory and convenient method of classification. In addition, this approach is so well established and understood that it undoubtedly would be difficult to change.

8.2   DUST, SMOKE, AND FUMES

Provide traditional definitions of dust, smoke, and fumes.

SOLUTION

Dust is typically formed by the pulverization or mechanical disintegration of solid matter into particles of smaller size by processes such as grinding, crushing, and drilling. Particle sizes of dusts range from a lower limit of about 1 μm up to about 100 or 200 μm and larger. Dust particles are usually irregular in shape, and particle size (to be discussed later) refers to some average dimension for any given particle. Common examples include fly ash, rock dusts, and ordinary flour.

Smoke implies a certain degree of optical density and is typically derived from the burning of organic materials such as wood, coal, and tobacco. Smoke particles are very fine, ranging in size from less than 0.01 μm (10 nm) up to 1 μm (1000 nm). They are usually spherical in shape if of liquid or tarry composition and irregular in shape if of solid composition. Owing to their very fine particle size, smokes can remain in suspension for long periods of time.

Fumes are typically formed by processes such as sublimation, condensation or combustion – generally at relatively high temperatures. They range in particle size from less than 0.1 μm to 1 μm. Similar to smokes, they settle very slowly.

8.3 MIST AND DRIZZLE

Briefly discuss mist and drizzle.

SOLUTION

Mists (or *fogs*) are typically formed either by the condensation of water or other vapors on suitable nuclei giving a suspension of small liquid droplets, or by the atomization of liquids. Particle sizes of natural mists and fogs usually lie between 2 and 200 μm and may also be classified as *drizzle* or *rain.*

8.4 CHANGING PROPERTIES

Explain why and how particulate properties can change when emitted to the atmosphere.

SOLUTION

When a liquid or solid substance is emitted to the air as particulate matter, its properties and effects may be changed. As a substance is broken up into smaller and smaller particles, more of its surface area is exposed to the air. Under these circumstances, the substance – whatever its chemical composition – tends to physically or chemically combine with other particulates or gases in the atmosphere. The resulting combinations are frequently unpredictable. Very small aerosol particles (from 0.001 to 0.1 μm, or 1 to 100 nm) can act as condensation nuclei to facilitate the condensation of water vapor, thus promoting the formation of fog and ground mist. Particles less than 2 or 3 μm in size – about half (by weight) of the particles suspended in urban air – can penetrate into the mucous and attract and convey harmful chemicals such as sulfur dioxide. By virtue of the increased surface area of the small aerosol particles, and as a result of the adsorption of gas molecules or other such properties that are able to facilitate chemical reactions, aerosols tend to exhibit greatly enhanced surface activity, a phenomenon well documented in the nanotechnology literature.

8.5 DUST EXPLOSIONS

Qualitatively and quantitatively describe dust explosions.

SOLUTION

There are seven key factors that are required for a dust explosion to take place [5].

1. Air (oxygen)
2. Fuel source (dust)

TABLE 8.1 Influence of the Type of Ignition Source and of the Ignition Energy on the Explosion Data of Combustible Dusts (1-m³ explosion chamber)

Type of Dust	Ignition Source	Ignition Energy	P_{max} (bar)	K_{st} (bar-m/s)
Lycopodium	D	10 000	8.2	186
	C	0.08	8.3	199
	S	10	8.4	153
Cellulose	D	10 000	9.7	150
	C	0.04	9.2	147
	S	10	8.2	63
2-Naphthol	D	10 000	8	100
	C	0.005	7.7	90
	S	10	7.9	90
2-Nitro-4-propionyl-aminoanisol	D	10 000	8.3	84
	C	16	8	95
	S	10	7.3	52
	D	10 000	6.4	47
Dibutyltin dioxide	C	8	6.6	55
	S	10	0	0

D = chemecal detonator, C = condenser discharge, S = permanent spark gap.

3. Mixture
4. Dry state
5. Minimum concentration
6. Ignition source (see Table 8.1)
7. Enclosure

All seven factors need to be present for an explosion to occur; thus, eliminating only one prevents the explosion.

An *explosion* is defined as an occurrence where energy is released over a sufficiently short period of time and in an enclosed (usually) volume to generate a pressure wave of finite amplitude traveling away from the source of the explosion. This energy may have been originally stored in the system as chemical, nuclear, electrical, or pressure energy. However, the release is not considered to be explosive unless it is rapid and concentrated enough to produce a pressure wave that can be heard.

An *explosion pressure*, P_{ex}, is the pressure in excess of the initial pressure at which the explosive mixture is ignited. The rate of pressure rise is represented by dP/dt, a pressure change with respect to time. This is a measure of the speed of the flame propagation, hence of the violence of the explosion. Typical values of maximum explosion pressures in a closed vessel range from 7 to 8 bar. The rate of pressure rise can vary considerably with the flammable gas. The influence of vessel volume on the maximum rate of pressure rise for a given flammable gas is characterized by the cubic law [3, 4]

$$(dP/dt)_{max} V^{1/3} = K_G = constant$$

where V = vessel volume (m^3), K_G = constant (bar-m/s), and $(dP/dt)_{max}$ = maximum pressure rise (bar/s). This latter term is given by

$$(dP/dt)_{max} V^{1/3} = K_{St} \qquad \text{for } V > 0.04 \, m^3$$

where K_{St} = constant (bar-m/s).

Table 8.1 provides some K_{St} values for fine dust [6].

As described above, many substances that oxidize slowly in their massive state will oxidize extremely fast or possibly even explode when dispersed as fine particles in air. Dust explosions are often caused by the unstable burning or oxidation of combustible particles, brought about by their relatively large specific surfaces.

8.6 ADSORPTION AND CATALYTIC ACTIVITY IN THE ATMOSPHERE

Discuss adsorption and catalytic activity relative to particulates.

SOLUTION

Adsorption and catalytic phenomena can be extremely important in analyzing and understanding particulate pollution problems. The conversion of sulfur dioxide to corrosive sulfuric acid assisted by the catalytic action of iron oxide particles, for example, demonstrates the catalytic nature of certain types of particles in the atmosphere. Finally, aerosols can absorb radiant energy and rapidly conduct heat to the surrounding gases of the atmosphere. These are gases that ordinarily would be incapable of absorbing radiant energy by themselves. As a result, the air in contact with the aerosols can become much warmer [7].

8.7 PARTICLE SIZE

Discuss particle size.

SOLUTION

Particle size is the single most important characteristic that affects the behavior of a particle. The range in sizes of particles observed in practice is remarkable. Some of the droplets collected in the demisters of wet scrubbers and the solid particles collected in large diameter cyclones are as large as raindrops. However, some if the particles created in high-temperature incinerators and metallurgical processes can consist of a few molecules clustered together. These particles cannot be seen by sensitive light microscopes because they are extremely small in size. In fact, particles composed of a few molecules clustered together can exist in a stable form. Some industrial processes such as combustion and metallurgical sources generate particles in the range of 10 to 100 nanometers. These sizes are approaching the size of individual gas molecules, which are in the range of 0.2 to 1.0 nanometers. However, particles in this size range can grow and agglomerate to yield particles in the greater than 100 nanometer range.

The general subject of particle size will be revisited in Chapter 10, "Particle Sizing and Measurement Methods".

8.8 PARTICLE VOLUME AND SURFACE AREA

A spherical particle has a diameter of 100 nanometers (nm). Calculate the volume (cm^3) and surface area (cm^2) of the particle.

SOLUTION

The reader is referred to Chapter 5 for additional details.

The volume (V) of the particle is

$$V = \pi D^3 / 6$$

$$= 0.524(100)^3$$

$$= 0.524 \times 10^6 \ nm^3$$

Since there are 10^7 nm/cm,

$$V = (0.524)(10^6)(10^{-7})^3 = 0.524 \times 10^{-15} \ cm^3$$

The surface area (A) is given by

$$A = \pi D^2$$

$$= 3.14 \times (100)^2$$

$$= 3.14 \times 10^4 \ nm^2$$

Converting leads to

$$A = (3.14)(10^4)(10^{-7})^2$$
$$= 3.14 \times 10^{-10} cm^2$$

Table 8.2 provides similar information for spherical particle size in the 1.0 nm to 1000 μm range.

8.9 VOLUME/SURFACE AREA RATIOS

Compare a 10 nm spherical particle to one with a diameter of 100 μm. Include the ratio of volumes and surface area in the calculation.

SOLUTION

Refer to Chapter 5 for additional details.
The ratio of diameters (RD) is

$$RD = \frac{100 \, \mu m}{10 \, nm} = \frac{(100)(1000)}{10}$$
$$= 10\,000 = 10^4$$

Since the volume and surface area are a function of the $(diameter)^3$ and $(diameter)^2$, respectively, the volume ratio (VR) and surface area ratio (SAR) are:

$$VR = (10^4)^3 = 10^{12}$$

TABLE 8.2 Spherical Particle Sizes, Volumes, and Surface Areas

Particle Size, nm (μm)	Particle Volume, cm^3	Particle Super Area, cm
1.0 (0.001)	5.24×10^{-22}	3.14×10^{-14}
10 (0.01)	5.24×10^{-19}	3.14×10^{-12}
100 (0.1)	5.24×10^{-16}	3.14×10^{-10}
1000 (1)	5.24×10^{-13}	3.14×10^{-8}
10^4 (10)	5.24×10^{-10}	3.14×10^{-6}
10^5 (100)	5.24×10^{-7}	3.14×10^{-4}
10^6 (1000)	5.24×10^{-4}	3.14×10^{-2}

and

$$SAR = (10^4)^2 = 10^8$$

respectively.

Thus 1 000 000 000 000 particles with a 10 nm diameter have the same mass as 1.0 (one) 100 μm particle.

The reader should note that the above two results could have been obtained directly from Table 8.2 in Problem 8.8. Referring to the previous problem,

$$VR = 5.24 \times 10^{-7}/5.24 \times 10^{-19} = 10^{12}$$

and

$$SAR = 3.14 \times 10^{-4}/3.14 \times 10^{-12} = 10^8$$

Also note that particle size itself is difficult to define in terms that accurately represent the types of particles of interest. This difficulty stems from the fact that particles exist in a wide variety of shapes, not just spheres as treated in this Problem. In the case of spherical particles, the definition of particle size is easy: it is simply the diameter. For the irregularly shaped particles, there are a variety of ways to define the size. This topic receives treatment in Chapter 10.

8.10   PARTICLE FORMATION

Provide a brief overview of particle formation.

SOLUTION

The range of sizes of particles formed in a process is largely dependent on the types of the particle formation mechanisms involved. It is possible to estimate the general size range simply by recognizing which of these mechanisms are important in the process being evaluated. The most important particle formation mechanisms include the following:

1. Physical attrition/mechanical dispersion;
2. Combustion particle burnout;
3. Homogeneous nucleation;
4. Heterogeneous nucleation;
5. Droplet evaporation.

For example, physical attrition occurs when two surfaces rub together; the grinding of a metal rod on a grinding wheel yields small particles, which break off the surface. The compositions and densities of these particles are generally identical to the parent materials.

9 Particle Size Distribution

Theodore et al [2] have defined particle size analysis as a determination of the particle size distribution of a sample, and particle-size distribution (PSD) as an equation, graph, or table which quantifies, usually by percent, the various sizes of particulates found in a sample.

Particle size distributions are often characterized by a "mean" particle diameter. Although numerous "means" have been defined in the literature, the most common are the arithmetic mean and the geometric mean. The arithmetic mean diameter is simply the sum of the diameters of each of the particles divided by the number of particles measured. The geometric mean diameter is the nth root of the product of the n number of particles in the sample. In addition to the arithmetic and geometric means, a particle size distribution may also be characterized by the "median" diameter. The median diameter is that diameter for which 50% of the particles are larger in size and 50% are smaller in size. Another important characteristic is the measure of central tendency. It is sometime referred to as the dispersion or variability. The most common term employed is the standard deviation.

There are various methods for expressing the results of particle size measurements; the most common method is by plotting either or both on (a) a frequency distribution curve, or (b) a cumulative distribution curve. These topics are considered in this Chapter.

Ultimately, the decision of how to represent particle size distribution information is left to the user and/or practitioner. Considerations include:

1. Choosing an approach;
2. Checking on the reasonableness of the choice; and
3. Ability to draw the proper conclusions from the approach selected

9.1 REPRESENTATIVE SAMPLING

Discuss the problems associated with the need to obtain representative samples for statistical analysis.

SOLUTION [8]

One of the main problems with sampling is the need to obtain samples that are representative of the population. This usually requires both taking a large enough sample

Nanotechnology: Basic Calculations for Engineers and Scientists, by Louis Theodore
Copyright © 2006 John Wiley & Sons, Inc.

and doing it randomly. To sample randomly is to select from a population such that each sample drawn has an equal chance of being chosen. This requires including the whole population when selecting and removing all possible bias(es) from the selection process. It also requires sampling from the population in such a manner that each removal of a sample is taken into account on a subsequent sample analysis.

As an example of what can happen in sampling, consider this notorious failure. A magazine took a telephone survey of eligible voters, and on the basis of this survey incorrectly predicted that Landon would beat Roosevelt in the 1936 U.S. presidential election. The problem was that the sample included only those voters who had telephones and was not representative because voters without telephones had no chance of being included. An analogy of this sampling error in a process application would be to draw samples only during the weekday shift and ignore the weekend period entirely.

It is also important to see how a sample compares with the overall population in terms of a measurable characteristic. If the sample matches the population on this characteristic, it is often valid to conclude that it matches the population on others. However, this assumption is not always valid, so that it is a weakness in the concept of sampling.

There are so many kinds of populations and samples in engineering and science applications that it may not be possible to guarantee a representative sample by any single method. The reader is cautioned on this matter and it is recommended that care be exercised.

9.2   TYPICAL PARTICLE SIZE RANGES

Provide size ranges for various classes of particulates.

SOLUTION

Traditional particle sizes of interest in science and engineering are generally in the range of $1.0-10^6$ nm for gas dispersoids (aerosols) and $10-10^5$ nm for liquid dispersoids (hydrosols). Colloid chemistry is concerned chiefly with hydrosols finer than 1000 nm. The approximate values given in Table 9.1 should assist in formulating a physical concept of such fine particles.

9.3   PARTICLE SIZE DISTRIBUTION AND CONCENTRATION FOR INDUSTRIAL PARTICULATES

Develop a table that contains particulate description, particle size distribution information, and typical particulate concentrations (loading) for the following industries:

1. Electric power;
2. Cement;

TABLE 9.1   Particle Size Ranges

Limit of resolution with naked eye	10 to 40 microns (μm)
Diameter of large molecules	5 μm
Wavelength of visible spectrum	0.4 to 0.7 μm (400–700 nm)
Diameter of human hair	50 to 200 μm
Flour	10 to 100 μm
Pigments	200 to 2000 μm
Sulfuric acid mist	0.5 to 15 μm (500–15 000 nm)
Tobacco smoke	200 μm
Talcum powder	2 to 20 μm

3. Steel;
4. Nonferrous smelters;
5. Chemical.

SOLUTION

The information requested is provided in Table 9.2 [7].

9.4   PARTICLE SIZE DISTRIBUTION

A typical particulate size analysis method of representation employed in the past is provided below. Explain the data.

	>5.0	nm	40%
<5	>2.5	nm	27%
<2.5	>1.5	nm	20%
<1.5		nm	13%
			100%

SOLUTION

These numbers mean that 40% of the particles by weight are greater than 5 nm in size, 27% are less the 5 nm but greater than 2.5 nm, 20% are less than 2.5 nm but greater than 1.5 nm, and the remainder (13%) are less than 1.5 nm.

Another form of representing data is provided in Table 9.3.

9.5   MEDIAN AND MEAN PARTICLE SIZE

The following particle sizes (in nm) were recorded.

$$22, \quad 10, \quad 8, \quad 15, \quad 13, \quad 18$$

TABLE 9.2 Particle Size and Concentration for Typical Industrial Particulates

Industry	Particulates	Particle Size by Weight (%)			Particle Loading (gr/ft³)
		0 to 1 μm	0 to 5 μm	0 to 10 μm	
1. Electrical	Fly ash from pulverized coal	1	25	50	3
2. Cement	Kiln dust	1	20	40	10
3. Steel	Blast furnace after dry dust catcher	5	30	60	3
	Open-hearth fume	90	98	99	1
4. Nonferrous smelters	Copper roaster dust			20	10
	Converter furnace dust			30	5
	Reverberatory furnace dust			60	3
5. Chemical	H$_2$SO$_4$ acid fume	99			0.05
	H$_3$PO$_4$ acid fume	15	99		20

TABLE 9.3 Size Ranges in Arithmetic Increments

Size Range (nm)	Percent in Size Range
0–2	10
2–4	15
4–6	30
6–8	30
8–10	10
>10	5

Find the median, the arithmetic mean, and the geometric mean of these particle sizes.

SOLUTION

One basic way of summarizing data is by the computation of a central value. The most commonly used central value statistic is the arithmetic average, or the mean. This statistic is particularly useful when applied to a set of data having a fairly symmetrical distribution. The mean is an efficient statistic in that it summarizes all the data in the set and because each piece of data is taken into account in its computation. The formula for computing the mean is

$$\bar{X} = \frac{X_1 + X_2 + X_3 + \cdots + X_n}{n} = \frac{\sum_{i=1}^{n} X_i}{n}$$

where $\bar{X}$ = arithmetic mean, X_i = any individual measurement, n = total number of observations, $X_1, X_2, X_3 \ldots$ = measurements 1, 2, and 3, respectively.

The *arithmetic mean* is not a perfect measure of the true central value of a given data set. Arithmetic means can overemphasize the importance of one or two extreme data points. Many measurements of a normally distributed data set will have an arithmetic mean that closely approximates the true central value.

When a distribution of data is asymmetrical, it is sometimes desirable to compute a different measure of central value. This second measure, know as the *median*, is simply the middle value of a distribution, or the quantity above which half the data lie and below which the other half lie. If n data points are listed in their order of magnitude, the median is the $[(n + 1)/2]$th value. If the number of data is even, then the numerical value of the median is the value midway between the two data nearest the middle. The median, being a positional value, is less influenced by extreme values in a distribution than the mean. However, the median alone is usually not a good measure of central tendency. To obtain the median, the particle size data provided (in nm) should first be arranged in order of magnitude:

$$8, \quad 10, \quad 13, \quad 15, \quad 18, \quad 22$$

Thus, the median is 14 nm, or the value halfway between 13 nm and 15 nm since this data set has an even number of measurements.

Another measure of central tendency used in specialized applications is the *geometric mean*, $\bar{X}_G$. The geometric mean can be calculated using the following equation:

$$\bar{X}_G = \sqrt[n]{(X_1)(X_2)\ldots(X_n)}$$

For the above particle sizes (substituting D for X),

$$\bar{D}_G = [(8)(10)(13)(15)(18)(22)]^{1/6} = 13.54\,\text{nm}$$

where the arithmetic mean, $\bar{D}$, is

$$\bar{D} = (8 + 10 + 13 + 15 + 18 + 22)/6$$
$$= 14.33\,\text{nm}$$

9.6   STANDARD DEVIATION

Refer to Problem 9.5. Calculate the standard deviation of the six particle sizes.

SOLUTION

The most commonly used measure of dispersion, or variability, of sets of data is the *standard deviation*, σ. Its defining formula is given by expression

$$\sigma = \sqrt{\frac{\sum (X_i - \bar{X})^2}{n - 1}}$$

where $\sigma =$ standard deviation (always positive), $X_i =$ value of the ith data point, $\bar{X} =$ mean of the data sample, and $n =$ number of observations.

The expression $(X_i - \bar{X})$ shows that the deviation of each piece of data from the mean is taken into account by the standard deviation. Although the defining formula for the standard deviation gives insight into its meaning, the following algebraically equivalent formula makes computation much easier (now applied to the particle diameter, D):

$$\sigma = \sqrt{\frac{\sum (D_i - \bar{D})^2}{n - 1}} = \sqrt{\frac{n \sum D_i^2 - (\sum D_i)^2}{n(n - 1)}}$$

The standard deviation may be calculated for the data at hand:

$$\sum D_i^2 = (8)^2 + (10)^2 + (13)^2 + (15)^2 + (18)^2 + (22)^2 = 1366$$

$$(\sum D_i)^2 = (8 + 10 + 13 + 15 + 18 + 22)^2 = 7396$$

Thus,

$$\sigma = \sqrt{\frac{6(1366) - 7396}{(6)(5)}} = 5.16 \, \text{nm}$$

The subject of the standard deviation as well as the mean (see previous Problem) is revisited in Part 4 in Problem 26.8.

9.7   THE FREQUENCY DISTRIBUTION CURVE

Describe the frequency distribution curve.

SOLUTION

Frequency distribution curves are usually plotted on regular coordinate (linear) paper. The curve describes the amount of material (particles) falling within each size range. When gas-borne particles produced in a given operation are measured, the data generally have a tendency to show a preferential (maximum amount of) particle size. A plot of percent mass versus particle size (d_p) on a linear scale gives a curve with a peak at the preferential size. Such a curve is shown in Figure 9.1.

The figure below shows a *normal* probability distribution (see also Problems 9.9–9.11) which is symmetrical about the preferential size. This curve is rarely encountered for particulates consisting of very fine particles. However, this curve may be approached for particles such as fumes formed by vapor phase reaction and condensation, or for tar and acid mists.

As decsribed above, frequency distribution curves are usually constructed on linear coordinate paper. The percentage by weight (or the number of particle) is plotted as the ordinate. The average particle size of each range (size fraction) is plotted on the abscissa. There are a number of methods for selecting size ranges for the construction of the frequency distribution curve. Two of the more common are:

1. Select equal arithmetic increments of size as shown in Table 9.3.
2. Choose size ranges bounded by sizes having the same ratio to each other as shown in Table 9.4.

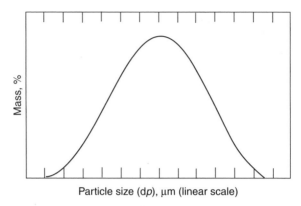

Particle size (dp), μm (linear scale)

Figure 9.1 Particle size distribution.

It is evident from this discussion that a large number of points are necessary to locate a position on the frequency curve. If this were the only method of representing the particle size data of a particulate sample, particle size analysis could be an extremely laborious and time-consuming procedure.

Another method of representing particle size distribution, which is generally more convenient in most applications, is the cumulative curve, in which the fraction of the total number or weight of particles that have a diameter greater (or less) than a given size is plotted against the size. This is essentially an integrated form of the frequency curve and is considered in the next Problem.

9.8 THE CUMULATIVE DISTRIBUTION CURVE

Describe cumulative distribution curves.

SOLUTION

Particle size data can also be plotted as a cumulative plot. Particle size for each size range is plotted on the ordinate. The cumulative percent by weight (frequency) is plotted on the abscissa. The cumulative percent by weight can be given as

TABLE 9.4 Size Ranges with the Same Ratio

Size Range (nm)	Percent in Size Range
0–5	15
5–10	10
10–20	30
20–40	20
40–80	20
>80	10

cumulative percent less than stated particle size (%LTSS) or cumulative percent larger than stated particle size (%GTSS). The cumulative percent by weight can be plotted on either a linear percentage or a probability percentage scale. The particle size range (ordinate) is usually a logarithmic scale.

If the particle size, d_p, of each size range is plotted versus the cumulative percent larger than d_p (linear scale), one would obtain a distribution curve as shown in Figure 9.2. In this figure, the distribution approaches the 0% and 100% values asymptotically. It is evident that the cumulative distribution is not a straight line for the entire range of particle size in the sample. The majority of the size ranges occur toward the 0% size and the 100% size.

More frequently the cumulative distribution is plotted on special coordinate paper called log probability paper. The particle size of each size range is plotted on a logarithmic ordinate. The percent by weight larger than d_p is plotted on the probability scale as the abscissa. This allows one to expand the cumulative distribution axis near 0% and near 100%. By expanding the axis the distribution may approach a straight line if the frequency distribution plot in the previous problem is *skewed*. If the distribution is log-normal (see Problem 9.10), the distribution curve plots out as a straight line. It should be noted that one can just as easily plot percent mass less than d_p (%LTSS) on the abscissa. In fact, %LTSS is preferred because the author's intials are carved in the acronym.

9.9   THE NORMAL DISTRIBUTION

Describe the normal distribution and briefly discuss its importance.

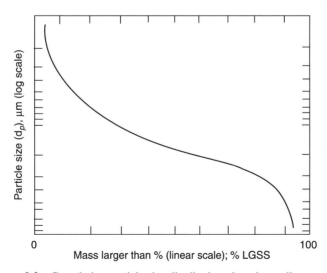

Figure 9.2   Cumulative particle size distribution plotted on a linear scale.

SOLUTION

One reason the normal distribution is so important is that a number of natural phenomena are normally distributed or closely approximate it. In fact, many experiments when repeated a large number of times, will produce values that approach the normal distribution curve. In its pure form, the normal curve is a continuous, symmetrical, smooth curve shaped like the one shown in Figure 9.3. Naturally, a finite distribution of discrete data can only approximate this curve.

The normal curve has the following definite relations to the descriptive measures of a distribution. The normal distribution curve is symmetrical; therefore, both the mean and the median are always to be found in the middle of the curve. Recall that, in general, the mean and median of an asymmetrical distribution do not coincide. The normal curve ranges along the x-axis from minus infinity to plus infinity. Therefore, the range of a normal distribution is infinite. The standard deviation, σ, becomes a most meaningful measure when related to the normal curve. A total of 68.2% of the area lying under a normal curve is included by the part ranging from one standard deviation below to one standard deviation above the mean. A total of 95.4% lies -2 to $+2$ standard deviations from the mean (see Figure 9.4). By using tables found in standard statistics texts and handbooks, one can determine the area lying under any part of the normal curve (see Problem 26.9 for additional details).

These areas under the normal distribution curve can be given probability interpretations. For example, if an experiment yields a nearly normal distribution with a mean equal to 30 and a standard deviation of 10, one can expect about 68% of a large number of experimental results to range from 20 to 40, so that the probability of any particular experimental result having a value between 20 and 40 is about 0.68.

Applying the properties of the normal curve to the testing of data and/or readings, one can determine whether a change in the conditions being measured is shown or whether only chance fluctuations in the readings are represented.

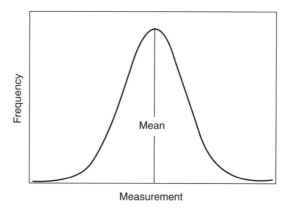

Figure 9.3   Gaussian distribution curve ("normal curve")

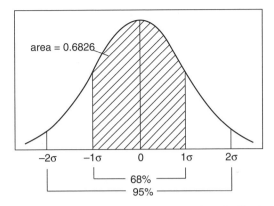

Figure 9.4 Characteristics of the Gaussian distribution.

For a well-established set of data, a frequently used set of control limits is ± 3 standard deviations. Thus, these limits can be used to determine whether the conditions under which the original data were taken have changed. Since the limits of three standard deviations on either side of the mean include 99.7% of the area under the normal curve, it is very unlikely that a reading outside these limits is due to the conditions producing the criterion set of data. The purpose of this technique is to separate the purely chance fluctuations from other causes of variation. For example, if a long series of observations of a measurement yield a mean of 50 and a standard deviation of 10, then control limits can be set up as 50 ± 30, i.e., ± 3 standard deviations, or from 20 to 80. A value above 80 would therefore suggest that the underlying conditions have changed and that a large number of similar observations at this time would yield a distribution of results with a mean different (larger) than 50.

9.10 THE LOG NORMAL DISTRIBUTION

You have been requested to determine if a particle size distribution is log-normal. Data are provided in Table 9.5.

SOLUTION

When measuring particulate emissions from industrial sources, the graph of the particle size distribution often displays the logarithmic variation of the normal distribution. The normal distribution has the fundamental defect related to its use in particle sizing analysis. Implicit in the statement that a random variable is normally distributed is the concept that the values of particle size are at equal distances from the central tendency or preferential size. Suppose the mean particle size or central

TABLE 9.5   Data for Log–Normal Distribution

Particle Size Range d_p (μm)	Distribution (μg/m³)
<0.62	25.5
0.62–1.0	33.3
1.0–1.2	17.85
1.2–3.0	102.0
3.0–8.0	63.75
8.0–10.0	5.1
>10.0	7.65
Total	255.0

tendency of the distribution were 20 μm. It would be equally probable to find either a 15 μm or 25 μm particle. One might also find a particle the size of 50 μm in the distribution. Were the distribution normal, it would be equally likely to find a particle the size of minus 10 μm.

However, if it can be assumed that the logarithm of the particle size is randomly distributed, then this problem is avoided. The ratios of particle size about a central tendency are equally probable, and the ratios are bounded on the lower end by zero.

As described earlier, the usefulness of the log normal distribution is more evident when the frequency distribution curve is characteristically *skewed*. The data can then be plotted as a cumulative plot on log probability paper. If the distribution follows the log-normal relationship, then the plot will result in a straight line. The linearity of the relationship allows one to describe the distribution statistically with a minimum of two parameters: the geometric mean, d_{gm}, and the geometric standard deviation, σ_{gm}.

The geometric mean value of a log-normal distribution can be read directly from a plot similar to that represented in Figure 9.5. The geometric mean size is the 50% size on the plot.

As discussed previously, the *geometric standard deviation* is a good measure of the dispersion or spread of a distribution. The geometric standard deviation is the root-mean square deviation about the mean value and can be read directly from a plot such as shown in Figure 9.5. For a log-normal distribution (which plots d_p maximum in nm versus percent mass larger than d_p), the geometric standard deviation is given by:

$$\sigma_{gm} = \frac{50\% \text{ size}}{84.13\% \text{ size}}$$

or

$$\sigma_{gm} = \frac{15.87\% \text{ size}}{50\% \text{ size}}$$

All one must do is determine the 50% size and the 84.13% size from the plot and divide to determine the geometric standard deviation. Therefore, knowing any

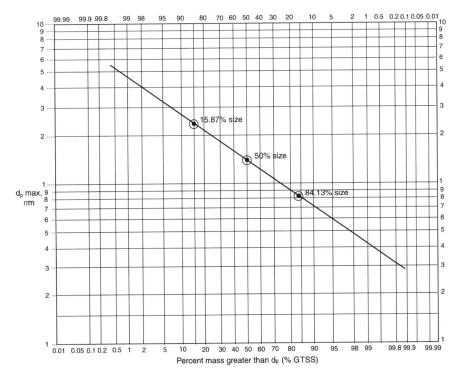

Figure 9.5 Geometric mean and standard deviation from a log-normal distribution plot.

combination of two of these parameters (15.87%, 50%, 84.13%, or σ_{gm}) describes the entire range of the particle size distribution.

Cumulative distribution information can be obtained from the Problem statement. Calculated results are provided in Table 9.6.

The cumulative distribution above can be plotted on log-probability graph paper. The cumulative distribution curve is shown in Figure 9.6. Since a straight line is obtained on log-normal coordinates, the particle size distribution is log-normal. Calculating the standard deviation is left as an exercise for the reader.

9.11 EFFECT OF SIZE DISTRIBUTION ON CUMULATIVE DISTRIBUTION PLOTS

Demonstrate the effect of size distribution on a log-probability plot.

SOLUTION

Refer to Figure 9.7 below. Note that the representation of particle-size analyses on probability coordinates has the advantage of simple extrapolation or interpolation

TABLE 9.6 Calculated Results from Cumulative Distribution Information

d_p (μm)	% Total	Cumulative %GTSS
<0.62	10	90
0.62–1.0	13	77
1.0–1.2	7	70
1.2–3.0	40	30
3.2–8.0	25	5
8.0–10.0	2	3
>10.0	3	0

from minimum data. The slope of the line is a measure of the breadth of the distribution of sizes in the sample. From Figure 9.7, curve A represents a batch of uniform-size particles; curve B has the same mean particle size with perhaps a 10-fold variation in diameter between the largest and smallest particles, whereas curve C may have a 20-fold variation. Curve D has the same 10-fold range of sizes as curve B, but in each weight percent category the particles are larger, so that curve D is a coarser grind than curve B, but equal in range of sizes.

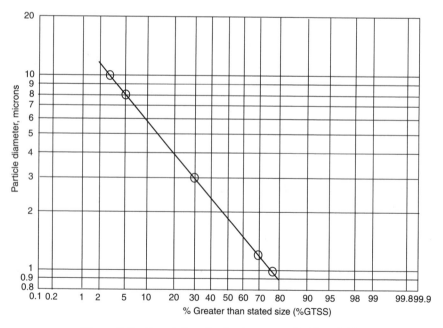

Figure 9.6 Cumulative distribution curve (Problem 9.10).

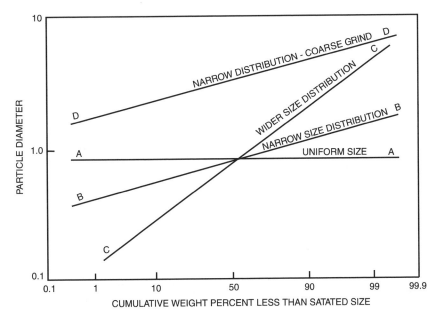

Figure 9.7   Effect of size distribution on relative probability plots.

9.12   NANOPARTICLE SIZE VARIATION WITH TIME

Consider the set of data in Table 9.7 which represents the average size of a nanoparticle emitted in the duct from a research laboratory for a given hour for 25 days. As a first step in summarizing the data, you are requested to form a frequency table, a frequency polygon, a cumulative frequency table, and a frequency distribution curve.

SOLUTION

Data are often unmanageable in the form in which they are collected. In this last Problem, the graphical techniques of summarizing such data are considered so that meaningful information can be extracted. Basically, there are two kinds of variables to which data can be assigned: continuous variables and discrete variables. A continuous variable is one that can assume any value in some interval of values. Examples of continuous variables are weight, volume, length, time and temperature. Most data are taken from continuous variables. Discrete variables, on the other hand, are those variables with possible values which are normally integers. Therefore, they involve counting rather than measuring. Examples of discrete variables are the number of sample stations, number of people in a room, and number of times a control standard is violated.

Since any measuring device is of limited accuracy, measurements in real life are actually discrete in nature rather than continuous, but this should not keep one from

TABLE 9.7 Set of Data for Problem 9.12

Days	Nanoparticle Size (nm)
1	53
2	72
3	59
4	45
5	44
6	85
7	77
8	56
9	157
10	83
11	120
12	81
13	35
14	63
15	48
16	180
17	94
18	110
19	51
20	47
21	55
22	43
23	28
24	38
25	26

TABLE 9.8 Frequency Table

Class Interval (nm)	Frequency of Occurence
25–40	4
40–55	7
55–70	4
70–85	4
85–100	2
100–115	1
115–130	1
130–145	0
145–160	1

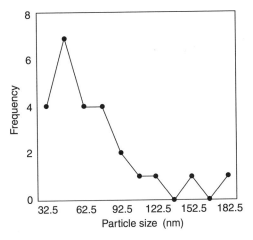

Figure 9.8 Size (midpoint of class interval) frequency polygon.

regarding such variables as continuous. As noted earlier when a size is recorded as 165 nm it is assumed that the actual size is somewhere between 164.5 and 165.5 nm.

A frequency table of the above data is first constructed. In constructing the frequency table (Table 9.8), it can be seen that the data have been divided into 11 class intervals with each interval being 15 units in length. The choice of dividing the data into 11 intervals was purely arbitrary. However, in dealing with data, it is a rule of thumb to choose the length of the class interval such that 8–15 intervals will include all of the data under consideration. Deriving the frequency column involves nothing more than counting the number of values in each interval. From observation of the frequency table, one can now see the data taking form. The

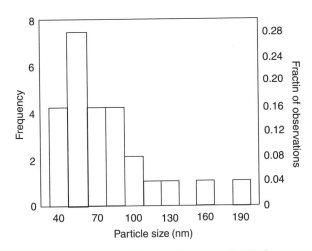

Figure 9.9 Histogram of percent frequency distribution curve.

TABLE 9.9   Cumulative Frequency Table

Level	Cumulative Frequency	Fractional Cumulative Frequency of Size Less Than Stated Value
≤40	4	0.16
≤55	11	0.44
≤70	15	0.60
≤85	19	0.76
≤100	21	0.84
≤115	22	0.88
≤130	23	0.92
≤145	23	0.92
≤160	24	0.96
≤175	24	0.96
≤190	25	1.00

values appear to be clustered between 25 and 85 nm. In fact, nearly 80% are in this interval.

As a further step, one can graph the information in the frequency table. One way of doing this would be to plot the frequency midpoint of the class interval. The solid line connecting the points of Figure 9.8 forms a frequency polygon.

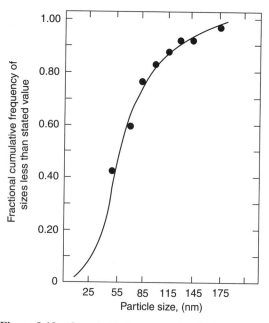

Figure 9.10   Cumulative frequency distribution curve.

Another method of graphing the information would be by constructing a histogram as shown in Figure 9.9. The histogram is a two-dimensional graph in which the length of the class interval is taken into consideration. The histogram can be a very useful tool in statistics, especially if one converts the given frequency scale to a relative scale so that the sum of all the ordinates equals one. This is also shown in Figure 9.9. Thus, each ordinate value is derived by dividing the original value by the number of observations in the sample, in this case 25. The advantage in constructing a histogram like this is that one can read probabilities from it, if one can assume a scale on the abscissa such that a given value will fall in any one interval is the area under the curve in that interval. For example, the probability that a value will fall between 55 and 70 is equal to its associated interval's portion of the total area of intervals, which is 0.16,

From the frequency table and histogram discussed above, one can also construct a cumulative frequency table (Table 9.9) and graph (Figure 9.10). These are shown on the previous page.

The cumulative frequency table gives the number of observations less than a given value. Probabilities can be read from the cumulative frequency curve. For example, to find the probability that a value will be less than 85, one should read the curve at the point 85 and read across to the value 0.74 on the y-axis.

10 Particle Sizing and Measurement Methods

Unfortunately, the term "measurement methods" has come to mean different things to engineers and scientists in the instrumentation field. Traditionally, this term has referred to sampling techniques employed for source characterization purposes. Source characterization includes either the determination of particle size distribution, concentration and flowrate, or chemical composition and other properties, or both. The equipment and procedures employed to measure particle data in a stream (usually air) represents the sampling part of the overall procedure. The analysis of the data is also important, usually providing the aforementioned information on particle size distribution, concentration, and flowrate. However, to complete the analysis may require information on shape, chemical composition, and other properties. Thus, it is important to differentiate between sampling and analysis, with analysis including several potential tasks. This Chapter primarily addresses the analysis of data for determining particle size distribution, concentration, and flowrate. Some material on equipment can be found in Problem 10.4.

An accurate quantitative analysis of the discharge of particulates from a process must be determined prior to the design and/or selection of recovery/control equipment. If the unit is properly engineered, utilizing the emission data as input to the device, most particulates can be successfully controlled by one or a combination of the methods to be discussed later in this Chapter.

There are many techniques available for measuring the particle-size distribution of particulates. The wide size range covered, from nanometers to millimeters, cannot always be analyzed using a single measurement principle. Added to this are the usual constraints of capital costs versus operating costs, speed of operation, degree of skill required, and, most important, the end-use requirement [9].

Physical sizing is traditionally one of the most common methods available for classifying (or sizing) particles. This is most often achieved presently by dry or wet screening with sieves, microscopic analysis, electric-grating techniques, and light-scattering methods.

More realistic determinations of particle behavior in any environment must consider the size, shape, and density. The technique best able to accomplish this is aerodynamic sizing. Only by knowing the aerodynamic size of particles is it possible to determine how they will behave in an air stream and the kind(s) of control equipment required to capture them. Consider, for example, a ping-pong ball and a golf ball. Under a microscope they will appear almost equal in size; however, if both were tossed into a moving air stream, they would behave quite differently. Even though the size and shape are similar, the density is quite different, and the behavior of the two objects is far from being similar aerodynamically. This is the primary fallacy in physical sizing.

The scientific community's ability to reliably observe and manipulate infinitesimally small matter has been a fundamental underpinning in the progress of nanotechnology-related research and development and commercial-scale development efforts. While nanotechnology concepts may have been evolving in the minds and imaginations of forward-thinking scientists and engineers for a long time, they did not emerge meaningfully until the 1980s, in large part because the scientific community had yet to develop adequate instruments, techniques, and methods to visualize and study materials and phenomena in the nanometer range. This changed in the mid-1980s, with the invention of two powerful microscopy techniques – atomic force microscopy (AFM; which is also referred to as scanning force microscopy, or SFM), scanning tunneling microscopy (STM), both of which permit accurate, atomic-scale measurements [1, 10].

Both of these relatively new techniques offer a radical departure from conventional types of microscopy, which work by reflecting either light (in the case of optical microscopes) or an electron beam (in the case of electron microscopes) off a surface and onto a lens. While conventional microscopy techniques are sufficient for detailed characterization and visualization of large macromolecular assemblies, no reflective microscope, not even the most powerful one, can provide precise imaging of individual atoms or nanometer-scaled structures [1]. These problems probably arise because the intensity of light scattered by a particle is not only a function of the particle size but also its index of refraction and shape.

Biswas and Wu [11] provide an excellent and detailed review of real-time nanoparticle instrumentation and characterization. Aerosol instrumentation reviewed include: real time vertical impactors, electrical instruments, and diagnostic lightscattering.

10.1   TYLER AND U.S. STANDARD SCREENS

The feedstock to a nano size reduction unit has its particle size specified as Tyler 8-14 screen mesh. Determine the size.

SOLUTION

A common method of specifying large particle sizes is to designate the screen mesh that has an aperture corresponding to the particle diameter. Since various screen scales are in use, confusion may result unless the screen scale involved is specified. The screen mesh generally refers to the number of screen openings per unit of length or area. The aperture for a given mesh will depend on the wire size employed. The

TABLE 10.1   Tyler and U.S. Standard Screen Scales

Tyler Mesh	Aperture Microns	U.S. Mesh	Aperture Microns
400	37	400	37
325	43	325	44
270	53	270	53
250	61	230	62
200	74	200	74
170	88	170	88
150	104	140	105
100	147	100	149
65	208	70	210
48	295	50	297
35	417	40	420
28	589	30	590
20	833	20	840
14	1168	16	1190
10	1651	12	1680
8	2362		
6	3327		
4	4699		
3	6680		

Tyler and the U.S. Standard Screen Scales (Table 10.1) are the most widely used in the United States. The screens are generally constructed of wire mesh cloth, with the diameters of the wire and the spacing of the wires being closely specified. These screens form the bottoms of metal pans about 8 in. in diameter and 2 in. high, whose sides are so fashioned that the bottom of one sieve nests snugly on the top of the next.

The clear space between the individual wires of the screen is termed the screen aperture. As indicated above, the term *mesh* is applied to the number of apertures per linear inch; for example, a 10-mesh screen will have 10 openings per inch, and the aperture will be 0.1 in. minus the diameter of the wire.

To calculate the size of the particle in the problem statement, refer to the Tyler screen information in Table 10.1.

8-mesh opening $= 2362$ μm
14-mesh opening $= 1168$ μm

Since an 8 by 14 mesh particle size indicates that the particle will pass through the 8-mesh screen but not pass (be captured) through the 14-mesh screen, *one* size cannot be specified for the particle in question.

The particle is in the size range 1168–2362 μm. The average arithmetic size is 1765 μm.

10.2   EQUIVALENT DIAMETER TERMS

Define various equivalent diameters that are employed in practice.

SOLUTION

Particle size is uniquely defined by particle diameter only for the case of spherical particles. Unfortunately, except for liquid droplets, certain metallurgical fumes, and combustion emissions, particles are usually not spherical. This may also be the case with nanoparticles. To deal with nonspherical particles it becomes necessary to define an *equivalent diameter* term that depends upon the various geometrical and/or physical properties of the particles.

Some of the methods used to express the size of a nonspherical particle measured by microscopy are illustrated in Figure 10.1. With reference to this figure, Ferret's diameter is the mean length between two tangents on opposite sides of the particle perpendicular to the fixed direction of the microscopic scan. Martin's diameter measures the diameter of the particle parallel to the microscope scan that divides the particle into two equal areas. The diameter of a circle of equal area is obtained by estimating the projected area of the particle and comparing it with a sphere that approximates its size.

The most popular choice is that sphere diameter (of the same density) that will settle with the same velocity as the particle in question under the influence of gravity. This is discussed later in this Chapter. Other diameters that are occasionally/rarely employed are listed in Table 10.2.

This general topic is revisited in Chapter 11.

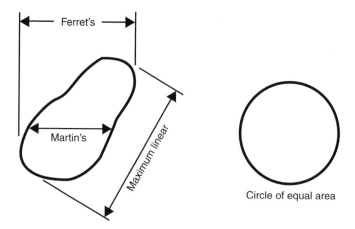

Figure 10.1   Diameters of nonspherical particles.

TABLE 10.2   Equivalent Diameters of Particles

Name	Definition
Surface diameter	The diameter of a sphere having the same surface area as the particle
Volume diameter	The diameter of a sphere having the same volume as the particle
Drag diameter	The diameter of a sphere having the same resistance to motion as the particle in a fluid of the same viscosity and at the same velocity
Specific surface diameter	The diameter of a sphere having the same ratio of surface area to volume as the particle

10.3   AERODYNAMIC DIAMETER

Calculate the aerodynamic diameter (μm) for the following three particles:

1. Solid sphere, equivalent diameter = 1.4 μm, specific gravity = 2.0;
2. Hollow sphere, equivalent diameter = 2.8 μm, specific gravity = 0.5;
3. Irregular sphere, equivalent diameter = 1.3 μm, specific gravity = 2.35.

SOLUTION

The aerodynamic diameter is defined as the diameter of a sphere of unit density (specific gravity = 1.0) having the same falling speed in air as the particle (Figure 10.2). It is most useful in evaluating particle motion in a fluid. The aerodynamic diameter is a function of the physical size, shape, and density of the particle. The aerodynamic diameter is useful when designing certain recovery control devices and is usually measured by a device called an impactor (see last Problem in this Chapter).

The aerodynamic diameter ($d_{p,a}$) is defined by:

$$d_{p,a} = d_p \sqrt{e_p c_c}$$

where $d_{p,a}$ = aerodynamic diameter, consistent units; d_p = actual (equivalent) diameter, consistent unit; ρ_p = particle specific gravity, dimensionless; and C_c = Cunningham correction factor, dimensionless (see Chapter 11, Problems 11.9 and 11.10). For purposes and analyses, assume C_c = 1.0.

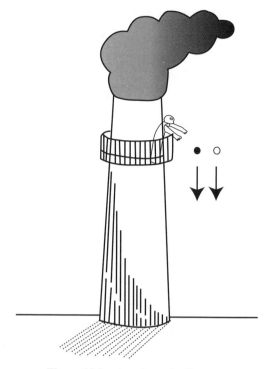

Figure 10.2 Aerodynamic diameter.

1. For the solid sphere,

$$d_{p,a} = 1.4(2.0)^{0.5}$$
$$= 1.98 \, \mu m$$

2. For the hollow sphere,

$$d_{p,a} = 2.80(0.51)^{0.5}$$
$$= 2.0 \, \mu m$$

3. For the irregular shape,

$$d_{p,a} = 1.3(2.35)^{0.5}$$
$$= 1.99 \, \mu m$$

Conversely, particles with different specific gravity but the same equivalent size, have different aerodynamic diameters. For example, if $d_p = 2$ μm, the reader is left the exercise of showing that the aerodynamic diameter for particles with specific gravity 1.0, 2.0, 4.0, and 8.0 is 2.00 μm, 2.83 μm, 4.00 μm, and 5.66 μm, respectively. Thus, particles of different size and shape can have the same aerodynamic diameter while particles of the same size can have different aerodynamic diameters.

10.4 SIZING DEVICES

Describe various particle sizing devices.

SOLUTION

If one were able to design an ideal particle measuring device, the device should be able to:

1. Measure the exact size of each particle.
2. Report data instantaneously without averaging data over some specified time interval.
3. Determine the composition of each particle including shape, density, chemical nature, and so on.

It would be an extremely difficult task to produce such an instrument. At this time there are devices that incorporate only one or two of these ideal functions. Various sizing techniques are examined and compared to such an ideal device, listing advantages and disadvantages of each. While this is not intended to be exhaustive, the more commonly employed methods for sizing of fine and nanoparticulates are reviewed below. For larger particles, sizing measures include: gravitational sedimentation, sedimentation balance, centrifugal sedimentation, sieving, and elutriation. Details are available in the literature [9].

10.4.1 Microscopy

The traditional microscope is one common instrument used in particle size analysis. The microscope measures the geometric diameter of each individual particle. The determination of particle size analysis is carried out by measuring the size of a number of particles. Particles are sized as they are traversed past the eyepiece micrometer. Each particle, presented in a fixed area of the eyepiece, is sized and tallied into a number of size classes. The number of particles sized may range from 100 to several thousand depending on the accuracy desired. This method can be time consuming and extremely tedious. The results are generally biased in favor of the larger particles.

10.4.2   Optical Counters

Optical particle counters have been used for particle sizing despite the fact that they cannot be directly applied to stack exhaust gas streams. The sample must be extracted, cooled, and diluted before entering the counter. This procedure must be done with extreme care to avoid introducing serious errors in the sample. The major advantage of the counter is its capability of observing emission (particle) fluctuations on an instaneous level. One can size particles as small as 0.3 μm (300 nm) with the optical counter.

10.4.3   Electrical Aerosol Analyzer

The electrical aerosol analyzer (EAA) is an aerosol size distribution measuring device that was commercially developed at the University of Minnesota. The EAA uses an electric field (which is set at an intensity dependent upon the size and mass of the particle) to measure the mobility of a charged aerosol. The analyzer operates by first placing a unipolar charge on the aerosol being measured and measuring the resulting mobility distribution of the charged particles by means of a mobility analyzer. The major advantage of the EAA is that this instrument can measure particles from 0.003 to 1.0 μm (3 to 1000 nm) in diameter.

10.4.4   Bahco Microparticle Classifier

The Bahco is a versatile particle classifier used for measuring powders, dust, and other finely divided solid materials. The Bahco's working range is approximately 1 to 60 μm. Developed in the 1950s, the Bahco has lost some of its initial appeal to more recently developed techniques. The Bahco uses a combination of elutriation and centrifugation to separate particles in an air stream. Particles can be collected onto a filter by using an EPA Method 5 sampling train. The collected particles are subsequently analyzed in the lab. Some of the major drawbacks of the Bahco are that:

1. The working size range is limited to between 1 and 60 μm;
2. Care must be exercised when measuring certain types of particles, especially those that are not friable, or hygroscopic;
3. The sample may not be representative due to particle agglomeration either in the stack or during the transfer of the collected sample to the lab;
4. The length of time required for analysis is several hours or more; and
5. The size grading regimes are not as sharp as with newer devices.

10.4.5   Impactors

Inertial impactors are commonly used to determine the particle size distribution of exhaust streams from industrial sources. Inertial impactors measure the aerodynamic diameter of the particles. The inertial impactor can be directly attached to an EPA Method 5 sampling train and easily inserted into the stack of an industrial source. The impactor is constructed using a succession of stages, each containing orifice openings with an impaction slide or collection plate behind the openings. In each stage, the gas stream passes through the orifice opening and forms a jet that is directed towards the impaction plate. The larger particles will impact on the plate if their momentum is large enough to overcome the drag of the air stream as it moves around the plate. Since each successive orifice opening is smaller that those on the preceding stage, the velocity of the air stream, and therefore that of the dispersed particles, is increased as the gas stream advances through the impactor. Consequently, smaller particles eventually acquire enough momentum to separate from the gas streamlines to impact on a plate. A complete particle size classification of the gas stream can therefore be achieved. More details are available in Problem 13.10.

10.4.6 Photon Correlation Spectroscopy (PCS)

The size distribution of particles ranging from a few nanometers to a few micrometers can be determined from their random motion due to molecular bombardment. This technique involves passing a laser beam into a suspension and measuring the Doppler shift of the frequency of light scattered at an angle (usually $90°$) with respect to the incident beam. The Doppler shift is related to particle velocity, which in turn, is inversely related to their size. Multiangle instruments are also available to generate the angular variation of scattered light intensity for the determination of molecular weight, radius of gyration, translational and rotational diffusion coefficients, and other molecular properties [9].

The discussion above describes the six major instruments that have been traditionally used for particle sizing purposes. However, other than the PCS, these devices are not suitable for measuring particles in the nanosize range. Theodore and Kunz [1] describe two measuring devices that are currently available for nanoparticles. Background material and a short description of the equipment is provided in the Introduction to this Chapter.

10.5 RECTANGULAR CONDUIT SAMPLING

Develop a sampling procedure for a rectangular parallepiped.

SOLUTION

As described in the Introduction to this Chapter, sampling is the keystone of source analysis. Sampling methods and tools vary in their complexity according to the specific task; therefore, a degree of both technical knowledge and common sense

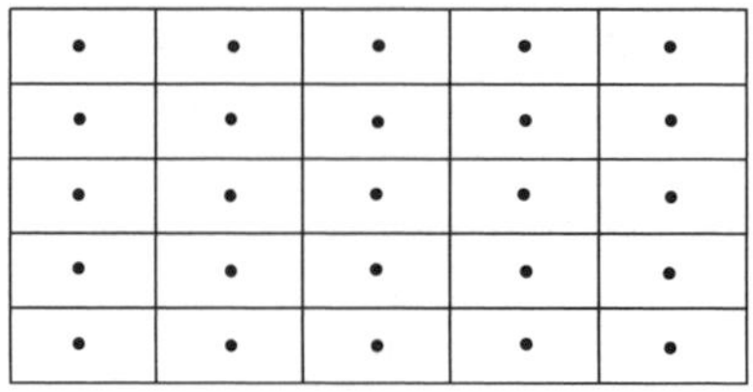

RECTANGULAR STACK
(MEASURE AT CENTER OF AT LEAST 12 EQUAL AREAS).

Figure 10.3   Traverse point locations for velocity measurement or for multipoint sampling in rectangular ducts (25 traverse points).

is needed to design a sampling function. Sampling is performed to measure quantities or concentrations of particulates in effluent gas streams, to measure the efficiency of a control process device, to guide the designer of recovery/control equipment and facilities, and/or to appraise contamination from a process or source. A complete measurement requires determination of the concentration and particulate characteristics, as well as the associated gas flow. Most statutory limitations require mass rates of emission; both concentration and volumetric flow rate data are, therefore, required.

The selection of a sampling site and the number of sampling points are based on attempts to obtain representative samples. To accomplish this, the sampling site should be at least eight stack or duct diameters downstream and two diameters upstream from any bend, expansion, contraction, valve, fitting, or visible flame. For a rectangular duct cross section, the equivalent diameter is determined from:

$$\text{Equivalent diameter} = 2\left[\frac{(\text{length})(\text{width})}{\text{length} + \text{width}}\right]$$

Once the sampling location is chosen, the duct cross-section is laid out in a number of equal areas, the center of each being the point where the measurement is to be taken. For rectangular stacks, the cross-section is divided into equal areas of the same shape, and the traverse points are located at the center of each equal, as shown in Figure 10.3.

10.6   VOLUMETRIC FLOW RATE CALCULATION

Calculate the volumetric rate of a fluid flowing through a 2 foot by 4 foot rectangular parrallepiped. The velocity $v(i,j)$ in ft/s passing each of the equal areas is provided as follows (see Figure 10.4)

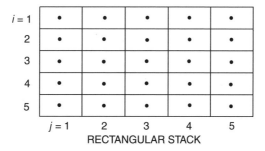

Figure 10.4   Probe location.

$v(1,1) = 14.2 \quad v(1,2) = 16.8 \quad v(1,3) = 22.7 \quad v(1,4) = 15.7 \quad v(1,5) = 16.2$
$v(2,1) = 26.1 \quad v(2,2) = 28.2 \quad v(2,3) = 30.7 \quad v(2,4) = 27.1 \quad v(2,5) = 25.0$
$v(3,1) = 26.9 \quad v(3,2) = 29.2 \quad v(3,3) = 30.9 \quad v(3,4) = 27.7 \quad v(3,5) = 25.1$
$v(4,1) = 24.3 \quad v(4,2) = 28.1 \quad v(4,3) = 30.3 \quad v(4,4) = 27.0 \quad v(4,5) = 24.7$
$v(5,1) = 17.3 \quad v(5,2) = 18.9 \quad v(5,3) = 24.0 \quad v(5,4) = 19.1 \quad v(5,5) = 17.0$

SOLUTION

Since the velocity probes are located at the center of equal areas, the volumetric flow rate (Q) is given by:

$$Q = \sum_{i=1}^{5} \sum_{j=1}^{5} v(i,j)A(i,j)$$

$$= \sum_{i=1}^{5} \sum_{j=1}^{5} v(i,j)A$$

where $A = A(i,j) = (2/5)(4/5) = 0.32 \text{ ft}^2$
   Therefore

$$Q = [14.2 + 26.1 + \cdots + 24.7 + 17.0]0.32$$
$$= (593.2)(0.32)$$
$$= 189.8 \text{ ft}^3/\text{s}$$

Also note that

$$Q = \sum_{i=1}^{5} \sum_{j=1}^{5} q(i,j)$$

where $q(i,j)$ is the volumetric flow rate passing location (i,j).

The average velocity ($\bar{v}$) is

$$\bar{v} = Q/A_T$$
$$= 189.8/(2 \times 4)$$
$$= 23.725 \text{ ft/s}$$

10.7 PARTICLE MASS FLOW RATE CALCULATION

Refer to Figure 10.4. The concentration $c(i, j)$ in mg/m^3 of particulates in the fluid passing each of the equal areas is provided as follows:

$c(1,1) = 201$ $c(1,2) = 222$ $c(1,3) = 222$ $c(1,4) = 219$ $c(1,5) = 198$
$c(2,1) = 213$ $c(2,2) = 227$ $c(2,3) = 231$ $c(2,4) = 226$ $c(2,5) = 213$
$c(3,1) = 214$ $c(3,2) = 233$ $c(3,3) = 240$ $c(3,4) = 229$ $c(3,5) = 216$
$c(4,1) = 214$ $c(4,2) = 230$ $c(4,3) = 233$ $c(4,4) = 229$ $c(4,5) = 212$
$c(5,1) = 201$ $c(5,2) = 226$ $c(5,3) = 228$ $c(5,4) = 225$ $c(5,5) = 196$

Calculate the duct particulate flow rate in g/s, mg/s, μg/s, and ng/s.

SOLUTION

The particle mass flow rate $\dot{m}(i, j)$ using $A(i, j)$ is

$$\dot{m}(i, j) = v(i, j)A(i, j)c(i, j)$$
$$= q(i, j)c(i, j)$$

The total rate is

$$\dot{m} = \sum_{i=1}^{5} \sum_{j=1}^{5} v(i, j)A\, c(i, j); \quad A = 0.32 \text{ ft}^2$$

This is simply the area times the cross product of the velocity and concentration, so that

$$\dot{m} = [(14.2)(201) + (26.1)(213) + \cdots + (24.7)(212) + (17.0)(196)]0.32$$

This may be represented in tabular form as shown below:

$\dot{m}(1,1) = 2854$ $\dot{m}(1,2) = 3730$ $\dot{m}(1,3) = 5039$ $\dot{m}(1,4) = 3438$ $\dot{m}(1,5) = 3208$

$\dot{m}(2,1) = 5559$ $\dot{m}(2,2) = 6401$ $\dot{m}(2,3) = 7092$ $\dot{m}(2,4) = 6125$ $\dot{m}(2,5) = 5325$

$\dot{m}(3,1) = 5757$ $\dot{m}(3,2) = 6804$ $\dot{m}(3,3) = 7416$ $\dot{m}(3,4) = 6343$ $\dot{m}(3,5) = 5422$

$\dot{m}(4,1) = 5200$ $\dot{m}(4,2) = 6463$ $\dot{m}(4,3) = 7060$ $\dot{m}(4,4) = 6183$ $\dot{m}(4,5) = 5236$

$\dot{m}(5,1) = \underline{3477}$ $\dot{m}(5,2) = \underline{4271}$ $\dot{m}(5,3) = \underline{5472}$ $\dot{m}(5,4) = \underline{4298}$ $\dot{m}(5,5) = \underline{3332}$

   22,847   27,669   32,079   26,387   22,523

The particle mass flow ratio is therefore

$$\dot{m} = (22{,}847 + 27{,}669 + 32{,}079 + 26{,}387 + 22{,}523)(0.32)$$
$$= (131{,}505)(0.32)$$
$$= 42{,}081.6 \, (\text{mg/m}^3)(\text{ft}^3/\text{s})$$

The applicable conversion factor is $35.31 \, \text{ft}^3/\text{m}^3$. Therefore,

$$\dot{m} = 42{,}081.6/35.31$$
$$= 1191.8 \, \text{mg/s}$$
$$= 1.192 \times 10^3 \, \text{mg/s}$$
$$= 1.192 \, \text{g/s}$$
$$= 1.192 \times 10^6 \, \mu\text{g/s}$$
$$= 1.192 \times 10^9 \, \text{ng/s}$$

10.8   AVERAGE PARTICLE CONCENTRATION

Refer to Problem 10.7. Calculate the average concentration ($\bar{c}$) of the fluid passing the cross-section of the duct. Provide results in lb/ft^3, g/ft^3, g/m^3, $\mu\text{g/m}^3$, and ng/m^3.

SOLUTION

The key to answering this question is to note the term *passing* in the problem statement. One would be tempted to simply add all the concentrations and divide by 25, the number of segments, that is,

$$\bar{c} = \frac{\displaystyle\sum_{i=1}^{5}\sum_{j=1}^{5} c(i,j)}{25}$$

This produces the following result

$$\bar{c} = 5498/25$$
$$= 220 \, \text{mg/m}^3$$

However, in accordance with the definition of average values, an appropriate weighing factor needs to be included in this determination. For this calculation,

$$\bar{c} = \frac{\sum\limits_{i=1}^{5} \sum\limits_{j=1}^{5} c(i,j)v(i,j)A}{\sum\limits_{i=1}^{5} \sum\limits_{j=1}^{5} v(i,j)A}$$

which reduces to

$$\bar{c} = \frac{\dot{m}}{Q}$$

Substituting Q from Problem 10.6 and $\dot{m}$ from Problem 10.7 leads to

$$\bar{c} = \frac{1.192\,\text{g/s}}{189.8\,\text{ft}^3/\text{s}}$$
$$= 6.28 \times 10^{-3}\,\text{g/ft}^3$$

In alternate units,

$$\bar{c} = (6.28 \times 10^{-3})35.31$$
$$= 2.22 \times 10^{-1}\,\text{g/m}^3$$
$$= 2.22 \times 10^{+2}\,\text{mg/m}^3$$
$$= 2.22 \times 10^{+5}\,\mu\text{g/m}^3$$
$$= 2.22 \times 10^{+8}\,\text{ng/m}^3$$

This is in very close agreement with the average concentration value calculated without the appropriate weighing factors. The reader is left the exercise of explaining the close agreement.

10.9 EQUAL ANNULAR AREAS FOR CIRCULAR DUCTS

In order to ensure equal annular areas for each measurement point, determine the sampling locations in a 6-ft I.D. circular duct for an eight-point traverse.

SOLUTION

For circular stacks, the cross-section is divided into equal annular areas, and the traverse points are located at the centroids of each area, as shown in Figure 10.5. When

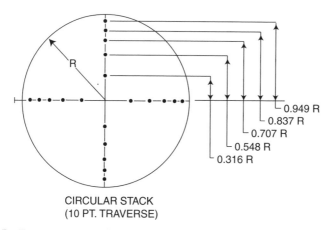

Figure 10.5   Traverse point locations for measurement or for multipoint sampling in circular stacks (10-point traverse, with a total of 20 traverse points).

the above sampling site criteria can be met, the minimum number of traverse points should be 12. Some sampling situations, however, render the above sampling site criteria impractical; in this case, one should choose convenient sampling locations. Under no condition should a sampling point be selected that is within 1 inch of the duct or stack wall.

The sampling locations are provided in Figure 10.6 [7].

10.10   TRAVERSE POINT LOCATION IN CIRCULAR DUCTS

Develop a chart that provides sampling locations in a duct as a function of the number of traverse points along a diameter.

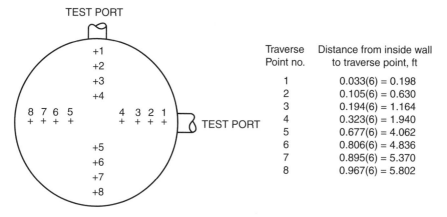

Figure 10.6   Sampling locations.

SOLUTION

The results are presented in Table 10.3 [7].

10.11 DUCT FLOW EQUATION DERIVATION

Derive the volumetric flow equation

$$Q = \pi R^2 \int_0^{1.0} v \, d\left(\frac{r}{R}\right)^2$$

for flow in a pipe of radius R and where v represents the local velocity at point r.

TABLE 10.3 Traverse Point Locations in a Circular Duct

Traverse Point Number on a Diameter	Number of Traverse Points on a Diameter									
	6	8	10	12	14	16	18	20	22	24
1	4.4	3.3	2.5	2.1	1.8	1.6	1.4	1.3	1.1	1.1
2	14.7	10.5	8.2	6.7	5.7	4.9	4.4	3.9	3.5	3.2
3	29.5	19.4	14.6	11.8	9.9	8.5	7.5	6.7	6	5.5
4	70.5	32.3	22.6	17.7	14.6	12.5	10.9	9.7	8.7	7.9
5	85.3	67.7	34.2	25	20.1	16.9	14.6	12.9	11.6	10.5
6	95.6	80.6	65.8	35.5	26.9	22	18.8	16.5	14.6	13.2
7		89.5	77.4	64.5	36.6	28.3	23.6	20.4	18	16.1
8		96.7	85.4	75	63.4	37.5	29.6	25	21.8	19.4
9			91.8	82.3	73.1	62.5	38.2	30.6	26.1	23
10			97.5	88.2	79.9	71.7	61.8	38.8	31.5	27.2
11				93.3	85.4	78	70.4	61.2	39.3	32.3
12				97.9	90.1	83.1	76.4	69.4	60.7	39.8
13					94.3	87.5	81.2	75	68.5	60.2
14					98.2	91.5	85.4	79.6	73.9	67.7
15						95.1	89.1	83.5	78.2	72.8
16						98.4	92.5	87.1	82	77
17							95.6	90.3	85.4	80.6
18							98.6	93.3	88.4	83.9
19								96.1	91.3	86.8
20								98.7	94	89.5
21									96.5	92.1
22									98.9	94.5
23										96.8
24										98.9

Note: Figures in body of table are percent of duct diameter from inside wall to traverse point.

SOLUTION [11]

Consider the differential element dA in Figure 10.7. The volumetric flow rate (dq) passing dA is given by

$$dq = vdA$$

since

$$dA = rd\, rd\phi \text{ (in cylindrical coordinates)}$$

then

$$dq = vrd\, rd\phi$$

This equation may be integrated across the entire area to obtain the total volumetric rate (Q).

$$Q = \int\int dq = \int_0^{2\pi} \int_0^R vr\, drd\phi$$

Since

$$\frac{dr^2}{2} = \frac{2rdr}{2} = rdr$$

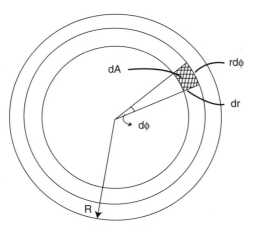

Figure 10.7

the above equation may be rewritten as

$$Q = \int_0^{2\pi} \left[\int_0^R v \, d\left(\frac{r^2}{2}\right) \right] d\phi$$

If the outside integral is evaluated,

$$Q = 2\pi \int_0^R v \, d\left(\frac{r^2}{2}\right)$$

Noting that R (or R^2) is a constant,

$$Q = (\pi R^2) \int_0^1 v \, d\left(\frac{r}{R}\right)^2; \quad r = R, \left(\frac{r}{R}\right)^2 = 1; \quad r = 0, \left(\frac{r}{R}\right)^2 = 0$$

The term in parentheses is the area while the integral is the average velocity. Thus, the area under a plot of v versus $(r/R)^2$ represents the average velocity. A smooth curve can be drawn from v versus r data to obtain the volumetric flow rate (Figure 10.8).

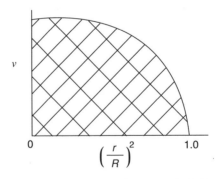

Figure 10.8   Volumetric flowrate calculation.

10.12   SOURCE CHARACTERISTICS AND VARIATIONS

Characterize sources and potential variations.

SOLUTION

Once a flow profile has been established, sampling strategy can be considered. Since sampling collection can be simplified and greatly reduced, depending on flow characteristics, it is best to complete the flow-profile measurement before sampling or measuring concentrations. Source characteristics may be placed in one of four categories.

1. No time variation, with the emission relatively uniform across the stack cross-section. In this case, only one concentration measurement is necessary for accurate results.

2. Steady generation of "contaminants," but nonuniform flow across the sampling location. In this case, a traverse is necessary to measure the average concentration. Typically, this is performed at the points selected for the velocity traverse. The time of sampling at each point should be the same in order to obtain a representative, composite sample.

3. Cyclical operation in which the actual sampling location is ideal and the variation across the stack is relatively uniform when the operation is running. Because the process involves time variation only, the sampling is conducted at one point for extended periods.

4. Nonuniform source and flow conditions. This case requires the most complicated procedure. If there is some measurable cycle related to the process, the sampling can be conducted over this period using simultaneously collected samples. One sample is collected at a reference point and the others at selected traverse points. This is repeated until a complete traverse is made. Results are corrected by using the reference point data as a measure of the time variation.

11 Fluid Particle Dynamics

The collection/recovery of solid or liquid particles in a fluid is based upon the movement of the particle in the fluid stream. In order to understand the mechanisms of particle dynamics it is necessary to examine the basic concepts of particle behavior in a fluid. To be "captured", the particle must be subjected to external forces large enough to separate the particle from the stream.

This Chapter examines the various forces acting on a particle and how they affect particle behavior. The problems that follow are primarily concerned with applications where the fluid is a gas, which is usually assumed to be air. However, the method provided can easily be extended to include other fluids, including liquids.

A number of forces act on a particle moving in a fluid. Three major ones are the *gravitational force*, F_G, the *buoyant force*, F_B, and the *drag force*, F_D. Others can include magnetic, inertial, electrostatic, and thermal forces. All forces must be considered when evaluating the behavior of a particle. This discussion of particle movement in a fluid will primarily examine the gravitational, buoyant, and drag forces. Electrostatic, centrifugal, inertial, and other forces are discussed superficially later in this Chapter.

Consider the following example. A parachutist relies on the phenomenon of terminal settling to assure a safe landing whether the jump is from 500 or 5000 feet. After the parachute is opened, there is a short period of deacceleration. In the absence of additional forces such as wind, a smooth descent at constant velocity results. With particles much smaller than human bodies, the settling velocity is much slower. This can be seen with smoke plumes, which remain aloft for long periods of time. As expected, this reduction in velocity with decreasing particle size continues into the micron size regime. Further reduction in size into the nanometer regime produces an increase in velocity that would not normally be expected. This strange and unique phenomenon is explained in this Chapter.

A total of 12 Problems follow. They vary from simple definitions to detailed particle behavior calculations. Included in the presentation is a discussion of the Cunningham Correction Factor and molecular diffusion – two effects that take on additional importance in the nanotechnology field.

Nanotechnology: Basic Calculations for Engineers and Scientists, by Louis Theodore

11.1   THE GRAVITATIONAL FORCE

Develop an equation describing the force due to gravity in terms of the diameter of a particle.

SOLUTION

All particles are subject to gravity. The gravitational force, F_G, which causes particles and masses to fall to the Earth can be expressed in the form of the general equation below. Assuming that there are no other forces acting on the particle, F_G is given as:

$$F_G = m_p g; \quad \text{consistent units}$$

where m_p = mass of the particle and, g = acceleration of the particle due to gravity.

The mass of the particle is equal to the particle density (ρ_p) multiplied by the particle volume (V_p).

$$m_p = \rho_p V_p$$

The volume of a spherical particle is equal to:

$$V_p = \frac{4}{3} \pi r_p^3 = \frac{\pi d_p^3}{6}$$

where d_p = particle diameter.

The above equation for the gravitational force can then be written as:

$$F_G = \frac{\rho_p \pi d_p^3 g}{6}; \quad \text{consistent units}$$

where ρ_p = density of the particle, d_p = particle diameter, and, g = acceleration of the particle due to gravity.

11.2   THE BUOYANT FORCE

Particles in a fluid can experience both a gravitational and buoyant force. Both act in the vertical direction; however, the gravity force is directed downward toward the center of the Earth while the buoyant force acts in the opposite direction. The buoyant force (F_B) can be written in the same form as the gravity equation presented in the previous problem, that is,

$$F_B = m_a g; \quad \text{consistent units}$$

where m_a = mass of air or fluid displaced, and g = acceleration of the particle due to gravity.

F_B can also be written as:

$$F_B = \rho_a V_a g = \rho_a \frac{\pi d_p^3}{6} g$$

where ρ_a = density of the (air) fluid.

Comparing the buoyant equation with the gravity equation, one can see that these equations are very similar except that ρ_p, and ρ_a represent the density of the particle and the density of the fluid, respectively. For "air" studies ρ_p is always much greater than ρ_a when air is the fluid in which the particle is suspended. Therefore, the term F_B can frequently be ignored. Note that it should not be ignored in "water" applications.

Develop an equation describing the buoyant force in terms of the diameter of a particle.

SOLUTION

Another force acting on a particle suspended in a fluid stream is the buoyant force, F_B. The buoyant force acting on a particle is equal to the weight of the displaced fluid. The concept of buoyant force can be seen from the following example. Consider two identical buckets, one containing water, the other containing air (Figure 11.1). A block of wood with identical size, shape, and density is placed in each bucket. The density of air is much less than the density of water. The buoyant force of the fluid (air) acting on the piece of wood in the bucket filled with air is not great enough to displace the weight of the object. However, the buoyant force of the fluid (water) in the second bucket is large enough to displace the wood. The object thus rises and floats on the fluid if the density (or specific gravity) is less than that of water.

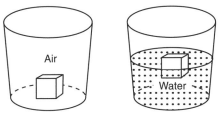

Figure 11.1 Identical objects in two different fluids; specific gravity of object greater than air but less than water.

11.3   THE DRAG FORCE

Define the drag force.

SOLUTION

Whenever a difference in velocity exists between a particle and its surrounding fluid, the fluid will exert a resistive force upon the particle. Either the fluid (gas) may be at rest with the particle moving through it or the particle may be at rest with the gas flowing past it. It is generally immaterial which phase (solid or gas) is assumed to be at rest; it is the *relative* velocity between the two that is important. This resistive force exerted on the particle by the gas is called the *drag*.

In treating fluid flow through pipes, a friction factor term is used in many engineering calculations. An analogous factor, called the drag coefficient, is employed in drag force calculations for flow past particles. The drag force expression, including the drag coefficient, is given by:

$$F_D = \left(\frac{\rho v^2}{2 g_c} \right) A_p C_D$$

For a sphere

$$A_p = \frac{\pi d_p^2}{4}$$

Substituting gives

$$F_D = \frac{\pi d_p^2 \rho v^2 C_D}{8 g_c}; \qquad \text{consistent units}$$

where C_D = drag coefficient, dimensionless; v = particle velocity; d_p = particle diameter; and ρ = density of the (air) fluid.

As indicated above, the drag force arises when a particle moves through a fluid. The particle clears or displaces the fluid immediately in front of it, imparting momentum to the fluid. The drag force produced is equal to the momentum (mv) per unit time imparted to the fluid by the particle. Since the moving particle has a velocity, v, a portion of the particle's velocity is transferred by momentum to the fluid as fluid velocity v_α. The amount of energy imparted from v to v_α is related to a friction factor which is called the drag coefficient, C_D. This coefficient is discussed in the next Problem.

11.4   THE DRAG COEFFICIENT

Define the drag coefficient and quantitatively discuss its effect on the drag force calculation.

SOLUTION

The drag coefficient (C_D) is related to the velocity of the particle and the flow pattern of the fluid around the particle. The Reynolds number (Re) of the particle is used as an indication of this flow pattern. The Reynolds number of the particle is a function of the fluid density, particle diameter, particle velocity, and fluid viscosity (and is dimensionless). The Reynolds number (developed by Osborne Reynolds in 1883) is defined as:

$$Re = \frac{\rho v d_p}{\mu}; \quad \text{dimensionless}$$

where ρ = density of the (air) fluid, v = particle velocity, d_p = particle diameter, and, μ = air (fluid) viscosity.

A mathematical expression describing C_D is very complex. Values of C_D can be estimated by using plots of C_D versus Reynolds number constructed from experimental data. It is essential to determine C_D so that one can solve for F_D, the drag force, as discussed in the previous Problem. From experiment it has been observed that three particle flow regimes exist. The three regimes are *laminar* (or *Stokes*), *transition* or *intermediate*, and *turbulent* (sometimes called the *Newtonian* regime). These regimes are related to the Reynolds number of the particle (Re).

For low values of the Reynolds number (Re < 2) the flow is said to be laminar. Laminar flow is defined as flow in which the fluid moves in layers, one layer sliding smoothly over an adjacent layer with only a molecular interchange of momentum between layers. For much higher values of the Reynolds number (Re > 500) the flow is turbulent. Turbulent flow is characterized by erratic motion of the fluid, with a violent chaotic interchange of momentum throughout the fluid. For Reynolds number values between 2 and 500 the flow pattern is said to be in the *transition* regime, where the flow can be either laminar or turbulent.

Mathematical expressions relating the values of C_D and Re have been estimated from experimental data. Equations for determining C_D in each flow regime are:

$$C_D = \frac{24}{Re_p}; \quad \text{Laminar (Re < 2.0)}$$

$$C_D = \frac{18.5}{Re_p^{0.6}}; \quad \text{Transtion (2 < Re < 500)}$$

$$C_D = 0.44; \quad \text{Turbulent (500 < Re < 200,000)}$$

A comparison between these three equations and experimental data produces results that are in reasonable agreement.

The drag force can be calculated by substituting the proper C_D expression into the earlier drag force equation. The equations for calculating F_D in all three flow regimes become

$$F_D = 3\pi\mu v_p d_p/g_c; \qquad \text{Laminar (Stokes)}$$

$$F_D = 2.31\pi(d_p v_p)^{1.4}\mu^{0.6}\rho^{0.4}/g_c; \qquad \text{(Intermediate)}$$

$$F_D = 0.055\pi(d_p v_p)^2\rho/g_c; \qquad \text{(Newtonian)}$$

Another empirical drag coefficient model is given in reference [12].

$$\log C_D = 1.35237 - 0.608\,10\,(\log \text{Re}) - 0.22961\,(\log \text{Re})^2$$
$$+ 0.098938\,(\log \text{Re})^3 + 0.041528\,(\log \text{Re})^4$$
$$- 0.032717\,(\log \text{Re})^5 + 0.007329\,(\log \text{Re})^6$$
$$- 0.0005568\,(\log \text{Re})^7$$

This is an empirical equation, which has been obtained by the use of a statistical fitting technique. This correlation gives excellent results over the entire range of Reynolds numbers. An advantage of using this correlation is that it is not partitioned for application to only a specific Reynolds number range.

Still another empirical equation [13] is

$$C_D = [0.63 + (4.80/\sqrt{\text{Re}})]^2$$

This correlation is also valid over the entire spectrum of Reynolds numbers. Its agreement with literature is generally good. However, in the range of $30 < \text{Re} < 10\,000$, there is considerable deviation. For $\text{Re} < 30$ or $\text{Re} > 10\,000$ the agreement is excellent.

11.5   EQUATION OF PARTICLE MOTION/BALANCE OF FORCES ON A PARTICLE

Develop the equation describing particle motion.

SOLUTION

If a particle is initially at rest in a stationary gas, and is then set in motion by the application of a constant external force or forces, the resulting motion can be considered to occur in two stages. The first period involves acceleration, during which the particle velocity increases from zero to some maximum velocity. The second stage occurs when the particle remains at this velocity. During the second stage, the particle is not accelerated. The final, constant, and maximum velocity is defined as the terminal settling velocity of the particle. Fortunately, in most studies and calculations, particles reach this terminal settling velocity almost instantaneously.

Newton's second law of motion states that the acceleration produced on a given mass by the action of a given force is proportional to the force and is in the direction of that force. The second law is simply a statement of the equation

$$F = \frac{ma}{g_c}.$$

The sum of the forces can be written as:

$$\sum F = \frac{ma}{g_c} = \frac{m}{g_c}\frac{dv}{dt}$$

where dv/dt = acceleration or change in velocity with respect to time.

Assuming gravity to be the external force, the sum of the forces is equal to a resultant force, F_R:

$$F_R = F_G - F_B - F_D = \frac{m}{g_c}\frac{dv}{dt}$$

As previously stated, the density of the particle is much greater than the density of air and the term F_B can be ignored. As the particle accelerates the velocity will increase. The drag force on the particle also increases with increasing velocity. At some point there will be a value of velocity where F_D will be as large as the other force(s). At this point the resultant force will be zero, and the particle will no longer accelerate. If the particle is not accelerating, then it must be at a constant velocity. This constant velocity, where all the forces balance out, is called the terminal settling velocity. The describing equation with $F_B = 0$ and $F_R = 0$ becomes

$$F_G = F_D; \qquad \text{at terminal settling velocity}$$

Keep in mind that another external force, e.g., centrifugal, electrical, and so on, can replace the gravity force, F_G, in the above development. The units, of course, must be consistent.

11.6 PARTICLE SETTLING VELOCITY EQUATIONS

Obtain the various equations describing the terminal settling velocity of spherical particles.

SOLUTION

The appropriate terminal particle settling velocity equation is derived by setting F_G equal to F_D (see previous Problem). For the three regimes one obtains

$$v_t = \frac{g\rho_p d_p^{\,2}}{18\,\mu}; \qquad \text{Laminar regime}$$

$$v_t = \frac{0.153\, g^{0.71}\, d_p^{1.14}\, e_p^{0.71}}{\mu^{0.43}\, e^{0.29}}\; ; \qquad \text{Transition regime}$$

$$v_t = 1.74 \left(\frac{g d_p e_p}{e} \right)^{0.5} ; \qquad \text{Turbulent regime}$$

Once again, the term g can be replaced by another external force (centrifugal, electrostatic, and so on), provided consistent units are employed.

11.7   DETERMINATION OF THE FLOW REGIME

Illustrate how the "K" factor is employed to determine the appropriate flow regime.

SOLUTION

As described earlier, the terminal velocity of a particle is a constant value of velocity reached when all forces (gravity, drag, buoyancy, and so on) acting on the particle balance; the sum of all the forces is then equal to zero (no acceleration). In order to calculate this velocity, a dimensionless constant K determines the appropriate range of the fluid–particle dynamic laws that apply.

$$K = d_p (g \rho_p \rho / \mu^2)^{1/3}$$

where $K = $ a dimensionless constant which determines the range of the fluid particle dynamic laws; $d_p = $ particle diameter; $g = $ gravity force; $\rho_p = $ particle density; $\rho = $ fluid (gas) density; and, $\mu = $ fluid (gas) viscosity. A consistent set of units that will yield a dimensionless K is: d_p in ft, g in ft/s^2, ρ_p in lb/ft^3, and μ in lb/ft-s. The numerical value of K determines the appropriate law.

$K < 3.3$; Stokes' law range: Re $= d_p v e / \mu \le 2.0$
$3.3 < K < 43.6$; Intermediate law range: $2.0 \le$ Re ≤ 500
$43.6 < K < 2360$; Newton's law range: Re > 500

For the Stokes' law range:

$$v = g d_p^2 \rho_p / 18\, \mu$$

For the Intermediate law range:

$$v = 0.153 g^{0.71}\, d_p^{1.14}\, \rho_p^{0.71} / \mu^{0.43} \rho^{0.29}$$

For Newton's law range:

$$v = 1.74(g d_p\, \rho_p/\rho)^{0.5}$$

Larocca [13] and Theodore [14] using the same approach employed above, defined a dimensionless term W that would enable one to calculate the diameter of a particle if the terminal velocity is known (or given). This particular approach has found application in industrial particle size calculations. The term W – which does not depend on the particle diameter, is given by:

$$W = v^3 \rho^2 / g \mu \rho_p$$

The two key values of W that are employed in a manner similar to that for K are 0.2222 and 1514; that is,

$$W < 0.2222;\ \text{Stokes' law}$$
$$0.2222 < W < 1514;\ \text{Intermediate law}$$
$$1514 < W;\ \text{Newton's law}$$

11.8 SETTLING VELOCITY APPLICATION

Calculate the settling velocity of a particle moving in a gas stream. Assume the following information:

$\rho_p = 0.899\,\text{g/cm}^3$
$\rho = 0.0012\,\text{g/cm}^3$
$\mu_{(AIR)} = 1.82 \times 10^{-4}\,\text{g/cm} \cdot \text{s}$
$g = 980\,\text{cm/s}^2$
$d_p = 45\,\mu\text{m}$

SOLUTION

Calculate the K parameter to determine the proper flow regime. Use the equation provided in the previous problem

$$K = d_p[g\rho_p\rho_a/\mu^2]^{\frac{1}{3}}$$

$$= 45 \times 10^{-4}\,\text{cm}\left[\frac{\left(980\,\frac{\text{cm}}{\text{s}^2}\right)\left(0.899\,\frac{\text{g}}{\text{cm}^3}\right)\left(0.0012\,\frac{\text{g}}{\text{cm}^3}\right)}{\left(1.82 \times 10^{-4}\,\frac{\text{g}}{\text{cm} \cdot \text{s}}\right)^2}\right]^{\frac{1}{3}}$$

$$= 1.43$$

Therefore, the flow regime is laminar; that is, Stokes' Law applies. The settling velocity is calculated using the equation for Stokes' Law:

$$v_t = \frac{g\rho_p d_p^2}{18\mu}; \qquad g_c = 1.0 \text{ (metric units)}$$

$$= \frac{\left(980 \frac{cm}{s^2}\right)\left(0.899 \frac{g}{cm^3}\right)(45 \times 10^{-4} cm)^2}{18\left(1.82 \times 10^{-4} \frac{g}{cm \cdot s}\right)}$$

$$= 5.45 \text{ cm/s}$$

11.9   THE CUNNINGHAM CORRECTION FACTOR

Describe the Cunningham Correction Factor.

SOLUTION

At very low values of the Reynolds number, when particles approach sizes comparable to the mean free path of the fluid molecules; the medium can no longer be regarded as continuous. For this condition, particles can fall between the molecules at a faster rate than predicted by the aerodynamic theories that led to the standard drag coefficients. To allow for this "slip", Cunningham [15] introduced a multiplying correction factor to Stokes' law. This alters the equation to the form

$$v = \frac{fd_p^2 \rho_p}{18\mu}\left(\frac{1 + 2A\lambda}{d_p}\right)$$

or

$$v = \frac{fd_p^2 \rho_p}{18\mu}(C); \qquad C = 1 + \frac{2A\lambda}{d_p}$$

where λ = mean free path of the fluid molecules; $A = 1.257 + 0.40 \exp(-1.10\ d_p/2\ \lambda)$; C = Cunningham Correction Factor.

The modified Stokes' law equation which is usually referred to as the Stokes–Cunningham equation, is then

$$F_D = 3\ \pi\mu d_p v/Cg_c$$

As shown in next problem, the correction factor should definitely be included in the drag force term when dealing with submicron and nanosize particles.

A word of interpretation is in order for λ, the mean free path of the fluid molecules (note that the values of λ is required in the calculation for C). Based on the kinetic theory of gases, λ is given by

$$\lambda = \frac{\mu}{0.499\rho\sqrt{8RT/PM}}; \qquad \text{consistent unit}$$

where μ = gas viscosity; ρ = gas density; R = ideal gas constant; T = absolute temperature of gas; P = absolute measure of gas; and M = molecular weight of gas.

For most gases, λ is approximately 100 nm. For air at 70°F and 1 atm, $\lambda = 653$ nm. As noted above, this correction is of importance only for particles smaller than 1000 nm (or 1.0 μm).

Finally, the CCF may be estimated from the following equation:

$$C = 1.0 + \frac{(6.21 \times 10^{-4})}{d_p} T$$

where T = temperature, K, and d_p = particle diameter, μm (microns).

11.10 CUNNINGHAM CORRECTION FACTOR VALUES FOR AIR AT ATMOSPHERIC PRESSURE

Calculate the Cunningham Correction Factor (CCF) for particle size variation from 1.0 nm to 10^4 nm at temperatures of 70°F, 212°F, and 500°F. Include a sample calculation for a particle diameter of 400 nm (0.4 μm) at 70°F, 1 atm.

SOLUTION

Employ the equations presented in Problem 11.9. The calculated results are provided in Table 11.1, along with a sample calculation.

TABLE 11.1   Cunningham Correction Factors

d_p (nm)	d_p (μm)	C (70°F)	C (212°F)	C (500°F)
1.0	0.001	216.966	274.0	405.32
10	0.01	22.218	27.92	39.90
100	0.1	2.867	3.61	5.14
250	0.25	1.682	1.952	2.528
500	0.5	1.330	1.446	1.711
1000	1	1.164	1.217	1.338
2500	2.5	1.066	1.087	1.133
5000	5	1.033	1.043	1.067
10 000	10	1.016	1.022	1.033

For a d_p of 0.4 μm, the Cunningham Correction Factor should be included. Employing the equation given in the previous problem,

$$A = 1.257 + 0.40e^{-1.10d_p/2\lambda}$$

$$= 1.257 + 0.40\exp -\left(\frac{1.10(0.4)}{(2)(6.53 \times 10^{-2})}\right)$$

$$= 1.2708$$

Therefore

$$C = 1 + \frac{2A\lambda}{d_p}$$

$$= 1 + \frac{(2)(1.2708)(6.53 \times 10^{-2})}{(0.4)}$$

$$= 1.415$$

The results clearly demonstrate that the CCF become more pronounced for nano-sized particles in the 10–1000 nm range. In addition, an increase in temperature also leads to an increase in this effect.

The reader should also note that a comparable effect does not exist for particles settling in liquids until the diameter become less than 10 nm (0.01 μm).

11.11   PARTICLE SETTLING VELOCITY – DIFFERENT REGIMES

Three different sized particles from a nano-operation settle through air. You are asked to calculate the particle terminal velocity and determine how far each will fall in 30 s. Also calculate the size of the particle that will settle with a velocity of 1.384 ft/s.

Assume the particles are spherical. Data are provided below.

Nano-operation particle diameters = 0.4, 40, 400 μm
Air temperature and pressure = 70°F, 1 atm
Specific gravity of particle = 2.31

SOLUTION

English units are employed in the solution to this Problem. As indicated in Problem 11.7, to calculate this velocity, a dimensionless constant K determines the appropriate range of the fluid–particle dynamic laws that apply:

$$K = d_p \left(\frac{g(\rho_p - \rho)\rho}{\mu^2} \right)^{1/3}$$

where K = dimensionless constant that determines the range of the fluid–particle dynamic laws; d_p = particle diameter; g = gravity force; ρ_p = particle density; ρ = fluid (gas) density; and, μ = fluid (gas) viscosity.

Once again, a consistent set of English units that will yield a dimensionless K is: d_p in ft, g in ft/s^2, ρ_p in lb/ft^3, and μ in lb/(ft-s). The numerical value of K determines the appropriate law:

$K < 3.3$; Stokes' law range

$3.3 < K < 43.6$; Intermediate law range

$43.6 < K < 2360$; Newton's law range

For the problem at hand, the particle density is calculated using the specific gravity given.

$$\rho_p = (2.31)(62.4)$$
$$= 144.14 \, \text{lb/ft}^3$$

The density of air is

$$\rho = P(M)/RT$$
$$= (1)(29)/(0.7302)(70 + 460)$$
$$= 0.075 \, \text{lb/ft}^3$$

The viscosity of air is

$$\mu = 0.021 \, \text{cP}$$
$$= 1.41 \times 10^{-5} \, \text{lb/ft·s}$$

The value of K for each fly ash particle size settling in air may now be calculated. Note that $\rho_p - \rho \approx \rho_p$.

For a d_p of 0.40 μm (400 nm)

$$K = \frac{0.4}{(25\,400)(12)} \left(\frac{(32.2)(144.1)(0.075)}{(1.41 \times 10^{-5})^2} \right)^{1/3} = 0.0158$$

For a d_p of 40 μm:

$$K = \frac{40}{(25\,400)(12)} \left(\frac{(32.2)(144.1)(0.075)}{(1.41 \times 10^{-5})^2} \right)^{1/3} = 1.58$$

For a d_p of 400 μm:

$$K = \frac{400}{(25\,400)(12)} \left(\frac{(32.2)(144.1)(0.075)}{(1.41 \times 10^{-5})^2} \right)^{1/3} = 15.8$$

Therefore,

For $d_p = 0.4$ μm; Stokes' law range
For $d_p = 40$ μm; Stokes' law range
For $d_p = 400$ μm; Intermediate law range

For a d_p of 0.4 μm (without the Cunningham Correction Factor):

$$v = \frac{g d_p^2 \rho_p}{18 \mu} = \frac{(32.2)[(0.4)/(25\,400)(12)]^2 (144)}{(18)(1.41 \times 10^{-5})}$$
$$= 3.15 \times 10^{-5} \text{ ft/s}$$

For a d_p of 40 μm:

$$v = \frac{g d_p^2 \rho_p}{18 \mu} = \frac{(32.2)[(40)/(25\,400)(12)]^2 (144)}{(18)(1.41 \times 10^{-5})}$$
$$= 0.315 \text{ ft/s}$$

For a d_p of 400 μm:

$$v = 0.153 \frac{g^{0.71} d_p^{1.14} \rho_p^{0.71}}{\mu^{0.43} \rho^{0.29}}$$
$$= 0.153 \frac{(32.2)^{0.71}[(400)/(25\,400)(12)]^{1.14}(144.1)^{0.71}}{(1.41 \times 10^{-5})^{0.43}(0.075)^{0.29}}$$
$$= 8.21 \text{ ft/s}$$

The distance that the particles will fall in 30 s may also be calculated.

For a d_p of 40 μm:

$$\text{Distance} = (30)(0.315)$$
$$= 9.45 \text{ ft}$$

For a d_p of 400 μm:

$$\text{Distance} = (30)(8.21)$$
$$= 246 \text{ ft}$$

For a d_p of 0.4 μm, the Cunningham Correction Factor should be included. For this diameter (see Problem 11.10), $C = 1.415$.

$$\text{Corrected } v = (3.15 \times 10^{-5})(1.415)$$
$$= 4.45 \times 10^{-5} \text{ ft/s}$$
$$\text{Distance} = (30)(4.45 \times 10^{-5})$$
$$= 1.335 \times 10^{-3} \text{ ft}$$

For the particle traveling with a velocity of 1.384 ft/s, first calculate the dimension-less number, W (see Problem 11.7):

$$W = \frac{v^3 \rho^2}{g \mu \rho_p}$$
$$= \frac{(1.384)^3 (0.075)^2}{(32.2)(1.41)(1.41 \times 10^{-5})}$$
$$= 0.2279$$

Since $W \approx 0.222$, assume Stokes' Law applies, and

$$d_p = \left(\frac{18 \, \mu v}{g \rho_p} \right)^{0.5}$$
$$= \left(\frac{(18)(1.41 \times 10^{-5})(1.384)}{(32.2)(144.1)} \right)^{0.5}$$
$$= 2.751 \times 10^{-4} \text{ ft}$$
$$= 77.6 \, \mu\text{m}$$

No correction factor is required because of the large size of the particle.

11.12   BROWNIAN MOTION/MOLECULAR DIFFUSION

Qualitatively describe Brownian motion.

SOLUTION

Brownian motion (also referred to as molecular diffusion) becomes the dominant collection mechanism for particles less than 500 nm (0.5 μm) and is especially significant as the particles become smaller. In a gas stream, very small particles deflect slightly when gas molecules impact on them. The transfer of kinetic energy from the fast moving gas molecules to the small particle causes this deflection, which is termed Brownian motion.

Diffusivity provides a measure of the extent to which molecular collisions cause very small particles to move in a random manner across the direction of gas flow. The diffusion coefficient in the equation below represents the diffusivity (D) of a particle at given gas stream conditions [16].

$$D = \frac{CKT}{3\pi d_{pa}\mu}; \qquad \text{consistent units}$$

where D = diffusion coefficient (diffusivity); C = Cunningham Correction Factor (dimensionless); K = Boltzmann constant; T = absolute temperature; d_{pa} = particle aerodynamic diameter; μ = gas viscosity.

Small particles obtain a high diffusion coefficient because the diffusion coefficient is inversely proportional to particle size.

Thus, small particles in a fluid are subject to a random displacement known as Brownian motion. This occurs in addition to the net motion in a given direction due to the action of any of the external forces mentioned earlier. This "chaotic" motion represents the statistical average displacement of particle in a given period of time. By definition, the average displacement of all particles over time is then zero. However, the actual path or trajectory taken by a given particle consists of an extremely large number of irregular and zig-zag jumps.

This topic is revisited in the next Chapter. Actual numerical details are provided at that time.

12 Particle Collection Mechanisms

The overall collection/removal process for particulates in a fluid essentially consist of four steps [14].

1. The application of an external force or forces from which the particle develops a velocity (as described in the previous Chapter) that displaces and/or directs the particle to a retrieval section/area.
2. The particle should be retained at the retrieval area with strong enough forces so that it is not re-entrained.
3. As collected/recovered particles accumulate, they are subsequently removed.
4. The ultimate disposition of the particles completes the process.

Obviously, the first is the most important step. The particle collection mechanisms discussed in this Chapter are generally applicable when the fluid is air; however, they may also apply if the fluid is water. In most commercial particulate systems, it is necessary to remove the particles from the gas stream and collect them in layers, dust cakes, or other forms. In the case of electrostatic precipitators, the dust is collected on a layer on a vertical collection plate. In the case of fabric filters [7], the particles collect as dust cakes on vertical bags. This is the initial collection step, and most particulate devices use one or a combination of collection mechanisms to accomplish this objective.

The forces listed below are basically the "tools" that can be used for particulate/recovery collection:

1. Gravity settling
2. Centrifugal action
3. Inertial impaction and interception
4. Electrostatic attraction
5. Thermophoresis and diffusiophoresis
6. Brownian motion

Note that all of these collection mechanism forces are strongly dependent on particle size.

Nanotechnology: Basic Calculations for Engineers and Scientists, by Louis Theodore
Copyright © 2006 John Wiley & Sons, Inc.

Each of these mechanisms receives treatment in this Chapter (Problems 12.1–12.6). Of these, Brownian motion (referred to by some as molecular diffusion) has the most significant effect on nanosized particles. Accordingly, it receives special attention. Other factors affecting collection mechanisms are considered in Problems 12.8–12.12. These effects include:

1. Nonspherical particles
2. Wall effects
3. Multiparticle effects
4. Multidimensional flow

12.1   GRAVITY

Describe the effect gravity force has on the collection process.

SOLUTION

All particulate devices collect/recover particles by a variety of mechanisms involving an applied force – the simplest of which is gravity. Large particles moving slowly (relatively speaking) in a gas stream can settle out and be collected/recovered. This mechanism is responsible for particle capture in the simplest of all devices – the settling chamber. Details of gravity forces can be found in Problem 11.1.

12.2   CENTRIFUGAL FORCE

Describe the effect of centrifugal force on the collection process.

SOLUTION

Centrifugal force is another collection mechanism used for particle capture. The shape or curvature of the collector causes the gas stream to rotate in a spiral motion. Large particles move toward the outside of the wall by virtue of their momentum (Figure 12.1). The particles lose kinetic energy there and are separated from the gas stream. Particles are then acted upon by gravitational forces and are collected. Centrifugal and gravitational forces are both responsible for particle collection in a cyclone.

In a conventional cyclone (Figure 12.2) the entire mass of the gas stream with the entrained particles is forced into a constrained vortex in the cylindrical portion of the cyclone. Upon entering the unit, a particle develops an angular velocity; due to its greater inertia, the particle tends to move outward across the gas streamlines in a

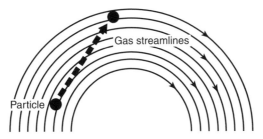

Figure 12.1   Centrifugal force.

tangential rather than a rotary direction, thus attaining a net outward radial velocity (v_r). By virtue of its rotation (v_ϕ) with the carrier gas around the axis of the tube and its higher density with respect to the gas, the entrained particle is forced toward the wall of the tube. Eventually, the particle may reach this outer wall where it is carried by gravity and/or secondary eddies toward the dust outlet at the bottom of the tube. The flow vortex is reversed in the lower (conical) portion of the tube, leaving most of

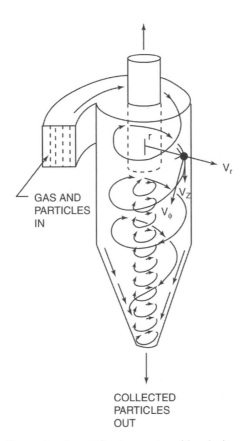

Figure 12.2   Conventional centrifugal separator with velocity vectors shown.

the entrained particles behind. The cleaned gas then passes through the central exit tube (inner vortex) and out of the collector.

For a particle in a centrifugal field, the centrifugal force per unit mass (m), is given by

$$f_c = r\omega^2/g_c = v_\phi^2/rg_c$$

The centrifugal force (F_c), is then

$$F_c = mf_c$$

where r = radius of the path of the particle; ω = angular velocity; and, v_ϕ = tangential velocity.

12.3   INERTIAL IMPACTION AND INTERCEPTION

Describe inertial impaction and interception.

SOLUTION

Inertial impaction occurs when an object (for example, a fiber or liquid droplet), placed in the path of a particulate-laden gas stream, causes the gas to diverge and flow around it. Larger particles, however, tend to continue in a straight path because of their inertia; they may impinge on the obstacle and be collected (as in Figure 12.3). Since the trajectories of particle centers can be calculated, it is possible to theoretically determine the probability of collision. Direct interception also depends on inertia and is merely a secondary form of impaction. Note that the trajectory of a particle center can be calculated; however, even though the center may bypass the target object, a collision might occur since the particle has finite size (Figure 12.4). A collision occurs due to direct interception if the particle's center misses the target object by some dimension less than the particle's radius. Direct interception is, therefore, not a separate principle, but only an extension of inertial impaction.

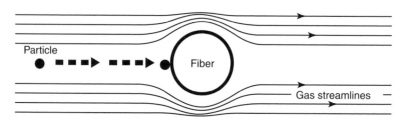

Figure 12.3   Impaction.

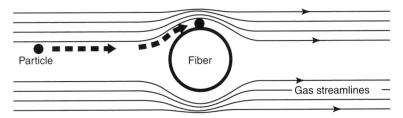

Figure 12.4 Direct interception.

Inertial impaction is analogous to a small car riding down an interstate highway at 65 mph and approaching a merge lane where a slow moving large truck has entered. If the car is not able to get into the passing lane to go around the merged truck, there could be an "impaction" incident. The faster the car is going, the more probable the impaction. The larger the car, the more difficulty it will have going around the truck. Conversely, if the truck enters the interstate road at the same speed as the car, the likelihood of impaction is reduced.

The efficiency of impaction can be related to the inertial impaction parameter shown in the equation below. As the value of this parameter increases, the efficiency of inertial impaction increases. Note that parameter is related to the square of the particle diameter.

$$N_I = \frac{C d_p^2 v \rho_p}{18 \mu d_c}$$

where N_I = impaction parameter, dimensionless; C = Cunningham Correction Factor, dimensionless; d_p = particle diameter; v = particle velocity relative to fluid; d_c = diameter of collector; μ = gas viscosity; and, ρ_p = particle density. Note that a coefficient of 9 (rather 18) is occasionally employed.

Consider the following example. Calculate the inertial impaction parameter for particles of 5 μm diameter and a specific gravity of 1.1 in an air stream at ambient conditions (viscosity = 1.21×10^{-5} lb/ft-s). The aerosol is flowing past a 256 μm diameter spherical collector at a relative velocity of 170 ft/s. Substituting the data into N_I gives

$$N_I = (170)(5.0 \times 3.28 \times 10^{-6})^2(1.1 \times 62.4)/(18)(1.21 \times 10^{-5})(256 \times 3.28 \times 10^{-6})$$

Solving gives

$$N_I = 17.16$$

This important dimensionless number is employed in the design of venturi scrubbers. Interestingly, it can be shown that this dimensionless number is the ratio of a spherical particle's "stopping" distance in the fluid to the collector radius. The stopping distance is defined as the distance a particle (moving with a given velocity) will come to rest in still air – a condition existing at the surface of the collector.

12.4   ELECTROSTATIC EFFECTS

Describe the effect electrostatic forces have on the collection process.

SOLUTION

Another primary particle collection mechanism involves electrostatic forces. The particles can be naturally charged, or, as in most cases involving *electrostatic attraction*, be charged by subjecting the particle to a strong electric field. The charged particles migrate to an oppositely charged collection surface (Figure 12.5). This is the collection mechanism responsible for particle capture in both electrostatic precipitators and charged droplet scrubbers. In an electrostatic precipitator, particle collection occurs due to electrostatic forces only. In a changed droplet scrubber, particle removal primarily occurs by the combined effects of impaction, direct interception, and electrostatic attraction. Particles are charged in these scrubbers in order to enhance other collection mechanisms that may be present.

The electrostatic force, F_E, experienced by a charged particle in an electric field is given by

$$F_E = qE_p; \qquad \text{consistent units}$$

where q = particle charge; and E_p = the collection field intensity (electric field). Theodore and Buonicore [7] provide illustrative examples detailing this calculation.

The forces created in an electric field can be thousands of times greater than gravity. The velocity with which particles in an electric field migrate can be determined in a manner similar to that shown for gravity settling. Setting the electrostatic force equal to the drag force and solving for the terminal (electrostatic) velocity is the usual procedure [7].

Particles in a precipitation will eventually reach a maximum or saturation charge, which is a function of the particle area. The saturation charge occurs when the

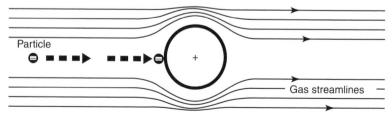

Figure 12.5   Electrostatic attraction.

localized field created by the captured ions is sufficiently strong to deflect the electrical field lines. Particles can also be charged by the diffusion of ions in the gas stream. The strength of the electrical charge imposed on the particles by both mechanisms is
particle size dependent.

Finally, *resistivity* is a measure of the ability of the particle to conduct electricity and is expressed in units of ohm-cm. Particles with low resistivity have a greater ability to conduct electricity (and higher electrostatic attraction) than particles with high resistivity. The following factors influence resistivity:

1. Chemical composition of the gas stream;
2. Chemical composition of the particle;
3. Gas stream temperature.

When a particle is intentionally charged by passing it through a strong electric field, it is called field charging. Very highly charged particles can be created in this manner. As these highly charged particles move through an electric field, they move along the field lines and are collected.

Most aerosols carry some electrical charge, positive or negative, which is continually transferred between randomly colliding particles. This phenomenon is referred to as diffusion charging. Diffusion charging is an important collection mechanism with particles in the nanometer range, and it is important up to 2 microns. For example, particles of 0.4 microns (400 nm) can hold 1 to 200 charges. Alternately, field charging predominates with larger particles.

12.5   THERMOPHORESIS AND DIFFUSIOPHORESIS

Briefly discuss thermophoresis and diffusiophoresis.

SOLUTION

These are two relatively weak forces that can affect particle collection. Thermophoresis is particle movement caused by thermal differences on two sides of the particle. The gas molecule kinetic energies on the hot side of the particle are higher. Therefore, collisions with the particle on this side transfer more energy than the molecular collisions on the cold side. Accordingly, the particle is deflected toward the cold area.

Diffusiophoresis is caused by an imbalance in the kinetic energies being transmitted to the particles by the surrounding molecules. When there is a strong difference in the concentration of molecules between two sides of the particle, there is a difference in the number of molecular collisions. The particle moves toward the area of lower concentration. Diffusiophoresis can be important when the evaporation or concentration of water is involved since these conditions often create substantial concentration gradients.

12.6 ACCELERATION EFFECTS

Discuss and describe acceleration effects.

SOLUTION

As described in Problem 11.5, if a particle is initially at rest in a stationary gas, and is then set in motion by the application of a constant external force or forces, the resulting motion can be considered to occur in two stages. The first period involves acceleration in line with Newton's Law. During this period the particle velocity increases from zero to some maximum velocity. The second stage occurs when the particle has achieved and remains at this velocity. During the second stage, the particle naturally is not accelerated. This final, constant, and maximum velocity is defined as the terminal settling velocity of the particle. As noted earlier, in most air/particulate studies and calculations, particles achieve this terminal settling velocity almost instantaneously. Thus, acceleration effects can be neglected in most real-world applications. The period of acceleration is comparatively short, usually on the order of 100th of a second or less. Nevertheless, this phenomenon should be well understood by the individual striving to achieve a thorough and fundamental understanding of fluid–particle dynamics.

12.7 BROWNIAN MOTION/MOLECULAR DIFFUSION EFFECTS

Discuss Brownian motion and compare its effect to that of gravity settling.

SOLUTION

Strictly speaking, the describing equations and calculations presented earlier in this and the previous Chapter are valid only under restricted conditions. The equations are *not* strictly valid if:

1. The particle is "very" small.
2. The particle is not a smooth rigid sphere.
3. The particle is located "near" the surrounding walls containing the gas.
4. The particle is located "near" one or more other particles.
5. The motion of the fluid and particle is multidimensional.

Only topic (1) is treated in this Problem. Topics 2–5 are examined in Problems 12.8–12.11.

Despite the above limitations, it should be noted that these latter four effects are rarely included in any analysis of a fluid–particle system. It is more common to use an empirical constant or factor that would account for all of these various effects.

Very small particles deflect slightly when they are struck by gas molecules. The deflection is caused by the transfer of kinetic energy from the fast moving gas molecule to the small particle. As the particle size, mass, and density increase, the extent of the particle movement decreases. The extent of this effect is indicated by the diffusivity rate parameter [16]:

$$D_p = \frac{CKT}{3\pi\mu d_p}; \qquad \text{consistent units}$$

where D_p = diffusivity of particles; K = Boltzman constant; T = absolute temperature; C = Cunningham Correction Factor; and μ = gas viscosity.

This deflection begins to be effective as a capture mechanism for particles less than approximately 0.3 microns, and it is significant for particles less than 100 nanometers.

As discussed earlier, due to the bombardment by fluid molecules, particles suspended in a fluid will be subjected to a random and chaotic molecular motion known as Brownian motion. This movement arises in addition to any net motion in a given direction due to the action of other external forces such as gravity. The following relationship expresses the average amplitude or displacement of a spherical particle due to Brownian movement [17]:

$$\Delta s = \sqrt{4RT\ Ct/3\ \pi^2\mu N D_p}$$

where N = molecules per unit volume.

The value Δs is a quantity representing the statistical average linear displacement or amplitude of a particle in a given direction (regardless of sign) in time t. The algebraic average displacement of all particles must be zero at all times. However, the actual path taken by a given particle consists of a very large number of irregular and zig-zag jumps.

Lapple [17] provides a comparison (Table 12.1) of the magnitude of Brownian movement displacement with that due to gravity settling. As expected, it is apparent that Brownian motion becomes appreciable compared to gravity settling for particles smaller than 3 microns in diameter and is entirely predominant for particles smaller than 100 nm. The effects are comparable in magnitude in the 0.5–1.0 micron range.

TABLE 12.1   Comparison of Brownian and Gravitational Displacements for Spherical Particles Suspended in Air and Water

Particle Diameter (microns)	In Air at 70°F, 1 atm		In Water at 70°F	
	Due to Brownian Movement[a]	Due to Gravitational Settling[b]	Due to Brownian Movement[a]	Due to Gravitaitonal Settling[b]
0.1	29.4	1.73	2.36	0.005
0.25	14.2	6.30	1.49	0.0346
0.5	8.92	19.9	1.052	0.1384
1.0	5.91	69.6	0.745	0.554
2.5	3.58	400.0	0.471	3.46
5.0	2.49	1550.0	0.334	13.84
10.0	1.75	6096.0	0.236	55.4

Displacement in 1.0 s (microns)

[a]This is the mean displacement along an axis for $t = 1$ s.
[b]This is the distance settled in 1 s by a particle with specific gravity 2.0, including the Cunningham Correction Factor for settling in air.

12.8   NONSPHERICAL PARTICLES

Discuss the effect nonspherical particles can have on the collection process.

SOLUTION

For particles having shapes other than spherical it is necessary to specify the size and geometric form of the body and its orientation with respect to the direction of flow of the fluid. One major dimension is chosen as the characteristic length, and other important dimensions are given as ratios to the chosen one. Such ratios are called shape factors. Thus, for short cylinders, the diameter is usually chosen as the defining dimension, and the ratio of length to diameter is a shape factor. The orientation between the particle and the stream also should be specified. For a cylinder, a sufficient angle can be formed by the axis of the cylinder and the direction of flow. The projected area is then determine and may be calculated. For a cylinder oriented so that its axis is perpendicular to the flow, $A_p = (l)(d_p)$, where l is the length of the cylinder. For a cylinder with its axis parallel to the direction of flow, A_p is $(\pi/4)$ d_p^2, the same as for a sphere of the same diameter.

Nonspherical bodies generally tend to orient in a preferred direction during the "settling" process. For example, at high Reynolds numbers a disk always falls horizontally, with its flat face perpendicular to its motion; a streamlines shape, on the other hand, falls nose down, in its position of least resistance. At low Reynolds numbers a particle such as a disk or ellipsoid, with three perpendicular symmetrical planes, will fall in any position [18]; theory predicts that such a particle maintains the orientation that it acquired by chance at the start of its fall. The general tendency is for the shape and surface of irregular particles to influence the rate of fall so that the particle falls at a lower velocity than a sphere of equivalent weight.

As mentioned previously, other factors must be introduced into the fluid resistance equations for spheres if equations of the same form are to be used for nonspherical particles. These include a linear dimension equivalent to the diameter of the sphere and a correction factor based on the surface area of the particle. The drag incorporates both of these factors on the aerodynamic behavior of the particles. This has been defined by some as the diameter of a sphere having the same resistance to motion as the particle in a fluid of the same viscosity and at the same velocity.

For isometric particles (cubes, spheres, tetrahedrons, and octahedrons), Pettyjohn and Christiansen [19] found a correlation between the drag coefficient and Reynolds number using the nominal diameter in both terms plus a parameter for the resulting family of curves. The parameter was equal to the ratio of the surface area of a sphere of equal volume to the actual particle surface.

12.9 WALL EFFECTS

Qualitatively discuss wall effects.

SOLUTION

In most particle collection equipment (i.e., settling chambers, cyclones, or electrostatic precipitators) the particles are negligibly small when compared to the dimensions of the unit; therefore, wall effects can usually be neglected. However, in certain other types of collection equipment, wall effects can be pronounced. When the fluid is of finite extent there are two effects. The fluid pulled along by the particle must produce a return flow since it cannot pass through the walls of the containing vessel. Also, since the fluid is stationary at a finite distance from the particle, there is a distortion of the flow pattern, which reacts back on the particle.

When particles are not spherical, the correction factor to be used is the same as for spheres with equivalent diameters. It should be emphasized again that the wall-effect correction factors are valid only at very low values of the Reynolds number, probably less than 1.0. Correction factors corresponding to velocities greater than those occurring in the streamline (low Re) region become much more complex, and the reader is referred to the literature [20–22]. At very high velocities, wall effects are negligible.

12.10 MULTIPARTICLE EFFECTS

Qualitatively discuss multiparticle effects.

SOLUTION

In the removal of particles from gas streams it is almost inevitable that large numbers of particles will be involved. It is also very likely that the particles will influence one another. Therefore, equations for the fluid resistance to the movement of single particles have to be modified to account for such interactions between particles. Particle interactions can become appreciable, even at very low concentrations. Even a particle-volume concentration (the ratio of particle volume to total volume) of 0.2% will increase the fluid resistance to particle movement by about 1% [23]. In general, for volume concentrations below 1%, the effect of particle interactions may be neglected.

If two identical particles separated by only a few diameters move through a viscous fluid, the fluid flows around the particles so that the resulting viscous force is greater than that acting on a single particle; thus, the terminal settling velocity is smaller than that predicted by Stokes' Law. In such hindered settling, the particles are sufficiently close together so as to cause the velocity gradients surrounding each particle to be affected by the presence of the neighboring particles. Also, in settling, the particles displace fluid and generate an appreciable upward velocity. The fluid velocity is, then greater with respect to the particle than with respect to the apparatus. The effective density of the fluid can be taken as that of the fluid–particle system itself and can be calculated from the composition of the fluid and the densities of the particles and the fluid.

12.11 MULTIDIMENSIONAL FLOW

Discuss multidimensional flow.

SOLUTION

Previous discussions on particle motion were limited to the unidimensional case, i.e., the parallel movement of a particle relative to the fluid. However, this is not always the case. This situation is defined as multidimensional flow. Equations must then be developed to describe each of the velocity components of the particle. The main complication arises with the drag if more than one relative velocity component exists.

Although a two-dimensional particle flow problem may exist, only one-dimensional drag effects are considered in this text. For example, consider a particle discharged to the atmosphere from a tall stack. The particle, while settling (vertically) under the influence of gravity, may also have a horizontal velocity component due to atmospheric motion (wind). No slip between the particle and air in the horizontal

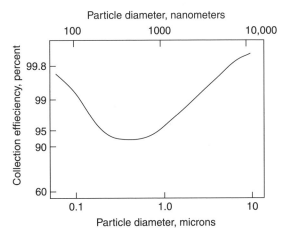

Figure 12.6 Effect of particle size on efficiency.

direction is normally assumed for this situation; i.e., the particle and air are moving with the same velocity in the specified direction. Only drag effects in the vertical direction need to be considered.

12.12 COLLECTION EFFICIENCY FOR NANOSIZED/SUBMICRON PARTICLES

Explain why nearly all particle size collection efficiency curves for high-efficiency recovery/control devices roughly take the form shown in Figure 12.6.

SOLUTION [8]

As illustrated in Figure 12.6, the collection efficiency for particulate control increases with increasing particle size over nearly the entire particle size range. However, for particles less than approximately 1.0 μm, the trend reverses and efficiency increases with decreasing size. This phenomenon is experienced by almost all particulate control devices, for example, baghouses, venturi scrubbers, electrostatic precipitators, and so on, and arises primarily because of molecular diffusion effects discussed in Problem 12.6. The random, chaotic motion of nanosized and submicron particles, similar to that predicted by the kinetic theory of gases, becomes more pronounced as the particle size decreases and approaches the molecular diameter of gases, resulting in higher efficiencies. This becomes an important consideration for systems requiring extremely high efficiencies, for example, in excess of 99.5%.

13 Particle Collection Efficiency

CHAPTER CONTRIBUTOR: STACY SHAFER

As described earlier, recovery/control equipment for particulates fall into five general classes: (1) gravity settlers, (2) centrifugal separators (cyclones), (3) electro-static precipitators, (4) fabric filters, and (5) wet scrubbers. Preliminary knowledge of the particulates is required in order to select and design compatible equipment, especially from a collection efficiency viewpoint. The particulate properties most fundamental to the performance, efficiency, and the choice of equipment are, par-ticle-size distribution, shape, structure, density, composition, electrical conductivity, abrasiveness, corrosiveness, flammability, hygroscopic properties, flowability, tox-icity, and agglomeration tendencies. Also important is knowledge of the gas stream properties including temperature, pressure, humidity, density, viscosity, dew point for condensable components, electrical conductivity, corrosiveness, toxicity, com-position, and flammability. The process conditions include the allowable pressure drop, electrical power requirements, particle concentration, and gas volumetric flow rate. The plant factors include maintenance, space limitations, availability of utilities, safety and health protection, disposal facilities, and materials of construc-tion. Finally, knowledge of auxiliary equipment is also required, including pumps, fans, compressors, motors, ducts, valves, control instrumentation, storage facilities, and conveying equipment [7].

Historically, there have been relatively little or few changes in traditional particu-late equipment in the past century. Unless some new control methodology is devel-oped in the near future, it is safe to assume at this time that similar type equipment will probably be employed for most, if not all, nanoemissions. In effect, it is antici-pated that "new" gaseous and particulate (no matter how small) emissions arising from nanoapplications will probably be "managed" by employing technology as it exists today. The reduction and control of these discharges will be difficult, but not impossible. Thus, it can be concluded that the equipment presently in place and available will probably see extensive use in the control of emissions from nano-technology processes.

The reader should note that the calculation of collection efficiencies is currently that employed for larger than nanosized particles. The applicability of these pro-cedures to submicron sized particles is probably valid, although the strange and unpredictable nature of nanoparticles may invalidate the present methodology.

Nanotechnology: Basic Calculations for Engineers and Scientists, by Louis Theodore
Copyright © 2006 John Wiley & Sons, Inc.

Thus, particle collection efficiency is certain to become a major concern for nanoemissions in the near future. This last Chapter of Part 2 addresses that issue. Ultimately, the trade-off between collection efficiency with capital plus operating costs and environmental concerns will have to be made by the engineer and/or scientist. Although the problems to follow are primarily concerned with particle capture efficiency, the reader should note that the last Problem (13.10) provides an overall review of material presented in Part 2.

13.1   COLLECTION EFFICIENCY: LOADING DATA

Given the following inlet loading and outlet loading (grains/ft^3) of a particulate control unit, determine the collection efficiency of the unit.

Inlet loading $= 2 \text{ gr/ft}^3$
Outlet loading $= 0.1 \text{ gr/ft}^3$

SOLUTION

Collection efficiency is a measure of the degree of performance of a control device; it specifically refers to the degree of removal of an agent and may be calculated through the application of the conservation law for mass. *Loading* refers to the concentration of the agent, usually in grains (gr) per cubic feet of gas stream.

The equation describing collection efficiency (fractional), E, in terms of inlet and outlet loading is

$$E = \frac{\text{Inlet loading - Outlet loading}}{\text{Inlet loading}}$$

Calculate the collection efficiency of the control unit in percent for the data provided.

$$E = \frac{2 - 0.1}{2} 100 = 95\%$$

The term η is also used as a symbol for efficiency E.

The reader should also note that the collected amount by the control unit is the product of E and the inlet loading. The amount discharged to the atmosphere is given by the inlet loading minus the amount collected.

13.2   COLLECTION EFFICIENCY: MASS RATE

The hazardous waste flow rate into a treatment device is 100 lb/h. Calculate the waste rate leaving the unit to achieve a collection efficiency of

1. 95%
2. 99%

3. 99.9%

4. 99.99%

5. 99.9999%

SOLUTION

The definition of collection efficiency, E, in terms of mass flowrate in $\dot{m}_{in}$ and mass flowrate out, $\dot{m}_{out}$, is

$$E = \left(\frac{\dot{m}_{in} - \dot{m}_{out}}{\dot{m}_{in}}\right)(100)$$

The above equation may be rewritten for $\dot{m}_{out}$:

$$\dot{m}_{out} = \dot{m}_{in}(100 - E)100$$

1. The mass flowrate out, $\dot{m}_{out}$, for an E of 95% is

$$\dot{m}_{out} = 100(100 - 95)100$$
$$= 5 \text{ lb/h}$$

2. The mass flowrate out, $\dot{m}_{out}$, for an E of 99% is

$$\dot{m}_{out} = 100(100 - 99)100$$
$$= 1 \text{ lb/h}$$

3. The mass flowrate out, $\dot{m}_{out}$, for an E of 99.9% is

$$\dot{m}_{out} = 100(100 - 99.9)100$$
$$= 0.1 \text{ lb/h}$$

4. The mass flowrate out, $\dot{m}_{out}$, for an E of 99.99% is

$$\dot{m}_{out} = 100(100 - 99.99)100$$
$$= 0.01 \text{ lb/h}$$

5. The mass flowrate out, $\dot{m}_{out}$, for an E of 99.9999% is

$$\dot{m}_{out} = 100(100 - 99.9999)100$$
$$= 0.0001 \text{ lb/h}$$

13.3 EFFICIENCY OF MULTIPLE COLLECTORS

A cyclone is used to collect particulates with an efficiency of 60%. A venturi scrubber is used as a second downstream control device. If the required overall efficiency is 99.0%, determine the minimum operating efficiency of the venturi scrubber.

SOLUTION

Many process systems require more than one piece of equipment to accomplish a given task. The efficiency of each individual collector or equipment may be calculated using the procedure set forth earlier. The overall efficiency of multiple collectors may be calculated from the inlet stream to the first unit and the outlet stream from the last unit. It may also be calculated by proceeding sequentially through the series of collectors.

Calculate the mass of particulate leaving the cyclone using a rate basis of 100 lb of particulate entering the unit. Use the efficiency equation:

$$E = (\dot{m}_{in} - \dot{m}_{out})/(\dot{m}_{in})$$

where E = fractional efficiency and $\dot{m}$ = mass rate.
Rearranging the above gives:

$$\dot{m}_{out} = (1 - E)(\dot{m}_{in}) = (1 - 0.6)(100) = 40 \, \text{lb}$$

Calculate the mass of particulate leaving the venturi scrubber using an overall efficiency of 99.0% (0.99, fractional basis):

$$\dot{m}_{out} = (1 - E)(\dot{m}_{in}) = (1 - 0.99)(100) = 1.0 \, \text{lb}$$

Calculate the efficiency of the venturi scrubber using $\dot{m}_{out}$ from the cyclone as $\dot{m}_{in}$ for the venturi scrubber. Use the same efficiency equation above and convert to percent efficiency:

$$E = (\dot{m}_{in} - \dot{m}_{out})/(\dot{m}_{in}) = (40 - 1.0)/(40) = 0.975 = 97.5\%$$

13.4 PENETRATION

Define and discuss penetration.

SOLUTION

An extremely convenient efficiency-related term employed in particulate control calculations is the penetration, P. By definition:

$$P = 100 - E, \qquad \text{percent basis}$$
$$P = 1 - E, \qquad \text{fractional basis}$$

Note that there is a 10-fold increase in P as E goes from 99.9% to 99%. For a multiple series of n collectors, the overall penetration is simply given by:

$$P = P_1 P_2 \ldots P_{n-1} P_n$$

For particulate control, penetrations and/or efficiencies can be related to individual size ranges. The overall efficiency (or penetration) is then given by the contribution from each size range; that is, the summation of the product of mass fraction and efficiency for each size range. This is examined in more detail later in this Chapter.

13.5 COLLECTION EFFICIENCY: NUMBERS BASIS

Consider a volume of aerosol which contains one hundred 1 μm particles and one hundred 100 μm particles. The efficiency of separation is 90% for 1 μm particles and 99% for 100 μm particles, calculate the overall efficiency on

1. A numbers basis, E_N;
2. A mass basis, E.

SOLUTION

Efficiency warrants further discussion. The efficiency of a particulate control device is usually expressed as the percentage of material collected by the unit compared with that entering the unit. It may be calculated on a particle number basis:

$$E_N = \left(\frac{\text{Particles collected}}{\text{Particles entering}} \right) \times 100$$

or (as shown earlier) on a total weight basis:

$$E = \left(\frac{\text{Inlet loading} - \text{Outlet loading}}{\text{Inlet loading}} \right) \times 100$$

It is extremely important to distinguish between the two. Larger particles, which possess greater mass and are more easily removed in a control device, will contribute much more to the efficiency calculated on a weight basis. Thus, if on a particle count basis, ninety 1 μm and ninety-nine 100 μm particles will be removed out of a total of 200. This gives a particle count efficiency of:

$$E_N = \left(\frac{189}{200}\right) \times 100 = 94.5\%$$

On a weight basis, however, if a 1 μm particle has unit mass, a 100 μm particle has 10^6 mass units. The weight efficiency is then given by:

$$E = \left(\frac{90(1) + 99(10^6)}{100(1) + 100(10^6)}\right) \times 100 = 99\%$$

Any expression of the efficiency of a particulate removal device is therefore of little value without a careful description of the size spectrum of particles involved.

It is interesting to note certain observations. The smaller particles, equivalent in mass to a considerably smaller number of large particles, have a much greater impact on visibility, health, and water droplet nucleation than the larger particles. When large tonnages are involved, the high mass efficiencies often reported for particle collection may lead to overly optimistic conclusions about emissions. The small weight percentages of particles that pass through the collector can still represent large *numbers* of particles escaping, usually to the atmosphere. It would then seem that tonnage-collection figures and weight-removal efficiencies may really not be that adequate to delineate the entire particulate emission problem.

The reader should also note that this analysis can also be applied to nanosized particles.

13.6   PARTICLE SIZE–COLLECTION EFFICIENCY RELATIONSHIPS

Describe the relationship between particle size and collection efficiency.

SOLUTION

Owing to the combined action of the various collection mechanisms described in Chapter 12, the performance of particulate control devices often has the particle size–efficiency relationship shown in Problem 12.10. Above 100 microns, particles are collected with very high efficiency by inertial impaction, electrostatic attraction, and even gravity settling. Efficiency remains high throughout the range of 10 to 100 microns due to the everpresent inertial and/or electrostatic forces (depending on type of collector), both of which are approximately proportional to the square of the particle diameter. For particles less than 10 microns, the limits of inertial forces and electrostatic forces begin to become apparent, and the efficiency drops. Efficiency due to these collection mechanisms reaches negligible levels between 3 and 0.3 microns depending on factors such as gas velocities (inertial forces) and electrical field strengths (electrostatic attraction).

Below 0.3 microns (300 nm), Brownian motion (molecular diffusion) begins to become more pronounced. Accordingly, the overall efficiency curve begins to rise in the very small size range.

The result of these various collection mechanisms is a low collection efficiency particle size range of 0.1 to 10 microns. In many control recovery devices, none of the collection mechanisms is highly efficient for particles in this range. These particles can be classified as "difficult-to-control" due to the inherent limitations of the collection mechanisms.

The above relationship, which has been reported in a number of studies of actual sources, indicates that sources generating high concentrations of particles in the 0.1 to 1.0 micron range may pose an especially challenging control problem. The data also suggest that particles in the nanosize range (<100 nm) are captured with relative ease. However, no one can say with certainty what the behavior of particles < 50 nm will be.

13.7 COLLECTION EFFICIENCY: SURFACE AREA BASIS

The size–collection efficiency information for a control device is provided in Table 13.1.

1. Calculate the overall efficiency for the unit on a number basis.
2. Calculate the overall efficiency for the unit on a mass basis.
3. Calculate the overall efficiency for the unit on a volume basis.
4. Calculate the overall efficiency for the unit on a surface area basis.

SOLUTION

The reader is referred to earlier Problems in this Chapter, particularly Problem 13.5.

TABLE 13.1 Size–Collection Efficiency Data

d (μm)	Collection Efficiency (%)	Number of Particles
1.0	0	10000
100	100	1

1. On a number basis,

$$E_N = \frac{(10\,000)(0) + (1)(1.0)}{10\,000 + 1}$$

$$= \frac{1}{10\,000 + 1} = \frac{1}{10\,001}$$

$$= 0.0001 = 0.01\%$$

2. On a mass basis,

$$E = \frac{(1)(100)^3(1.0) + (10\,000)(1.0)^3(0)}{(1)(100)^3 + (10\,000)(10)^3}$$

$$= \frac{10^6 + (0)}{10^6 + 10^4}$$

$$= 0.99 = 99\%$$

3. On a volume basis,

$$E_V = 0.99 = 99\%$$

since volume is proportional to mass.

4. On a surface area basis, one notes that the surface area of the particle is proportional to the square of the diameter. Therefore,

$$E_{SA} = \frac{(1)(100)^2(1.0) + (10\,000)(1.0)^2(0)}{(1)(100)^2 + (10\,000)(1.0)^2}$$

$$= \frac{10^4}{10^4 + 10^4}$$

$$= 0.5 = 50\%$$

13.8 PARTICLE SIZE DISTRIBUTION/SIZE–EFFICIENCY CALCULATION

The following data are provided in Table 13.2. Calculate the overall collection efficiency.

TABLE 13.2 Particle Size Distribution and Efficiency Information

PSR (μm)	% in PSR	% LTSS	% GTSS	$\bar{d}_p$ (μm)	E_i (%)
0–5	13	13	87	2.5	0.6
5–10	12	25	75	7.5	7.7
10–15	9	34	66	12.5	19.2
15–20	8	42	58	17.5	38.3
20–30	9	51	49	25.0	56.2
30–50	11	62	38	40.0	79.1
50–100	13	75	25	75.0	98.9
100+	25	—	—	100.0+	100

PSR = particle size range; % in PSR = percent in particle size range; % LTSS = percent less than stated size; % GTSS = percent greater than stated size; $\bar{d}_p$ = average particle size in PSR; E_i = mass efficiencies in PSR.

Note that columns 1–4 represent particle size distribution data (see Chapter 9). Columns 5 and 6 provide size-efficiency information.

SOLUTION

Refer to Table 13.2. The overall collection efficiency is obtained from the cross product of columns 2 and 6 for each PSR and summing the results, i.e.,

$$E = \sum_{i=1}^{m} [(2) \times (6)]/100; \qquad \text{percent basis}$$

The calculation is provided below and in Table 13.3.

From Table 13.3,

$$E = \frac{5741}{100} = 57.41\%$$

$$= 0.5741; \qquad \text{fractional basis}$$

TABLE 13.3 Calculation of Overall Collection Efficiency

PSR (μm)	2×6
0–5	7.8
5–10	92.4
10–15	172.8
15–20	306.4
20–30	505.8
30–50	870.1
50–100	1285.7
100+	2500
	5741

13.9 CHECK FOR EMISSION STANDARDS COMPLIANCE: NUMBERS BASIS

As a consulting engineer, you have been contracted to modify an existing control device used in nanofeedstock byproduct emission removal. The federal standards for emissions have been changed to a total numbers basis. Determine if the unit will meet an effluent standard of $10^{5.7}$ particles/acf (actual cubic foot). Data for the unit are given below.

> Average particle size, $d_p = 10 \, \mu m$; assume constant
> Byproduct specific gravity $= 2.33$
> Inlet loading $= 3.0 \, gr/ft^3$
> Efficiency (mass basis), $E = 99\%$

SOLUTION

The outlet loading (OL) is

$$OL = (1.0 - 0.99)3.0$$
$$= 0.030 \, gr/ft^3$$

Assume a basis of $1.0 \, ft^3$:

$$Particle \ mass = \rho_p V_p = \rho_p \frac{\pi d_p^3}{6}$$
$$= \left(\frac{\pi[(10 \, \mu m)(0.328 \times 10^{-5} \, ft/ \, \mu m)]^3 (2.33)(62.4 \, lb/ft^3)}{6} \right)(7000 \, gr/lb)$$
$$= 1.880 \times 10^{-8} \, gr/particle$$

$$Number \ of \ particles = (0.03 \, gr)/(1.880 \times 10^{-8} \, gr/particle)$$
$$= 1.596 \times 10^6 \ particles \ in \ 1 \ ft^3$$

$$Allowable \ number \ of \ particles/ft^3 = 10^{5.7}$$
$$= 5.01 \times 10^5$$

Therefore, the unit will not meet a numbers standard.

13.10 ANDERSON 2000 SAMPLER

Given Anderson 2000 sampler data from a particulate sidestream emission from a nanoprocess, you have been requested to plot a cumulative distribution curve on log-probability paper and determine the mean diameter and geometric standard deviation of the particulate emission. Pertinent data are provided in Table 13.4.

$$\text{Sampler volumetric flow rate, } q = 0.5 \, \text{cfm}$$

See also Figure 13.1 for aerodynamic diameter vs. flow rate data for an Anderson sampler.

SOLUTION [8, 24]

As described in Chapter 9, the Anderson sampler consists of a series of stacked stages and collection surfaces. Depending on the calibration requirements, each stage contains from 150 to 400 precisely drilled jet orifices, identical in diameter in each stage but decreasing in diameter on each succeeding stage. A constant flow of air is drawn through the sampler so that as the air passes from stage to stage through the progressively smaller holes, the velocity increases as the air stream makes a turn at each stage; thus, the particle gains enough inertia to lose the aerodynamic drag. It is "hurled" from the air stream and impacted on the collection surface. The particle is considered aerodynamically sized the moment it leaves the turning air stream. Adhesive, electrostatic, and Van der Waal's forces hold the particles to each other and to the collection surface.

TABLE 13.4   Anderson 2000 Sampler Data

Plate Number	Tare Weight (g)	Final Weight (g)
0	20.48484	20.48628
1	21.38338	21.38394
2	21.92025	21.92066
3	21.55775	21.55817
4	11.40815	11.40854
5	11.61862	11.61961
6	11.76540	11.76664
7	20.99617	20.99737
Backup filter	0.20810	0.21156

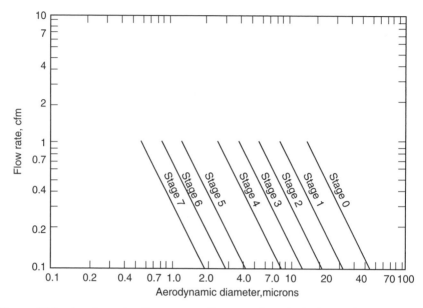

Figure 13.1 Aerodynamic diameter vs. flow-rate through Anderson Sampler for an impaction efficiency of 95%.

Table 13.5 provides the net weight (in mg), percent of total weight, and cumulative percent for each plate. A sample calculation (for plate 0) follows:

$$\text{Net weight} = \text{Final weight} - \text{Tare weight}$$
$$= 20.48628 - 20.48484$$
$$= 1.44 \times 10^{-3}\,\text{g}$$
$$= 1.44\,\text{mg}$$
$$\text{Percent of total wt.} = (\text{net wt.}/\text{total net wt.})(100)$$
$$= (1.44/10.11)(100\%)$$
$$= 14.2\%$$

Calculate the cumulative percent for each plate. Again for plate 0,

$$\text{Cumulative \%} = 100 - 14.2$$
$$= 85.8\%$$

TABLE 13.5   Anderson 2000 Sampler Data

Plate	Number Net Weight (mg)	Percent of Total Weight
0	1.44	14.2
1	0.56	5.5
2	0.41	4.1
3	0.42	4.2
4	0.39	3.9
5	0.99	9.8
6	1.24	12.3
7	1.20	11.9
Backup filter	3.46	34.2
Total	10.11	100.0

For plate 1,

$$\text{Cumulative } \% = 100 - (14.2 + 5.5)$$
$$= 80.3$$

Table 13.6 shows the cumulative percent for each plate.

Using the Anderson graph in Figure 13.1 determine the 95% aerodynamic diameter at $q = 0.5$ cfm for each plate or stage (see Table 13.7).

The cumulative distribution curve is provided on log-probability coordinates in Figure 13.2 (see Chapter 9).

The mean particle diameter is the particle diameter corresponding to a cumulative percent of 50%.

$$\text{Mean particle diameter} = Y_{50} = 1.6 \, \mu\text{m}$$

TABLE   13.6   Cumulative   Percent for Each Plate

Plate No.	Cumulative Percent
0	85.8
1	80.3
2	76.2
3	72.0
4	68.1
5	58.3
6	46.0
7	34.1
Backup filter	—

TABLE 13.7   95% Aerodynamic Diameters

Plate No.	95% Aerodynamic Diameter (μm)
0	20.0
1	13.0
2	8.5
3	5.7
4	3.7
5	1.8
6	1.2
7	0.78

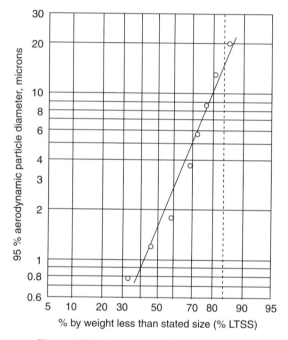

Figure 13.2   Cumulative distribution curve.

The distribution appears to approach log-normal behavior. The particle diameter at a cumulative percent of 84.13 is

$$Y_{84.13} = 15\ \mu m$$

Therefore, the geometric standard deviation is

$$\sigma_G = Y_{84.13}/Y_{50}$$
$$= 15/1.6$$
$$= 9.4$$

References: Part 2

1. L. Theodore and R. Kunz, *Nanotechnology: Environmental Implications and Solutions*, Hoboken NJ: John Wiley & Sons, 2005.

2. L. Theodore, J. Reynolds, and K. Morris, *Concise Dictionary of Environmental Terms*, Boca Raton FL: CRC/Lewis Publishers, 1997.

3. W. Bartknecht, Brenngas und Staufexplosemen, Forsehungebericht F 45 des Bundesinstitutes fur Arbeilssckutz, Kobleng, Germany, 1971.

4. W. Barlknecht, *Explosions*, New York City: Springer-Verlag, 1981.

5. L. Theodore, personal notes, 2000.

6. A. Flynn and L. Theodore, *Health, Safety and Accident Management in the Chemical Process Industries*, New York City: Marcel Dekker, 2002

7. L. Theodore and A. Buonicore, *Industrial Control Equipment for Gaseous Pollutants*, Boca Raton FL: CRC Press, 1976.

8. J. Reynolds, L. Theodore, and J. Jeris, *Handbook of Chemical and Environmental Engineering Calculations*, Hoboken NJ: John Wiley & Sons, 2002.

9. R. Perry and D. Green (editors), *Perry's Chemical Engineers' Handbook*, 7th edition, New York City: McGraw-Hill, 1997.

10. A. Gosh, *Nanotechnology will initiate a quantum leap in manufacturing efficiency*, ARC Insights, #202-49M, ARC Advisory group, Dedham, MA, October 23, 2002.

11. P. Biswas and W. Chang-Yu, *J. Air & Waste Man. Assoc.*, "Nanoparticles and the Environment", Vol. 55, 708–746, June 2005.

12. L. Theodore and A. Buonicore, personal notes, 1975.

13. S. Barnea, I. Mizraki, Ph.D. thesis, Haifa University, Haifa, 1972.

14. L. Theodore, personal notes, 1984.

15. E. Cunningham, *Proc. R. Soc. London Ser. A*, 83, 357 (1910).

16. Einstein, A., *Ann. Physik*, 19, 289 (1906).

17. C. Lapple, *Fluid and Particle Mechanics*, University of Delaware, Newark, Delaware, 1951.

18. J. F. Heiss, and J. Coult, *Chem. Eng. Prog.*, 48, 133 (1952).

19. E. Pettyjohn and E. Christiansen, *Chem. Eng. Prog.*, 44, 157 (1948).

20. P. G. W., Hawksley, *B. C. V. R. A. Bull.*, 15, 105 (1951).

21. H. Faxen, *Ann. Phys.* (Leipzig), 68, 89, (1922).

22. H. Liebster, *Ann. Phys.* (Leipzig), 82, 541 (1927).

23. W. Strauss, *Industrial Gas Cleaning*, New York: Pergamon Press, 1967.

24. L. Theodore and A. Buonicore, *Air Pollution Control Equipment: Volume I, Particulates*, Boca Raton FL: CRC Press, 1988.

PART 3
Applications

There are a host of topics that can be included in and fall under the "Applications" umbrella. This arises because of the interdisciplinary nature of nanotechnology. The six topics chosen for study include material that one could define not only as engineering/science related but also management policies, societal concerns, and ethics. In effect there is more contained in this Part than just the traditional engineering (including chemical, mechanical, and electrical), chemistry, physics, material science, and mathematics.

Nanoparticles that are small enough to enter human cells have the opportunity to interact (in a positive or negative sense) with the cells' biochemical functions. In addition, nanoparticles can absorb ultraviolet light and trigger chemical reactions. Scientists want to engineer these properties to produce new drugs, deliver drugs, clean up pollution, develop warfare armament, and improve consumer products, like dental products, sunscreen, and so on. Since nanoparticles can have unexpected effects on the environment and human health, especially if they are used in medicine, the next Part reviews environmental concerns.

Despite some recent modest successes in the nanotechnology field, most concede, however, that real, commercial-scale success in this arena is still years, if not decades, away. Nonetheless, despite the fact that the quickly evolving field of nanotechnology integrates the well-established disciplines of engineering and science, it also represents a brave new world – a futuristic, imaginative journey that some might say is worthy of a Jules Verne novel.

At a time when most industrialized nations are investing heavily in nanotechnology research and development, a plethora of potential applications – both practical and fantastic – are being considered. As indicated above, cautious observers note that the potential environmental, health, and safety risks associated with nanotechnology are not being adequately studied and that the complex ethical, legal, and societal implications of this powerful new technological paradigm are not being explored or debated on a large scale [1, 2]. Some of these issues are discussed in the Problems that follow.

While some expectations from nanotechnology may be highly hyped and overestimated in the short term, many feel that the long-term implications for healthcare, productivity, and the environment, among others, are underestimated when one

Nanotechnology: Basic Calculations for Engineers and Scientists, by Louis Theodore
Copyright © 2006 John Wiley & Sons, Inc.

considers the depth and breadth of technological breakthroughs recorded to date, and the pace at which further research and applications are being undertaken [1, 3]. Only time will tell.

As is usually the case in preparing Problems involving applications, the problem of what to include and what to omit has been particularly difficult. However, every attempt has been made to offer material to individuals with a limited technical background at a level that should enable them to better cope with some of the complex and real-world applications that may arise in the future. After some thought it was decided to include the following 6 topics:

1. Legal considerations
2. Size reduction
3. Industrial processes
4. Ventilation
5. Atmospheric dispersion
6. Ethics

It is hoped that a review of these areas will prove useful to interested readers.

14 Legal Considerations [1]

Technology affects almost every area of human activity in one way or another. And so, one can expect that legal relations between people will have to be taken into account. Even though anticipated developments in nanotechnology may yet be in the realm of speculation, those involved with nanotechnology and those involved with law should consider how nanotechnology and law might interact.

It is incumbent upon those engaged in any area of technological development to acquire a basic understanding of patent law because the patent portfolio of a company, particularly one focused on research and development, may represent its most valuable asset(s). Certain activities, such as premature sale or public disclosure, can jeopardize one's right to obtain a patent. Patents are creatures of the national law of the issuing country and are enforceable only in that country. Thus, a U.S. patent is enforceable only in the United States. To protect one's invention in foreign countries, one must apply in the countries in which protection is sought. One can obtain a general idea of the developmental progress of a new technological field by monitoring the number of patents issued in that field.

The readers should note that Problems 14.1–14.5 presented in this Chapter were drawn from material prepared by A. Calderone on Legal Considerations, which appears in reference [1].

14.1 INTELLECTUAL PROPERTY LAW

Provide a general introduction to intellectual property law.

SOLUTION

One area of law with which the nanotechnologist must be concerned is intellectual property law. Technological development is all about ideas. Ideas have commercial value only if they can be protected by excluding others from exploiting those ideas. Typically, the way to protect ideas is through intellectual property rights such as patents, trademarks, copyrights, trade secrets, and maskworks. Patents can be used to protect useful inventions, ornamental designs, and even botanical plants. The patent allows the owner of the patent the right to prevent anyone else from making, using, or selling the "invention" covered by the claims of the patent. Trademarks are distinctive marks associated with a product or service (these are usually referred to as service marks), which the owners of the mark can use exclusively to identify themselves as the source of the product or service. Copyrights protect the expression of an idea, rather than the idea itself, and are typically used to protect literary works and visual and performing arts, such as books, photographs, paintings and drawings, sculptures, movies, songs, and the like. Trade secret law protects technical or business information that a company uses to gain a competitive business advantage by virtue of the secret being unknown to others. Customer or client lists, secret formulations, or methods of manufacture are typical business secrets.

14.2   PATENTS

Discuss patents.

SOLUTION

Of all the intellectual property rights, the most pertinent for nanotechnological inventions are patents. (Recent patent activity is discussed in a later Problem.) A patent can protect, for example, a composition of matter, an article of manufacture, or a method of doing something. Patent rights are private property rights. Infringement of a patent is a civil offense, not criminal. The patent owner must come to his or her own defense through litigation, if necessary. And, this is a very expensive undertaking. Lawsuits costing more than a million dollars are not unusual. But at stake can be exclusive rights to a technology worth a hundred times as much.

14.3   CONTRACT LAW

Discuss contract law.

SOLUTION

Another legal area that is and will be relevant to nanotechnology is contract law. Any time two or more parties agree upon something, the principles of contract law come into play. The essential components of a contract are that parties be competent to enter into a contractual agreement, subject matter (what the contract is about), legal consideration (the inducement to contract such as the promises or payment exchanged, or some other benefit or loss or responsibility incurred by the parties), mutuality of agreement, and mutuality of contract. While oral contracts can be legally binding, in the event of a dispute it may be difficult to establish in court who said what. It is far better to memorialize the agreement in the form of a written contract.

One of the basic principles of contract law is that the parties should have a meeting of minds. In effect, they should have a common understanding of what the terms of the contract mean. Sometimes it is not so clear what particular terms mean, or the meaning or its implications may change in time. What, for example, qualifies as "nanotechnology"? Not only is nanotechnology not well defined now, it may encompass things in the future that are not even imagined today.

Generally, contracts are employed with the sale and licensing of exclusive rights to a technology. Also, there are agreements to fund technological research and development.

Who are the entities engaged in the contract? Typically, these are business entities. So one must also consider whether there may be some peripheral issues of corporation or partnership law.

14.4 TORT LAW

Briefly discuss tort law.

SOLUTION

There is a branch of law that can retrospectively address certain situations in which property or people are harmed. That is tort law. A tort is a civil wrong, other than a breach of contract, for which the law provides a remedy. One can recover damages under tort law if a legal duty has been breached that causes foreseeable harm. These duties are created by law other than duties under criminal law, governmental regulations, or those agreed to under a contract. Tort law can be very encompassing.

Nanotechnologists have to consider the possibilities of reasonably foreseeable harm arising from their developments and take prudent precautions to avert such harm. In the event that a technology is inherently dangerous, nanotechnologists may be held to a standard of strict liability for any harm caused by the technology regardless of whether an accident was foreseeable.

Most nanotechnologists are not at all interested in deliberately causing harm. But some will be. Nanotechnology can have military applications, and governments may be interested in developing nanoweapons. Suppose that nanoproducts, perhaps nanomachines, are developed that can invade the human body and do harm. Is a cloud of such nanomachines to be considered a poison gas? Or is it a collection of antipersonal mechanical implements, like shrapnel? How will nanomachines as weapons be treated under the Geneva convention? And suppose such a cloud of nanoweapons drifts over, or is released over, a civilian population? The devastating effect of land mines, which remains lethal long after hostilities are ended and which wreak havoc upon unsuspecting civilians wandering into mine fields, has been amply documented. Will nanoweapons remain harmful years after their deployment? What responsibilities do government have morally, and under international law?

But suppose these nanoproducts are not designed to cause harm but simply to obtain and transmit information. For example, suppose such nanoproducts, if ingested, provide information about bodily functions. Or suppose they enable a person to be tracked wherever he or she goes. Larger devices are already known that enable a person to be tracked by global positioning satellites. These devices are worn voluntarily. These are also implantable devices (the author of this text has one). But nanomachines would be undetectable and could very well be implanted in a person without his or her knowledge or consent. Under what circumstances should such invasions of privacy be allowed or forbidden?

Then there is the matter of the criminal use of nanotechnology. The past decade has already seen the growth of a new area of crime: computer crime. For example, in addition to conventional theft, law enforcement agencies must now become technologically proficient to handle computer fraud, identify theft, theft of information, embezzlement, copyright violation, computer vandalism, and like activities, all accomplished over the computer network under conditions such that not only is the criminal hard to trace but even the crime may go undetected. The computer criminals are technologically very savvy and willing to exploit the potentials of any new technology. About the only thing one can expect is that if nanotechnology provides great new potentials, someone will use those potentials for criminal purpose, and the laws will again be forced to play catch-up in response to the crimes, after the harm has occurred.

14.5 RECENT PATENT ACTIVITY

Discuss recent patent activity.

SOLUTION

In terms of issued patents alone, the (USPTO) reports that those involving nanotechnology have increased by over 600% in the 1997–2002 time period; from 370 in 1997 to 2650 in 2002. While these patents made up only 2% of all patents issued in 2002, this compares to a figure of 0.3% in 1997. New filings of nanotechnology-related patent applications are evenly split between process inventions and product inventions, as is typical for all patent applications. Most of these applications (approximately 90%) come from private corporations, with Hewlett-Packard, Texas Instruments, 3M, and Motorola filing the largest number. Universities are filing approximately 7%, with the University of California and Stanford University in the lead. About 3% are being filed by agencies of the U.S. government and collaboration research centers such as the Department of Energy, the Department of Defense, and Sandia National Laboratories. Most of the inventions in these applications are refinements to known technology, but a significant number can be considered "revolutionary" or pioneering in nature.

14.6 CONSERVATION LAW FOR MASS

An inventor claims to have devised a new steady-state process that produces nano-sized particles at a rate of 175 lb/hr. The flow diagram for the process is given in Figure 14.1 with line nano flow rates. Does the inventor have a legitimate claim?

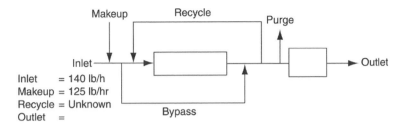

Figure 14.1 Flow diagram of new process (Inlet = 140 lb/hr; Makeup = 125 lb/hr; Recycle = Unknown; Bypass = unknown; Purge = 0.0 lb/day; Outlet = 175 lb/day).

SOLUTION

The *conservation law* for mass can be applied to any process or system. The validity of this law has been confirmed by countless experiments at the macroscopic level. The general form of this law is given by

Mass in − mass out + mass generated = mass accumulated

or on a time rate basis by

Rate of mass in − rate of mass out + rate of mass generated
= rate of mass accumulated

This equation may be applied either to the total mass involved or to a particular species, on either a mole or mass basis. This law can be applied to steady-state or unsteady-state (transient) processes and to batch or continuous systems.

In order to isolate a system for study, it is separated from the surroundings by a boundary or envelope. This boundary may be real (e.g., the walls of a size reduction unit) or imaginary. Mass crossing the boundary and entering the system is part of the *mass in* term above, while that crossing the boundary and leaving the system is part of the *mass out* term. This equation may be written for any compound whose quantity is not changed by chemical reaction and for any chemical element whether or not it has participated in a chemical reaction. It may be written for one piece of equipment, around several pieces of equipment, or around an entire process. It may be used to calculate an unknown quantity directly, to check the validity of experimental data, or to express one or more of the independent relationships among the unknown quantities in a particular problem situation [4].

A *steady-state* process is one in which there is no change in conditions (pressure, temperature, composition, and so on) or rates of flow with time at any given point in the system. The accumulation term above is then zero. (If there is no chemical or nuclear reaction, the generation terms is also zero.) All other processes are *unsteady-state*. In a *batch* process, a given quantity of reactants is placed in a container, and by chemical and/or physical means, a change is made to occur. At the end of the process, the container (or adjustment containers to which material may have been transferred) holds the product or products. In a *continuous* process, reactants are continuously fed to a piece of equipment or to several pieces in series, and products are continuously removed from one or more points. A continuous process may or may not be steady state [4].

As indicated previously, the above may be applied to the total mass of each stream (referred to as an *overall* or *total material balance*) or to the individual component(s) of the streams (referred to as a *componential* or *component material balance*). Often the primary task in preparing a material balance is to develop the quantitative relationships among the streams.

Four important processing concepts are bypass, recycle, purge, and makeup. With *bypass*, part of the inlet stream is diverted around the equipment to rejoin the (main) stream after the unit. This stream effectively moves in parallel with the stream passing through the equipment. In *recycle*, part of the product stream is sent back to mix with the feed. If a small quantity of nonreactive material is present in the feed to a process that includes recycle, it may be necessary to remove the nonreactive material in a *purge* stream to prevent it building up above a maximum tolerable value. This can also occur in a process without recycle. If a nonreactive material is added in the feed and not totally removed in the products, it will accumulate until purged. The purging process is sometimes referred to as *blowdown*. *Makeup*, as its name implies, involves adding or making up part of a stream that has been removed from a process. Makeup may be thought of, in a final sense, as the opposite of purge and or blowdown [4].

Applying the conservation law for mass to the nanosized particles gives

$$\text{in} = \text{out}$$

$$140 + 125 \overset{?}{=} 175$$

$$265 \neq 175$$

Since the conservation law is not satisfied, the claim should be rejected. However, as noted above, the conservation law for mass has been validated *only* at the macroscopic level. It may not be applicable at the nanolevel (see Problem 14.8).

14.7   CONSERVATION LAW FOR ENERGY

An inventor claims to have devised a new steady-state flow process that delivers energy an enthalpy rate of 500 000 Btu/hr. If the input rate of energy (enthalpy) to the process is 400 000 Btu/hr, does the inventor have a legitimate claim?

SOLUTION

Thermodynamics is defined as that science that deals with the relationship among the various forms of energy. A system may possess energy due to its:

1. Temperature
2. Velocity
3. Position
4. Molecular structure
5. Surface, and so on

The energies corresponding to these states are, respectively:

1. Internal
2. Kinetic
3. Potential
4. Chemical
5. Surface

Engineering thermodynamics is founded on three basic laws. Energy, like mass and momentum, is conserved. Application of the conservation law for energy gives rise to the first law of thermodynamics. An application involving the second law of thermodynamics can be found in Problem 14.8. The first law has been verified by numerous experiments over time at the macroscopic level. This law, in steady-state equation form for batch and flow process, is presented here.

For *batch* processes:

$$\Delta E = Q - W$$

For flow process:

$$\Delta H = Q - W_s$$

where potential, kinetic, and other energy effects have been neglected and Q = energy in the form of heat transferred across the boundaries of the system; W = energy in the form of work transferred across the boundaries of the system; W_s = energy in the form of mechanical work transferred across the boundaries of the system; E = internal energy of the system; H = enthalpy of the system; $\Delta E, \Delta H$ = changes in the internal energy and enthalpy, respectively, during the process.

The internal energy and enthalpy as well as the other terms, may be on a *mass* basis (that is, for 1 kg or 1-lb of material), on a *mole* basis (that is 1 gmol or 1 lbmol of material) or represent the total inernal energy and enthalpy of the entire system. As long as these equations are dimensionally consistent, it makes no difference [5].

Based on the problem statement, one immediately notes that the first law of thermodynamics is *not* satisfied. The claim must therefore be rejected. As it was in the previous Problem, this law has been verified at the macroscopic level; the reader should note that it might not apply at the nanolevel (see Problem 14.8).

14.8   THE SECOND LAW OF THERMODYNAMICS

A patent has been submitted that claims that a newly developed flow process consisting of specialty nanomaterials has unique capabilities. The patent claim indicates that the process can take steam at 212°F and 1 atm and provide heat to a constant temperature reservoir at 300°C while discharging heat to a low-temperature reservoir at 60°F. Pertinent data is provided in the flow diagram of Figure 14.2.

SOLUTION

For any energy-producing process to be theoretically possible, it must meet the requirements of both the first and second laws of thermodynamics. As with the first law, the second law has been verified by experiments; its validity cannot be questioned at the macrolevel.

The first law was discussed in the previous Problem. The second law states that every naturally occurring process produces a total entropy change that is positive; the limiting value of zero is attained only for a reversible process. Thus, no process is possible for which the total entropy decreases. The reader should note that the entropy change of a system may be positive $(+)$, negative $(-)$, or zero (0); the entropy change of the surroundings during this process may likewise be positive, negative, or zero. However, the total entropy change, ΔS_T, must be

$$\Delta S_T \geq 0$$

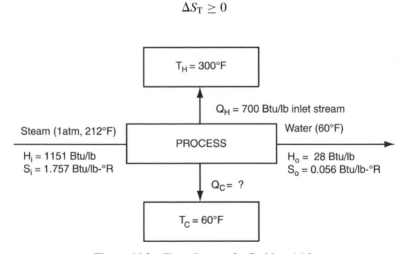

Figure 14.2   Flow diagram for Problem 14.8.

To satisfy the first law for the flow process above (assuming a basis of 1 lb of steam)

$$\Delta H - \sum Q_i = 0$$
$$1151 - 28 - 700 - Q_C = 0$$
$$Q_C = 423 \text{ Btu/lb}$$

Regarding the second law, the entropy change of steam is

$$\Delta S_S = 0.056 - 1.757$$
$$= -1.701 \text{ Btu/lb-}°\text{R}$$

The entropy gained by the reservoir is given by

$$\Delta S_i = \frac{Q_i}{T_i}$$

For the hot reservoir,

$$\Delta S_H = \frac{Q_H}{T_H} = \frac{700}{300 + 460}$$
$$= 0.921 \text{ Btu/lb-}°\text{R}$$

For the discharge to the cold reservoir,

$$\Delta S_C = \frac{Q_C}{T_C} = \frac{423}{60 + 460}$$
$$= 0.813$$

The total entropy change is therefore

$$\Delta S_T = \Delta S_S + \Delta S_H + \Delta S_C$$
$$= -1.701 + 0.921 + 0.813$$
$$= +0.033 \text{ Btu/lb-}°\text{R}$$

The calculations performed above automatically satisfied the first law since the energy lost by the conversion of the steam to water, that is, 1123 Btu/lb, equals the energy transferred to the hot and cold reservoir, that is, 700 and 423 Btu/lb, respectively. Interestingly, the second law is also satisfied, but only marginally. This suggests that the claim may not be possible.

It is important to note that Evans [6] conducted experiments at the nanolevel where a total entropy decrease was recorded for short periods of time. He concluded that the second law of thermodynamics was only suited for the macroscopic level and not a precise model at the molecular level. A fundamentally new approach could be needed in order to develop a precise thermodynamic description of the system suitable at the nanoscale level where classical thermodynamics may have been shown to fail.

14.9 ALLOWABLE PATENT APPLICATION CLAIMS

Researcher John Smith invents a catalyst for the conversion of certain organic compounds. The catalyst includes a nanosized catalytic metal deposited on a porous inorganic oxide support. The catalyst metal is preferably a lanthanide, europium being especially preferred. The catalyst support is preferably zirconium oxide having an average pore size ranging from 100 to 200 nm. Smith has a patent application prepared and submitted to the U.S. Patent and Trademark Office in 2004. Smith's application includes a specification describing the invention along with claims that define the limits as to what can be legally protected by the patent. Smith's claims include a broad claim to the chemical conversion process employing a catalyst composition, including a catalytic metal on a porous inorganic oxide support. Smith's application also includes narrower claims directed to the specified pore size, the use of zirconium oxide as the support, and the use of lanthanide metals as the catalytic metal, particularly europium. During prosecution of the application two prior art patent references were found. Reference #1, a patent that was issued in 1973, discloses and claims the use of platinum on any porous inorganic oxide support for the same chemical conversion process contemplated by Smith. Reference #2, based on an application that was filed in 1996 and was issued in 1999, discloses and claims a porous zirconium oxide having an average pore size ranging from 100 to 250 nm. Reference #2 mentions in the specification that the material can be used in catalysts, but does not claim the use of the material in any chemical conversion process.

Which claims of Smith's patent application are likely to be allowed?

SOLUTION

Both References #1 and #2 are prior art to Smith's application. Smith's broad claim to the chemical conversion process using a catalyst composition including a porous inorganic oxide with a catalytic metal is anticipated by Reference #1 and will be rejected by the patent examiner.

Smith's claim to the chemical conversion process using a catalyst composition, including a catalytic metal on porous zirconium oxide is likely to be rejected as being obvious over References #1 and #2 combined. Since Reference #2 discloses the same porous zirconium oxide support and suggests its use as a catalyst, the patent examiner is likely to argue that one skilled in the art will find it obvious to combine the teachings of Reference #2 with the teachings of Reference #1 to reproduce Smith's claim. Smith can overcome the rejection by successfully arguing that there is no suggestion in the references that the specific zirconium oxide porous support of Reference #2 would be useful in the particular chemical conversion process of Reference #1, or that the use of the porous zirconium oxide support with the specified pore size range provides unexpected benefits (e.g., greater efficiency, selectivity, and so on) as opposed to the use of other supports.

Smith's narrower claims directed to the use of lanthanides, and particularly europium are allowable over References #1 and #2 since neither of the references disclose or suggest these features.

14.10 PRACTICING ONE'S OWN INVENTION

Refer to Problem 14.9. Can Smith practice his own invention?

SOLUTION

No. That is, not without a license from the owners of Patent #2. Patent #1 is covered by the old patent law in accordance with which a patent is considered to be in force for a period of 17 years from the time of issuance. Patent #1 expired in 1990 and is now part of the public domain. Patent #2 is entitled under the new law to a period of enforceability extending from date of issuance to a date 20 years from the time of filing. It is therefore still in force. Smith uses a material covered by the claims of Patent Reference #2. Smith will either have to get an agreement to make and use the particular zirconium oxide support, or simply purchase the material from the owners or legitimate suppliers of the material covered by the claims of Patent #2.

15 Size Reduction

Reduction in the size of solid particles is frequently required. . . and this is especially true in the nanotechnology field. Reduction in size involves the production of smaller mass units from larger mass units of the same material. Traditionally, it is an operation that will cause fracture to take place in the larger units. This fracturing or shattering of the larger mass units is often accomplished by the application of pressure.

Solid materials are crystalline in nature. As such, the atoms in the individual crystals are arranged in definite repeating geometric patterns, and there are certain planes in the crystal along which shear takes place more readily (see Chapter 6). The pressure applied must be sufficient to cause failure by shear along these planes.

In general, there are six widely used methods for producing nanoscaled particles – on the order of 1 to 100 nm in diameter – of various materials. These are listed below: [7, 8]

1. Plasma-arc and flame-hydrolysis methods (including flame ionization);
2. Chemical vapor deposition (CVD);
3. Electrodeposition techniques;
4. Sol-gel synthesis;
5. Mechanical crushing;
6. Promising technologies.

These methods are detailed in the Problems in this Chapter. Emphasis is placed on the fifth method: mechanical crushing.

The ongoing challenge for the research community and industry is to continue to devise, perfect, and scale-up viable production methodologies that can cost-effectively and reliably produce the desired nanoparticles with the desired particle size, particle size distribution, purity, and uniformity in terms of both composition and structure [1].

15.1 SIZE REDUCTION OBJECTIVES

Briefly describe the objectives of particle size reduction.

Nanotechnology: Basic Calculations for Engineers and Scientists, by Louis Theodore
Copyright © 2006 John Wiley & Sons, Inc.

SOLUTION

The purpose of size reduction is not only to make "little ones out of big ones" when the effectiveness (from a nanoperspective) can be measured by the degree of fineness of the product but also to produce a product of the desired size and/or size range. The size requirements for various products may vary widely, and hence different equipment and procedures are employed. A size range entirely satisfactory for one purpose may be highly undesirable for another, even when the same material is involved.

In some instances, it is necessary to use a product with rather narrow limits in size variations. It is usually impossible to accomplish this by size reduction only. For mechanical crushing, separation and classification by various means may be required to secure the desired limitation in size and/or size range. The two key operating variables are:

1. Reduction ratio;
2. Moisture content.

The *reduction ratio* is the ratio of the average diameter of the feed to the average diameter of the product. The *moisture content* of solids to be reduced in size can also be important. If it is below 3 or 4% by weight, no particular difficulties are usually encountered; indeed, it appears that the presence of this amount of moisture is of real benefit in size reduction if for no other reason than for dust control. When the moisture content exceeds about 4%, most materials become sticky or pasty, with a tendency to clog the equipment and agglomerate.

15.2   PLASMA-BASED AND FLAME-HYDROLYSIS METHODS

Describe high-temperature processes such as the plasma-based and flame-hydrolysis methods that are employed for size reduction.

SOLUTION

These two production routes involve the use of a high-temperature plasma or flame ionization reactor. As an electrical potential difference is imposed across two electrodes in a gas, the gas, electrodes, or other materials ionize and vaporize if necessary and then condense as nanoparticles, either as separate structures or as surface deposit. An inert gas or vacuum is used when volatilizing the electrodes.

During flame ionization, a material is sprayed into a flame to produce ions [7]. Using flame hydrolysis, highly dispersed oxides can be produced via high-temperature hydrolysis of the corresponding chlorides. Flame hydrolysis produces extremely fine, mostly spherical particles with diameters in the range of 7 to 40 nm and high specific surface areas (in the range of 50 to 400 m^2/g [8].

In general, high-temperature flame processes for making nanoparticles are divided into two classifications – gas-to-particle or droplet-to-particle methods – depending on how the final particles are made [9].

In gas-to-particle processes, individual molecules of the product material are made by chemically reacting precursor gases or rapidly cooling a superheated vapor. Depending on the thermodynamics of the process, the molecules then assemble themselves into nanoparticles by colliding with each other or by repeatedly condensing and evaporating into molecular clusters. Gas-to-particle processes involve the use of flame, hot-wall, evaporation–condensation, plasma, laser, and sputtering-type reactors.

In droplet-to-particle processes, liquid atomization is used to suspend droplets of a solution or slurry in a gas at atmospheric pressure. Solvent is evaporated from the droplets, leaving behind solute crystals, which are then heated to change their morphology. Spray drying, pyrolysis, electrospray, and freeze-drying equipment are typically used in the droplet-to-particle production process.

15.3   CHEMICAL VAPOR DEPOSITION AND ELECTRODEPOSITION

Briefly describe the following two processes:

1. Chemical vapor deposition
2. Electrodeposition

SOLUTION

1. *Chemical Vapor Deposition.* During CVD, a starting material is vaporized and then condensed on a surface, usually under vacuum conditions. The deposit may be the original material or a new and different species formed by chemical reaction.
2. *Electrodeposition.* Using this approach, individual species are deposited from solution with an aim to lay down a nanoscaled surface film in a precisely controlled manner.

15.4   SOL-GEL PROCESSING

Describe the sol-gel process.

SOLUTION

This is a wet-chemical method that allows high-purity, high-homogeneity nanoscale materials to be synthesized at lower temperatures compared to competing high-temperature methods. A significant advantage that sol-gel science affords over more conventional materials-processing routes is the mild conditions that the approach employs.

Two main routes and chemical classes of precursors have been used for sol-gel processing [8].

1. The inorganic route ("colloidal route"), which uses metal salts in aqueous solution (chloride, oxychloride nitrate) as raw materials. These are generally less costly and easier to handle than the precursors discussed below for the metal-organic route, but their reactions are more difficult to control and the surfactant that is required by the process might interfere later in downstream manufacturing and end use.

2. The metal-organic route ("alkoxide route") in organic solvents. This route typically employs metal alkoxides $M(OR)Z$ as the starting materials, where M is Si, Ti, Zr, Al, Sn, or Ce; OR is an alkoxy group, and Z is the valence or the oxidation state of the metal. Metal alkoxides are preferred due to their commercial availability. They are available for nearly all elements, and cost-effective synthesis from cheap feedstocks have been developed for some. The selection of appropriate OR groups (bulky, functional, fluorinated) allows developers to fine-tune the properties. Other precursors are metal diketonates and metal carboxylates. A larger range of mixed-metal nanoparticles can be produced in mild conditions, often at room temperature, by mixing metal alkoxides (or oxoalkoxies) and other oxide precursors.

In general, the sol-gel process consists of the following five steps:

1. Sol formation;
2. Gelling;
3. Shape forming;
4. Drying;
5. Densification.

First, after mixing the reactants, the organic or inorganic precursors undergo two chemical reactions: hydrolysis and condensation or polymerization, typically with an acid or base as a catalyst, to form small solid particles or clusters in a liquid (either organic or aqueous solvent).

The resulting solid particles or clusters are so small (1 to 1000 nm) that gravitational forces are negligible and interactions are dominated by van der Waals, coulombic, and steric forces. These sols – colloidal suspensions of oxide particles – are stabilized by an electric double layer, or steric repulsion, or a combination of both. Over time, the colloidal particles link together by further condensation and a dimensional network occurs. As gelling proceeds, the viscosity of the solution increases dramatically.

The sol-gel can then be formed into three different shapes: thin film, fiber, and bulk. Thin (100 nm or so) uniform and crack-free films can readily be formed on various materials by lowering, dipping, spinning, or spray coating techniques.

Sol-gel chemistry is promising, yet it is still in its infancy and a better understanding of the basic inorganic polymerization chemistry has to be reached [8]. Some drawbacks include the high cost for the majority of alkoxide precursors, relatively long processing times, and high sensitivity to atmospheric conditions. And, the batch nature of present sol-gel processing leaves cost and scale-up issues associated with the development of viable production routes.

15.5 MECHANICAL CRUSHING

Describe the mechanical crushing process.

SOLUTION

Traditional methods employed to achieve coarse size reduction include: jaw crushers, gyratory crushers, the Bradford breaker, toothed roll crusher, hammer mill, squirrel-cage disintegrating cone crusher, crushing rolls, gravity stamps, and the various mills, including the roller, ball, grate, compound, rod cage, rumbling, pebble, vibratory and tube mills. Of these, progressive particle size reduction or pulverization using a conventional ball mill is one of the primary methods for preparing nanoscaled particles of various metal oxides. High-energy ball milling is in use today, but its use is considered by some to be limited because of the potential for contamination problems. Today, the availability of tungsten carbide components and the use of inert atmosphere and high-vacuum processes has helped operators to reduce impurities to acceptable levels for many industrial applications [8]. Other common drawbacks, however, include the highly polydisperse size distribution and partially amorphous state of nanoscaled powders prepared by pulverization.

15.6 PROMISING TECHNOLOGIES

Briefly discuss some of the promising technologies that may some day be employed for size reduction.

SOLUTION

In addition to the proven techniques discussed in Problems 15.2–15.5, additional, promising technologies are emerging to produce nanoscaled particles of various materials [8]. These include the following five processes:

1. Flame or jet flame reactors that introduce an additional flame behind the reaction zone in order to transform the aggregates into spherical particles more effectively.

2. Plasma processes that are designed to promote more rapid cooling in order to produce fewer agglomerates.

3. Sonochemical processing routes, in which an acoustic cavitation process generates a transient localized hot zone with an extremely high temperature gradient and pressure. Sudden changes in temperature and pressure assist in the destruction of the precursor material (e.g., organometallic solution) and promote the formation of nanoparticles.

4. Hydrodynamic cavitation processes in which nanoparticles are generated through the creation and release of gas bubbles inside a sol-gel solution. Erupting hydrodynamic bubbles are responsible for nucleation, growth, and quenching of the nanoparticles. Particle sizes can be controlled by adjusting the pressure and the solution retention time in the cavitation chamber.

5. Microemulsion techniques, which show promise for the synthesis of metallic semiconductor silica, barium sulfate, magnetic, and superconductor nanoparticles.

15.7 ENERGY AND POWER REQUIREMENTS

Discuss energy and power requirements for size-reduction operations.

SOLUTION

Obviously, the energy and power requirements for any particle size reduction process is an extremely strong function of the method employed to achieve the size reduction. There are some data for crushing operations but little to none for the other processes. However, one can generally say that these requirements are minimal for all but crushing operations. For this reason, only crushing (and grinding) operations are considered below.

Although most of the power required for driving crushers and grinders is used in overcoming mechanical friction, the actual energy used in size reduction is theoretically proportional to the new surface produced, since there is no change in the material except size and the creation of new surface. This principle was first recognized by Rittinger [10].

Rittinger's number designates the new surface produced per unit of mechanical energy adsorbed by the material being crushed. The values vary for different solids, depending on the elastic constants and their relation to the ultimate strength and on the manner or fate of the applied force. These numbers are available in the literature, but it would be unlikely that they could be applied to nanoprocesses.

Three laws have been proposed to relate size reduction to a single variable, the energy input to the mill. These laws are encompassed in a general differential equation [11].

$$dE = -CdX/X^n$$

where E = work, X = particle size, and C and n = constants. For $n = 1$ the solution is Kick's law [12]. This can be written as:

$$E = C \log (X_F/X_P)$$

where X_F = feed particle size, X_P = product size, and, X_F/X_P = reduction ratio. For $n > 1$ the solution to the above equation becomes:

$$E = \left(\frac{C}{n-1}\right)\left(\frac{1}{X_P^{n-1}} - \frac{1}{X_F^{n-1}}\right)$$

For $n = 2$ this becomes *Rittinger's law* [10], which states that the energy is proportional to the new surface produced. The *Bond law* [13] corresponds to the case in which $n = 1.5$:

$$E = 100E_i\left(\frac{1}{\sqrt{X_P}} - \frac{1}{\sqrt{X_F}}\right)$$

where E_i is the Bond work index, or work required to reduce a unit weight from a theoretical infinite size to 80% passing 100 μm.

Consider the following application. The feedback to a certain mill with an average particle size of 2.5 μm has its size reduced to 450 nm. The power required to process 2600 lb/day is 1.3 hp. Calculate a revised power requirement if the feed rate is increased to 3200 lb/day and the product size is 660 μm. Assume Rittinger's law to apply. For this example, a simple ratio is generated for both process conditions so that:

$$P = 1.3\left(\frac{3200}{2600}\right)\frac{\left(\frac{1}{660}\right) - \left(\frac{1}{2500}\right)}{\left(\frac{1}{450}\right) - \left(\frac{1}{2500}\right)}$$

$$= 1.6(0.00112/0.00182)$$

$$= 0.98 \, \text{hp}$$

The reader should note that the feedstock can often be neglected in the above calculation as the size reduction ratio approaches infinity (or becomes large). For the above case, the approximation reduces to:

$$P = 1.3(3200/2600)(450/660)$$
$$= 1.09 \, \text{hp}$$

15.8 POTENTIAL DUST EXPLOSIONS WITH SIZE REDUCTION

Explain why dust explosions may arise during particle size reduction operations.

SOLUTION

Some of the basics and principles of dust explosions were reviewed in Problem 8.5. The seven key factors listed at that time for a dust explosion to occur included:

1. Air (oxygen);
2. Fuel source (the dust itself);
3. Mixture;
4. Dry;
5. Minimum concentration;
6. Enclosure; and
7. Ignition source.

Note once again that *all* factors need to be present for an explosion to occur.

Concentration of the dust in air and its particle size and size distribution are important factors that determine explosibility. Below a lower limit of concentration, no explosion can result because the enthalpy of combustion is insufficient to propagate it. Above a maximum limiting concentration, an explosion cannot be produced because insufficient oxygen is available. The finer the particles, the more easily is ignition accomplished and the more rapid is the rate of combustion [14].

Finely divided nonmetallic (coal, plastics, cereal, flour, wood, and so on) and metals suspended in the atmosphere are potential explosion hazards. The explosion possibility is enhanced during crushing (and grinding) operations since a significant portion of the energy requirement is released as heat, producing elevated temperatures that are conducive to combustion. Explosions are initiated – see factor 7 above – from spontaneous combustion, hot surfaces, and sparks and flames.

15.9 MATERIAL BALANCE SIZE REDUCTION

A nano size-reduction unit (NSRU) with a feed rate of 100 g/min of 150 μm diameter solids is operated in conjunction with a particle classifier (PC). See Figure 15.1 and its accompanying notation. The percent of particles less than 20 nm for R, F, and P are r = 45, f = 55, and p = 95, respectively. Calculate all flow rates plus the percent of the feed, I, that is recirculated, R.

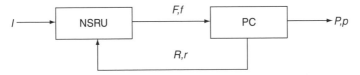

Figure 15.1   Size reduction unit (*I*, feed, coarse material; *F*, feed to particle classifies; PC, particle classifier; *P*, product, fines; *R*, recycle, less coarse material).

SOLUTION

An overall balance on the unit gives

$$I = P = 100 \text{ g/min}$$

An overall balance on the PC yields, in consistent units,

$$F = P + R; \quad P = 100$$

A "componential" balance on less than 20 nm size particles is

$$Ff = Pp + Rr$$
$$Ff = 100p + Rr$$

since

$$F = 100 + R$$
$$(100 + R)f = 100p + Rr$$

Rearranging and substituting

$$R(f - r) = 100(p - f)$$
$$R = 100\left(\frac{p - f}{f - r}\right) = 100\left(\frac{0.95 - 0.55}{0.55 - 0.45}\right)$$
$$= 100(0.4/0.1) = 400 \text{ g/min}$$

The percent of the feed that is recirculated is therefore $(0.4/0.1)100 = 400\%$. Completing the overall balance gives

$$F = 100 + R$$
$$= 100 + 400 = 500 \text{ g/min}$$

15.10   SIZE REDUCTION SURFACE AREA INCREASE

Refer to Problem 15.9. Estimate the increase in surface area that results from the size reduction process.

SOLUTION

First note that insufficient information is available for an exact calculation since a complete particle size distribution of the product stream P is not provided. However, in line with an overall mass balance (and as shown in the previous Problem), the input (I) and product (P) streams of necessity are equal. In effect, the inlet mass must equal the outlet mass.

Choose as a basis one 150 μm particle entering the NSRU. Its surface (S) is

$$S_{150} = \pi d^2$$
$$= \pi(150)^2$$
$$= 7.07 \times 10^4 \ (\mu m)^2$$
$$= 7.07 \times 10^{10} \ (nm)^2$$

The volume of the same particle is

$$V_{150} = \pi d^3/6$$
$$= \pi(150)^3/6$$
$$= 1.77 \times 10^6 \ (\mu m)^3$$
$$= 1.77 \times 10^{15} \ (nm)^3$$

Since 95% of the product stream has a size less than 20 nm, assume the average particle size is 10 nm for estimation purposes. For this assumption,

$$S_{10} = \pi(10)^2$$
$$= 3.14 \times 10^2 \ (nm)^2$$

and

$$V_{10} = \pi(10)^3/6$$
$$= 0.5236 \times 10^3 \ (nm)^3$$

Since the mass balance must be satisfied, the number of 10 nm particles is then

$$N_{10} = \frac{1.77 \times 10^{15}}{0.5236 \times 10^3}$$
$$= 3.380 \times 10^{12}$$

The total surface area, TS, of the 10 nm particle is

$$TS_{10} = (S_{10})(N_{10})$$
$$= (3.14 \times 10^2)(3.380 \times 10^{12})$$
$$= 10.62 \times 10^{14} \ (\text{nm})^2$$

The ratio of the surface areas, R_s, is

$$R_s = 10.62 \times 10^{14}/7.07 \times 10^{10}$$
$$= 1.5 \times 10^4$$

The ratio, R_p, of the 150 μm particles to 10 nm particles is

$$R_p = 7.07 \times 10^{10}/3.14 \times 10^2$$
$$= 2.25 \times 10^8$$

The reader should note that the result could have been deduced from the results presented in Problem 5.9. R_s is given by the ratio of the two diameters while R_p is given by this ratio squared.

15.11 FINES EDUCTOR APPLICATION

Refer to Problem 15.9. The output from the PC is fed to a fines eductor (FE). The output from the FE contains particles, 45% of which are less then 10 nm in size. The FE bottom stream contains 98% particles that are less 20 nm in size, with 56% of the particles less than 10 nm in size (Figure 15.2). Estimate the discharge particle flow stream from the FE if the flow rate factor (D/B) is $1.71/1.0$.

SOLUTION

As with Problem 15.9, another overall and componential balance is required. Applying an overall balance to the unit gives

$$P = 100 = D + B$$

with

$$D/B = 1.71$$

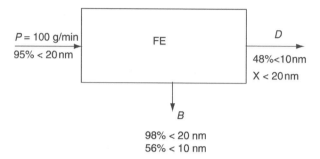

Figure 15.2   Flow diagram for Problem 15.11.

Since the eductor is essentially a separator, one may also write balances for the particles less than 20 nm and also for those less than 10 nm

<20 nm: $100(0.95) = D(X) + B(0.98);\ X = $ particles <20 nm in D
<10 nm: $100(Y) = D(0.48) + B(0.56);\ Y = $ particles <10 nm in P

Solving the first two equations leads to

$$B = 36.9$$
$$D = 63.1$$

The third equation gives

$$X = 0.932$$

while the fourth equation gives

$$Y = 0.5095$$

From these results, it appears that the FE has produced little in both size reduction and separation.

15.12   FINES EDUCTOR SIZE REDUCTION

Refer to Problem 15.11. Estimate the increase in surface resulting from the FE unit.

SOLUTION

Since some particle size distribution data is provided, a better estimate of the increase in surface area can be made. However, in examining the results of

Problem 15.11, it appears that the FE has served the purpose of separating two streams with little to no increase in surface area.

For *P*, nearly 100% of the particles have a size equal to or less than 20 nm and approximately 50% of the particles in *P* has a size equal to a less than 10 nm. However, the same can be said for stream *D* and *B*.

Thus, and as noted in the previous Problem, there is essentially no size reduction and no classification of the inlet stream.

16 Prime Materials

Over the last 20 years, the desire to produce and use matter whose particle size dimensions are so small that they were previously unimagined has been accelerated by the discovery and verification of a broad array of novel, size-dependent properties and phenomena that occur or are vastly improved in the nanometer range [15]. In this nanoscale range, it is not only the chemical composition but also the size, shape, and surface characteristics of ultrafine particles that determines the properties of various materials [8]. For instance, when they are produced at infinitesimally small particle sizes, materials as varied as metals, metal oxides, polymers, and ceramics (all discussed in greater detail in the Problems that follow), and carbon derivatives such as carbon nanotubes and fullerenes or "buckyballs" (discussed separately in the next Chapter), have extraordinary ratios of surface area to particle size.

Engineers and scientists in pursuit of nanoscaled materials that bring functional advantages to their end-use applications have left few stones unturned. Today, the range of elements and compounds that have been successfully produced and deployed as nanometer-sized particles includes:

1. Metals such as iron, copper, gold, aluminum, nickel and silver;
2. Oxides of metals such as iron, titanium, zirconium, aluminum, and zinc;
3. Silica sols and colloidal silicas;
4. Clays such as talc, mica, smectite, asbestos, vermiculite, and montmorillonite;
5. Carbon compounds, such as fullerenes, nanotubes, and carbon fibers.

Each of the first four types of materials is discussed in this Chapter. The manufacturing methods used to render them into nanoscaled particles, are discussed in the next Chapter.

Nanometer-sized particles of various oxides – including the iron oxides (Fe_2O_3 and Fe_3O_4), silica dioxide (SiO_2), titanium dioxide (TiO_2), aluminum oxide (alumina; Al_2O_3), zirconium dioxide (zirconia; ZrO_2), and zinc oxide (ZnO), among others, are already available in commercial quantites. These materials are presently being used in a wide range of existing applications and envisioned for use in many others. In addition to these widely used oxides, nanoparticle versions of other compounds, such as antimony (III) oxide, chromium (III) oxide, iron (III) oxide, germanium (IV) oxide, vanadium (V) oxide, and tungsten (VI) oxide, are also being developed and their possible uses explored [8].

Nanotechnology: Basic Calculations for Engineers and Scientists, by Louis Theodore
Copyright © 2006 John Wiley & Sons, Inc.

Much of the information presented in the Problems that follow was excerpted and/or adapted from an outstanding comprehensive report prepared by SRI Consulting [8]. Much of this material can also be found in Chapter 2 of reference [1], prepared by S. Shelley.

16.1 METALS

Discuss the use of metals in nanotechnology.

SOLUTION

While the use of metals in catalyst systems is not new, the ability to reliably render a large number of atomic species and alloys into particle sizes below 100 nm, with good control over the size distribution, dispersion, and surface characteristics, is a new and evolving field.

When metal particles are produced in this range, they exhibit properties not found in the standard particle sizes. This includes quanta effects, the ability to sinter at temperatures significantly below their standard melting points, increased catalytic activity due to higher surface area per weight, and more rapid chemical reaction rates. And, when nanoscaled particles of metals are consolidated into larger structures, they often exhibit increased strength, hardness, and tensile strength, compared to structures made from conventional micron-sized powders.

However, reducing some metals to nanoscaled powders presents problems, not the least of which is that when reduced to sufficiently small particle sizes, many metals become more reactive and subject to oxidation – often explosively. Copper and silver powders are among the few metallic nanoparticles that are not pyrophoric and thus can be handled in air. Many others need to be stabilized with a passivation layer or handled in an inert, blanketed environment. And in some applications, effort is required to minimize unwanted particle agglomeration.

Nanoscaled copper powders are not available yet in commercial-scale quantities. However, they are being pursued to take advantage of copper's superior ability to conduct electricity and heat.

A range of methods are available or under development to manufacture nanoparticle metals; while the details of many remain closely guarded proprietary secrets, the technologies generally fall into two classes – chemical methods or physical methods. These are detailed in the next Chapter.

16.2 IRON

Provide some details on iron nanoparticles.

SOLUTION

Needle-shaped iron particles (with diameters on the order of 50 to 100 nm and lengths of 20 nm or so) are already widely used in magnetic recording for analog and digital data. Finer size particles (on the order of 30 to 40 nm in length) are also available. Preliminary work suggests that such particles can provide magnetic recording media with 5 to 10 times greater recording capacity per unit space [8]. Meanwhile, when iron nanoparticles (100 to 200 nm in diameter) are coated with palladium, they show promise in the decontamination of ground-water, by reducing chlorinated hydrocarbons to nontoxic hydrocarbons and chlorides.

Iron–platinum alloy nanoparticles are available in particle sizes as low as 3 nm. Such nanoparticles are expected to be used in the magnetic storage media of the future. Because such particles have stronger magnetization than the currently used tiny crystals of cobalt–chromium alloys, a 10-fold increase in information density is predicted.

16.3   ALUMINUM

Provide some details on aluminum nanoparticles.

SOLUTION

Nanoaluminum powder, with a particle size range from 10 to 100 nm, can be made with a plasma reactor. In a few thousandths of a second, a rod of solid material with a massive pulse of electrical energy is pulverized, heating it to 50 000°C, followed by rapid cooling of the gas of vaporized material. This rapid cooling or quenching is how the size of the resulting nanoparticles is controlled.

Applications of nanoscaled aluminum powder include various electronics circuits, optical applications such as scratch-resistant coatings for plastic lenses, biomedical applications such as antimicrobial agents, as well as new tissue-biopsy tools. Energy-related applications include fuel cells, improved batteries, and solar energy applications.

16.4   NICKEL

Provide some details on nickel nanoparticles.

SOLUTION

Traditional nickel powders – which are valued for their high conductivity and high melting point characteristics – have an average diameter of 2 to 7 μm and are consumed in powder metallurgy, nickel–cadmium batteries, and welding rods. Today, finer-scale nickel powders (with a particle size range of 100 to 500 nm) are being produced commercially by various processes, including chemical vapor deposition, wet-chemical processes, and gas-phase reduction. Increasingly, multilayer ceramic capacitors have become the major market for nickel nanoparticles as a lower-cost alternative to conventional pastes made from fine powders of palladium and silver.

16.5 SILVER

Provide some information on silver nanoparticles.

SOLUTION

Silver is well known for its excellent conductivity and its antimicrobial effects, and silver powder is already approved for use as a biocide by the U.S. Food and Drug Administration. However, nanoscaled silver particles have a much larger surface area, offering an opportunity to gain higher efficacy while using less materials. To date, nanoscaled silver powders (with particle diameters of 10 to 90 mm) have been used as an ingredient in a biocide, in transparent conductive inks and pastes, and in various consumer and industrial products that need enhanced antimicrobial properties.

16.6 GOLD

Provide some information on gold nanoparticles.

SOLUTION

Production of gold nanoparticles is easier in comparison to other metal nanoparticles, primarily because of that element's inherent chemical stability. Colloidal gold has been used in medical applications for some time, and additional medical end uses for gold nanoparticles are under development.

Gold nanoparticles are also in commercial use in a wide array of catalytic applications (e.g., for low-temperature oxidation processes, including carbon monoxide oxidation in a hydrogen stream, selective oxidation of propene to propylene, and the oxidation of nitrogen-containing chemicals) and optical and electrical applications (as components in various probes, sensors, and optical devices).

16.7 IRON OXIDES

Provide some of the properties and applications of iron oxides.

SOLUTION

This is the first of five metal oxides to be reviewed. In general, particles of ferric-oxide (Fe_2O_3) are used in pigment applications, while the magnetic (Fe_2O_4) lend themselves to electromagnetic uses. Ferric oxide nanoparticles may be translucent to visible light but opaque to ultraviolet (UV) radiation, and the ultrasmall particle size allows for the creation of rugged yet ultrathin transparent coatings with enhanced UV-blocking capabilities.

When suspended in fluids, nanoparticles of magnetic magnetite create so-called ferrofluids, which can be made to respond to electromagnetic energy in many useful ways. Ferrofluids have been used in many industrial and medical applications for decades. Magnetite nanoparticles are also being used to improve various electromagnetic media for data storage, such as magnetic tapes and computer hard drives, and to produce advanced magnets and supercapacitors, and various medical diagnostic devices.

16.8 ALUMINUM OXIDE

Discuss some of the properties and applications of aluminum oxide.

SOLUTION

Nanosized powders of Al_2O_3 (also known as alumina) have a lower melting point, increased light absorption, improved dispersion in both aqueous and inorganic solvents, and ultrahigh surface area, compared to conventionally sized alumina crystals [8]. They are used in chemical mechanical planarization (CMP) slurries to polish semiconductor chips during manufacturing (nanoparticle use in CMP slurries is discussed in the next Chapter) and as components in both advanced ceramics (such as those used to make catalysts, refractory materials, ceramic filtration membranes, and substrates for microelectronic components) and advanced composites (nanocomposites are discussed in the next Chapter).

Today, a major end use for nanoscaled alumina is in the coating of lightbulbs and fluorescent tubes. The addition of aluminum oxide results in a more uniform emission of light and better flow ability of the fluorescent material mixture during the electrostatic coating [8]. Such nanoparticles are also used today as a component in clear coating that boast increased hardness and improved scratch and abrasion resistance, as a flame-retardant agent, as a performance filler in tires, as a rheology control agent, as a surface friction agent in vinyl flooring, as a detackifying agent in paints, and as a coating of high-quality inkjet papers.

16.9 ZIRCONIUM DIOXIDE

Discuss some of the properties and applications of zirconium dioxide.

SOLUTION

Nanoscaled particles of ZrO_2 (also called zirconia) are valued for their ability to impart improved fracture toughness and resistance to fracture and chipping. Such nanoparticles are already being used as a component in structural ceramics (in wear parts, extrusion dies, pump components, and cutting edges), in electronic ceramics (such as those used for oxygen sensors, dielectrics, and piezoelectric components), and in thermal spray coatings used to protect components in jet engines and combustion turbines from extreme heat. Such thermal spray coatings based on nanosized particles of zirconia have sintering temperatures that are as much as 400°C lower than comparable micrometer-sized coatings, with no sacrifice in toughness or flexibility [8].

Zirconia nanoparticle use is also being explored for use in fuel cell power generation, gas-stream purification applications, in optical connectors (ferrules), in high-end orthopedic and dental prostheses, and as carrier particles in pharmaceutical applications.

16.10   TITANIUM DIOXIDE

Discuss some of the properties and applications of titanium dioxide (TiO_2).

SOLUTION

TiO_2 is already the largest-volume inorganic pigment produced in the world, and micrometer-sized TiO_2 powders are widely used in surface coating, paper and plastic applications, and as a filler and whitening agent.

In applications where white pigmentation is not the aim, smaller, nanometer-scale particles of TiO_2 are finding a host of commercial uses. For instance, at diameters of 50 nm or less, TiO_2 particles still maintain a strong UV light-blocking capability, but such miniscule particles transmit rather than scatter visible light, so they do not impart any white pigmentation or opacity to the matrix or substrate into which they are incorporated. This had made such alumina nanoparticles a sought-after UV-blocking additive for sunscreens and cosmetics and even for varnishes for the preservation of wood, textile fibres, and packaging films.

And, because nanosized particles of TiO_2 demonstrate catalytic, photocatalytic, and electrical properties, their use is also being explored in novel applications, such as the development of self-sanitizing tiles for restaurants and hospitals, and catalytic coatings on glass that catalyze the decomposition of organic buildup (essentially giving glass, such as car windshields, a "self-cleaning" capability).

The use of nanosized TiO_2 powders is also being explored in photo electro-chemical solar cells and various types of improved thermal coating for corrosion protection, and as a component in various polymer composites to yield a product with a tunable refractive index and improved mechanical properties for photonic and electronic applications. Since light scattering is significantly reduced in such advanced nanocomposites, they are considered an attractive building block for optical network component applications [8]. Once again, more information on nanocomposites can be found in the next Chapter.

16.11 ZINC OXIDE

Discuss some of the properties and applications of zinc oxide.

SOLUTION

This is the last oxide to be reviewed. ZnO, is valued for its UV opacity and fungicidal action. As is the case with the other metal oxides, nanoparticles of zinc oxide are differentiated from larger particles by their increased surface area and transparency to visible light, making them essentially invisible when added to other matrices, such as cosmetics, sunscreens, and antifungal foot powders. And, whereas TiO_2 blocks only UV-B radiation from the sun, zinc oxide particle absorb both UV-A and UV-B radiation, so they offer broader protection and improved aesthetic appeal when formulated into sunscreens and cosmetic, and into clear-coat film polymers to protect automobiles, furniture, and fabrics from sun damage.

An evolving use for ZnO is in the field of photoelectronics, where nanowires of ZnO are being developed for use in UV nanolasers [8]. Meanwhile, ZnO nanoparticles are also being used in ceramics and rubber processing, where they are said to improve elastomeric toughness and abrasion resistance.

16.12 SILICA PRODUCTS

Discuss the various forms of silica, and some of their properties and applications.

SOLUTION

The main forms of nanosized silica are precipitated silica, silica gels, colloidal silica, silica sols, and fumed or pyrogenic silica. Today, precipitated silicas and silica gels are considered commodity chemicals, so they are not discussed further here, but ample discussion can be found in the literature [8].

Colloidal silica or silica sol products are stable suspensions of independent, non-agglomerated, and nonporous spherical SiO_2 particles. They can be obtained by a liquid-phase process that involves passing a solution of sodium silicate through an ion exchange column to partially remove the sodium ions. Under alkaline conditions, silica particles start to grow, forming a stable and homogenous silica suspensions. The pH is adjusted to control particle size, and another counter-ion, such as an ammonium ion, may be introduced to stabilize the suspension. The resulting particle size range is very narrow and the process is said to have excellent batch-to-batch consistency [8]. Most commercial sols consist of discrete spheres with a diameter range between 5 and 200 nm.

The stability of colloidal silicas is improved via the addition of bases that generate negative charges on the particle surface. When the particles are forced to repel each other, a stable sol is formed. Silica sols are considered to be either anionic (when they are stabilized with cations, such as Na^+, K^+, NH_4^+) or cationic (when the stabilizer is an aluminum derivative). Meanwhile, by lowering the pH value, the stability decreases and the particles react to form a gel. The gelling rate between pH 4 and 7 is very high and acidification must take place very quickly.

Chemical mechanical planarization (CMP), a precision polishing technique used during the production of semiconductor chips, is the largest market application that is already in use of nanosized silica, in particular, fumed silica and silica sols. A detailed discussion of CMP can be found in the next Chapter.

17 Production Manufacturing Routes

As discussed in the previous Chapter, when materials such as metals, metal oxides, ceramics, and polymers are manipulated into increasingly smaller and smaller particle sizes, the resulting increase in available surface area leads to a direct improvement in a variety of material properties, including thermal and electrical conductivity, surface chemistry (which affects particle dispersibility and reactivity), photonic behavior (which is a functionality that changes in the presence of light of varying wavelengths), and catalytic conversion rates, among others. Today, nanoscaled particles are already in use in a diverse array of viable commercial applications, while many other promising laboratory- and pilot-scale developments are being groomed for commercial-scale production.

Ultimately, the high cost associated with nanoscaled particles will have to be justified in terms of enhanced performance. Nanotechnology researchers are under increasing pressure to not only devise viable manufacturing methods that bring down production costs, but also to exploit to the fullest extent possible the broadest array of functional advantages associated with a given type of nanoscaled material.

Today, nanoparticle use is the most robust in the manufacture of semiconductor devices and in the production of both advanced composite materials for a wide variety of end uses, including improved consumer products such as cosmetics and sunscreen. In addition, nanoparticle-related developments are being actively pursued to improve fuel cells, batteries and solar devices, advanced data storage devices such as computer chips and hard drives plus magnetic audio and videotapes, and sensors and other analytical devices. Meanwhile, nanotechnology-related developments are also being hotly pursued in various medical applications, such as the development of more effective drug-delivery mechanisms and improved medical diagnostic devices, to name just a few.

Much of the information presented in the Problems that follow was excerpted and/or adapted from an outstanding comprehensive report prepared by SRI Consulting [8]. Much of the material can also be found in Chapter 2 reference [1], prepared by S. Shelley.

Nanotechnology: Basic Calculations for Engineers and Scientists, by Louis Theodore
Copyright © 2006 John Wiley & Sons, Inc.

17.1 CARBON NANOTUBES AND BUCKYBALLS

Provide background material on carbon nanotubes.

SOLUTION

One particular type of nanometer-scaled structure that is generating considerable interest is the carbon nanotube. Today, carbon nanotubes are among the most hotly pursued type of nanoparticle, and in some applications they are leading the charge in terms of nanoscaled particles that are making their way into commercial-scale use.

Carbon nanotubes are seamless cylinders composed of carbon atoms in a regular hexagonal arrangement, closed on both ends by hemispherical endcaps. They can be produced as single-wall nanotubes (SWNT) or multiwall nanotubes (MWNT). While carbon nanotubes can be difficult and costly to produce, and challenges remain when it comes to distributing and incorporating them homogeneously within the matrices of other materials, they have been an object of desire among researchers working in the nanosphere since their discovery in 1991 due to their excellent intrinsic physical properties and promising potential applications in a wide array of fields [16].

The following list summarizes some of the unprecedented structural, mechanical, and electronic properties that are exhibited by carbon nanotubes [8, 16–18].

1. SWNT can have diameters ranging from 0.7 to 2 nm, while MWNT can have diameters ranging from 10 to 300 nm. Individual nanotube lengths can be up to 20 cm.

2. Carbon nanotubes have a surface area of upto 1500 m^2/g and a density of 1.33 to 1.40 g/cm^3.

3. Depending on their structure, nanotubes can function as either conductors or semiconductors of electricity and heat; thermal conductivity has been demonstrated to achieve values of 2000 W/mK (with some published work suggesting values as high as 6000 W/mK).

4. They exhibit extremely high thermal and chemical stability.

5. They are extremely elastic, with a modulus of elasticity on the order of 1000 gigapascals (Gpa).

6. They have many times the tensile strength greater than 65 Gpa, with predicted value as high as 200 Gpa; a widely cited comparison is that carbon nanotubes have a tensile strength 100 times that of steel, but at only one-sixth the weight of steel.

7. Some SWNTs can withstand 10 to 30% elongation before breakage.

8. As nanocapillaries, they can accommodate guest molecules and form dispersions with surfactants in water.

9. They have the potential for further chemical functionalization via various types of surface treatments.

Today, a variety of production methods exist for making carbon nanotubes, although in many cases, the process specifics are closely guarded by companies hoping to retain a competitive advantage by developing and patenting more effective or less costly production routes. In general, most of the prevailing methods are variations of the following two basic process types [8, 16, 18, 19].

1. Carbon is subjected to high heat (such as through the use of a 3000 K electric arc), and the resulting carbon nanotubes and other carbon nanostructures are isolated from the soot mixture that is formed. Arc discharge processes can make either SWNT or MWNT using a plasma-based process whose feedstock carbon typically comes from burning either a solid carbon electrode (to produce MWNT) or a composite of carbon and a transition-metal catalyst (to produce SWNT). Another high-temperature route involves the use of pulsed Nd : YAG laser aimed at powdered graphite that is loaded with metal catalyst (a layer variation of this process uses a continuous, high-energy CO_2 laser, focused on a composite, carbon–metal feedstock source).

2. Alternately, carbon-containing compounds (such as graphite rods) are decomposed using heat or catalysis to yield carbon nanotubes and other structures. This latter method, which typically uses chemical vapor decomposition (CVD), is increasingly used since it offers better control over the type of product manufactured. CVD-based processes can be used to make either SWNT or MWNT, and the technique produces them by heating a precursor gas and flowing it over a metallic or oxide surface with a prepared catalyst (typically a nickel, iron, molybdenum, or cobalt catalyst).

Still another nanoscaled carbon derivative has been making headlines in recent years. Until the late 1980s, graphite and diamond were the only known and macroscopically available modifications of carbon [17]. This changed when chemistry professor Richard Smalley and his colleagues at Houston's Rice University found that by vaporizing carbon and allowing it to condense in an inert gas, it formed highly stable crystals, composed of 60 atoms of carbon apiece (C-60).

Since the novel, cagelike molecule was suggestive of the soccer-ball structure used by the architect R. Buckminster Fuller in his futuristic geodesic buildings, the researchers named their discovery "buckminsterfullerene", which was quickly shortened to fullerene or buckyball. The buckyball remains nanotechnology's most famous discovery. It not only earned Smalley and his colleagues the 1996 Nobel Prize in Chemistry but it also cemented nanotechnology's reputation as a cutting-edge research field [20]. Additional details are available in the literature [1].

17.2 SEMICONDUCTOR MANUFACTURING

Discuss semiconductor manufacturing.

SOLUTION

Perhaps considered the grandfather of all nanoparticle-related applications is chemical mechanical planarization (CMP), a precision polishing process that is used during the production of integrated electonic circuits on semiconductor chips. Today's semiconductor chips involve increasingly fine and exacting geometries, and increased numbers of metal layers are used to create the integrated circuits. According to one report, circuits designed at 25 nm can incorporate as many as 10 million transistors and 50 million connections [8].

During CMP, nanoscaled particles of abrasive materials – typically oxides of aluminum and zirconium, and collodial or fumed silica, but increasingly cerium as well – with diameters of 20 to 300 nm are formulated into a polishing slurry. Slurries for polishing silicon chip surfaces are typically made from a fumed or collodial silica suspension stabilized with dilute potassium or ammonium hydroxide. Slurries for treating metal surfaces typically use aluminum oxide suspensions plus an oxidizer such as ferric nitrate, potassium iodate, or hydrogen peroxide [8]. This slurry combines both mechanical and chemical action to smooth the topography of the semiconductor device, reduce height variations across the dielectric region, and scavenge metal deposits and excess oxide that is used during fabrication. In doing so, the slurry ensures that the metal and dielectric layers on the silicon wafers are smooth and essentially defect free.

As the drive toward miniaturization continues to spur the development of smaller and smaller chip geometries, the exacting demands placed on CMP slurries will continue to grow [21]. For instance, while line widths of 130 nm are common in today's semiconductor devices, 90 nm line widths will be the standard for next-generation devices [16]. Such intricate chip assemblies have no tolerance for even the smallest defects.

Today, CMP is considered the "gold standard" for polishing and cleaning complex semiconductor chip geometries, and this has created a viable commercial-scale market for related nanoparticles. In the coming years, growth in demand for various oxide nanoparticles for CMP will be driven primarily by three factors: an overall increase in the number of semiconductor wafers produced, a rise in the percentage of wafers that are processed using CMP, and an increase in the number of layers per wafer that must be planarized during manufacture. Because the technology is "absolutely critical during semiconductor manufacturing," nanoparticle use for CMP applications is projected by one market analyst to grow by 20%/year [16, 21].

17.3 ADVANCED COMPOSITES

Discuss advances composites.

SOLUTION

Nanocomposites are formed when nanometer-sized particles of useful additives are blended with a polymer matrix resin. The idea of blending functional additives with plastics is not new, but the ability to produce and use nanoscaled versions of the traditional additive materials brings many previously unknown benefits to this class of materials. Today, nanometer-sized clay platelets and carbon nanotubes are the most widely used additives incorporated into commercially available nanocomposites, along with various oxides. For instance, today, nanometer-sized oxides of alumina (aluminium oxide) are being used to improve scratch and abrasion resistance, nanoparticles of zinc oxide and cerium oxide to improve ultraviolet light attenuation, nanoscaled oxides of cerium to improve oxygen storage, antimony/tin oxides (ATO) to improve near-infrared light attenuation, and indium/tin oxides and ATO to improve conductivity.

One of the earliest nanocomposites was pioneered by Toyota Central R&D labs in 1987 [8]. By combining nylon-6 with just 5 wt% nanoscaled clay platelets, the nanocomposite had the following property improvements compared to the neat resin:

1. 40% higher tensile strength;
2. 68% higher tensile modulus;
3. 60% higher flexural strength;
4. 126% higher flexural modulus; and,
5. 65 to 125°C higher heat distortion temperature.

Today, many different types of nanocomposites are commercially available. An exhaustive range of polymers has been used in various nanocomposite applications, including but not limited to nylon (6, 6-6, 11, 12, and aromatic), polystyrene, polypropylene, polyethylene, polyimides, ethylene vinyl acetate copolymers, poly(styrene-*b*-butadiene) copolymers, poly(dimethyl-siloxane) elastomers, ethylene propylene diene monomer, acrylonitrile butadiene rubber, polyurethane polypyrrole, polyethylene oxide, polyvinyl alcohol, polyaniline, epoxies, and phenolic resins [8].

In general, using much smaller loadings of nanoscaled additives, today's nanocomposites demonstrate the following improved properties compared with both the neat resin and conventional composites made using micron-sized additive particles [8]:

1. Better mechanical properties, such as higher tensile and flexural strength and moduli, improved toughness, improved scratch resistance, and surface appearance even at low additive loading levels;
2. Better barrier properties in terms of decreased liquid and gas permeability, and improved resistance to organic solvents;
3. Improved fire retardancy;
4. Better optical clarity;

5. Lower coefficient of thermal expansion and higher heat distortion temperature;
6. Improved electroconductivity and catalytic activity;
7. Self-cleaning capabilities and antimicrobial properties; and,
8. The ability to make lower-weight plastic parts due to the lower density of the nanoscaled fillers.

By offering an excellent combination of improved mechanical properties, heat resistance, optical properties, and barrier properties, nanocomposites are being used in both flexible and rigid film and sheets for various food-packaging applications. For instance, nanoscaled clay platelets (measuring 1 nm by 100 nm) are impermeable to O_2 and CO_2, so when they are dispersed in a polymeric matrix, these nanoplatelets impart improved gas-barrier and oxygen-scavenging properties to the composite.

According to process developers, during the polymerization process, the nanoplatelets form an obstacle course in the solidified nylon around which gas molecules have difficulty maneuvering [22]. From a practical standpoint, this increased resistance to gas diffusion allows the food and pharmaceutical industry to design novel packaging schemes that, for example, keep unwanted O_2 out to reduce product spoilage, while keeping CO_2 inside the packaging where it belongs, to keep carbonated beer and soft drinks from going flat.

Meanwhile, in automotive and aviation applications, nanocomposites are valued not just for their improved physical properties but for their ability to produce parts with reduced weight (which leads to improved fuel efficiency) and increased recyclability. And, as detailed in the discussion on carbon nanotubes, the addition of nanometer-sized carbon nanotubes to inherently nonconductive polymeric materials allows the resulting nanocomposite parts to be painted using electrostatic spray painting. This is finding favor among automakers who are eager to phase out solvent-borne painting processes [23]. Similarly, nanotube-bearing nylon is being used to make fuel lines for cars, where it is critical to dissipate static charges that could spark an explosion [24]. Thanks to their improved fire retardancy, nanocomposites are also being widely used today in the electrical and electronics industries, and in the manufacture of wire and cable and TV housings.

As in the case when working with other types of nanoparticles, the ability to get nanoparticles, carbon nanotubes, and nanoplatelets of clays to disperse effectively in the polymer matrix is no easy task. Ongoing research and development efforts are geared toward perfecting various surface treatments to mitigate excessive surface charges in order to minimize agglomeration and improve dispersion, and to get the nanoparticles to bond to the polymer rather than to one another.

17.4 ADVANCED CERAMICS

Discuss advanced ceramics.

SOLUTION

In general, ceramics are inorganic, nonmetallic materials that are consolidated at high temperatures, usually starting as powder particles and ending as solid, usable forms [8].

Structural ceramics are defined as stress-bearing components or ceramic coatings, which show special resistance to corrosion, wear, and high temperature. Today, advanced ceramics based on aluminum oxide or alumina (Al_2O_3) and zirconium dioxide or zirconia (ZrO_2) are used for many industrial applications, such as cutting tools, bearings and seals, filters and membranes, refractory materials, catalysts and catalyst supports, automotive engine components, and in bone implants used in dental and orthopedic prostheses. In addition, electronic uses of advanced ceramics include substrates and packaging for integrated circuits, capacitors, transformers, inductors, piezoelectric devices, and chemical and physical sensors.

Finished ceramic components are typically resistant to high temperatures and corrosion, but they are also brittle and hard to work with. Traditionally, high-performance ceramics have been made from powders whose constituent particles have a diameter of just under 1 μm, or 1000 nm. Increasingly, however, improvements have been demonstrated, by producing ceramics from powders consisting of much smaller particles that are 100 nm diameter or less [8].

The size and form of the particles have a big influence on the properties of the ceramic. In high-density components, for example, they determine the component's strength, while in porous ceramics, they can be used to deliberately control the size and form of the pores. In microelectronics applications, nanoceramics can be used to create structures whose dimensions are far smaller than the limit allowed by conventional techniques.

Nanocrystalline ceramics also allow a higher level of homogeneity to be attained, and more well-tailored grain sizes in the final sintered product, and this leads to enhanced properties such as increased fracture toughness, transparency for use in lamps and windows, superplasticity for near-set shaped parts, very high hardness and wear resistance, enhanced gas sensitivities, diffusion bonding for ceramic seals, improved tribologic properties such as higher resistance to friction and wear, and increased high-temperature resistance [8].

Meanwhile, since there is a strong relationship between sintering temperature and particle size, the ability to reduce the particle size of the initial material to about 20 nm allows the sintering temperature for zirconia to be lowered from 1400 to 1110°C.

As with other nanoparticles, the reduction in the size of the powder particles drastically increases the total particle surface area. For example (as demonstrated in analogous Problem earlier in the text), a 30 g sample of zirconium powder with a 700 mm diameter has a surface area of only 200 m^2, while the same amount of zirconium powder with 15 nm particles has 1500 m^2 of surface area [8]. One additional feature of nanoscale materials is the high fraction of atoms that reside at particle surfaces or grain boundaries. At a grain size of less than 20 nm, as much as 30% of the atoms can be present at the grain boundaries. This is important, since these grain-to-grain interfaces play a crucial role in determining the mechanical, optical, and electrical properties of the final material.

Ceramic nanopowders are typically synthesized using either precipitation from organometallic or aqueous salt solutions (including sol–gel processing) or condensation from the vapor phase, using spray pyrolysis and gas condensation.

However, the use of nanoscaled ceramic powders creates three sets of challenges that are not encountered when working with oxide powders at the micron or submicron scale [8]. These are detailed below.

1. Conventional ceramic powder is produced by progressively fine grinding of the material. However, nanocrystalline powder is typically chemically precipitated from solutions or condensed from gaseous states. Both of these alternative production methods are costly and more elaborate compared to traditional production routes. As a result, issues related to viable, cost-effective production capacities and batch-to-batch consistency remain with these nanoparticle based production routes for making ceramics.
2. Shaping nanoceramic powders into homogeneous solids remains another big challenge. Nanopowders tend to clump together unevenly because they have large surface areas with excessive surface charge.
3. During the sintering process, grain size may be difficult to control. The higher reactivity of the fine powder can lead to increased crystal growth, resulting in coarse and inhomogeneous grain structures.

17.5 CATALYTIC AND PHOTOCATALYTIC APPLICATIONS

Briefly discuss catalytic and photocatalytic applications.

SOLUTION

Researchers are using nanotechnology to harness the photocatalytic capabilities of certain nanoscaled substances and use these capabilities for novel commercial application. When exposed to UV light, photocatalytic substances, such as the anatase form of titanium dioxide (TiO_2; but not the rutile form), strongly absorb UV radiation. In the presence of water, oxygen, and UV light, such substances generate free radicals that can be used to decompose unwanted chemical substances and reduce the adhesive forces that bind dirt (both organic and inorganic matter) and algae to various surfaces.

The photocatalytic effect can be exploited for various commercial applications, such as water and air purification, or to impart self-cleaning, antimicrobial, and anti-algae properties to surfaces. Since all such reactions take place very close to the surface area it allows these reactions to proceed at orders-of-magnitude faster rates than ordinary grades of the oxide [25]. Such photocatalytic reactions can be used to treat, sanitize, and deodorize air and water plus various surfaces and fabrics.

In addition, the unique physical interaction of water droplets with photocatalytic surfaces has led to the development and marketing of TiO_2 films that impart "self-cleaning" or "antifogging" properties to windows, mirrors, and other substrates. During exposure to light, the contact angle of a water droplet on a photocatalytic surface is gradually reduced to zero; in other words, the water cannot maintain the form of a droplet on such a surface so that it spreads out in a sheetlike form. This results in a superhydrophilic or extremely water-loving surface [21]. Today, the use of photocatalytic substances is being explored in the development of antifogging bathroom mirrors and self-cleaning tiles for use in hospitals and restaurants, as well as air cleaners, water purifiers, and germicidal lamps, among others.

Conventional photocatalytic TiO_2 films can be produced using chemical vapor deposition (CVD) techniques, although more recently, nanotechnolgy researchers have devised ways to produce them from dispersions of nanoparticles. Anatase-phase TiO_2 is currently the most popular semiconducting oxide for photocatalysis, although zinc oxide, tin oxide, and other oxides also exhibit photocatalytic behavior, so their use in such applications is also being pursued.

Meanwhile, in search of improved catalysts to destroy air pollutants, one group of researchers is focusing its attention on gold nanoparticles layered on TiO_2. The group has shown that the addition of gold (a notoriously unreactive element) to TiO_2 (a widely used industrial catalyst) changes the electronic properties of both materials. The result is a catalyst that is said to be 5 to 10 times more reactive than ordinary TiO_2 in terms of dissociating sulfur dioxide (SO_2) [25].

17.6 GAS SENSORS AND OTHER ANALYTICAL DEVICES

Discuss gas sensors and other analytical devices.

SOLUTION

The research community is hard at work to exploit the extraordinary surface area and increased reactivity and catalytic properties of many nanoscaled materials and use these attributes to develop highly sensitive sensors and other analytical devices (such as the so-called lab-on-a-chip applications). Such devices are envisioned to improve industrial monitoring devices, check food quality, improve medical diagnostics such as disease detection, and improve the detection of chemical, biological, radiological, and nuclear hazards [23].

Resistive metal oxide gas sensors (using, for example nanoscaled oxides of zinc, tin, titanium, and iron) rely on a change in electrical conductivity at the surface of the sensor, as it comes in contact with the target gas. Thus, the reaction rate is dictated by the amount of surface area on the probe. Ongoing advances in nanoparticle engineering have given manufacturers the opportunity to greatly increase surface area on the sensor probe in order to improve the gas detection sensitivity and selectivity. For example, one company has patented a new fabrication process for preparing metal oxide sensors that may overcome some of the disadvantages of conventional thick-film fabrication techniques (including considerable labor, limited reproducibility, and high costs) [21].

Comparing prototypes of sensors made from conventional, coarse-grained oxides (>1 μm) with those produced using nanopowders (<100 nm), the nanostructured sensors demonstrate enhanced sensitivity and respond more quickly – which gives them a distinct advantage for certain monitoring applications. Using the new approach, different layers are deposited to form multiplayer, chip-style sensors with a series of internal electrodes.

17.7 CONSUMER PRODUCTS

From an application perspective, discuss consumer products.

SOLUTION

Makers of sunscreens, cosmetics, and other personal-care products have discovered that the use of nanometer-scaled versions of common additives can improve the effectiveness and aesthetic appeal of many products, compared to conventional formulations. For instance, the recent trend in sunscreen development has been to devise products that offer broader-spectrum protection against not just ultraviolet B (UV-B) radiation (which is to blame for sunburn), but shorter wavelength ultra violet-A (UV-A) radiation (the chief culprit in skin cancer and other skin damage) as well. With the advent of affordable methods to produce and use nanoscaled particles of the common UV blockers titanium oxide (TiO_2) and zinc oxide (ZnO), sunscreen manufactures can now use these broad-spectrum UV-blocking agents to produce transparent lotions that are aesthetically superior to the opaque white oxide creams that are the hallmark of surfers and lifeguards [21]. Because the wavelength of visible light is 400 to 800 nm, nanoscaled particles are able to transmit, rather than scatter light, thereby allowing transparent (rather than opaque) sunscreen, cosmetic, and other personal-care product formulations to be made.

17.8 DRUG DELIVERY MECHANISMS AND MEDICAL THERAPEUTICS

The intersection of nanotechnology with the life sciences has the potential to revolutionize medicine and health care. Today, a new era in biotechnology and biomedical engineering is emerging with the use of nanoscaled structures for disease diagnoses, gene sequencing, and improved drug delivery [23]. Meanwhile, a host of biocompatible nanomaterials are being pursued to allow for the development of more robust artificial tissues and organs, and novel materials such as alumina ceramic toughened with nanosized zirconia are being developed to extend the life of ceramic hip and knee replacement material [22]. Nanotechnology-based materials and devices are also being used to develop high-performance medical instruments and diagnostic devices.

One of the most promising nanotechnology-based developments in the medical arena is the use of nanoscaled materials and devices to improve the way toxic drugs often both reach their intended target within the human body and are sequestered preferentially in and around a tumor, while avoiding capture by the body's immune system or accumulation in healthy organs and tissues [21]. For example, the systemic, intravenous administration of chemotherapy drugs often causes devastating damage to healthy cells and severe side effects for the patient. One industry observer notes that to minimize the collateral damage associated with chemotherapy, lower than optimal drug doses are often used because administering many cancer drugs at the doses needed to induce the desired anticancer effect "could kill the patient" [21].

Some developmental efforts may sound like they are right out of the pages of a science fiction novel. For instance, one research team is developing superparamagnetic iron oxide, Fe_2O_3, as a potential tool for treating tumors. About 20 nm in diameter and dispersed within an amorphous silicon dioxide matrix, the tiny iron oxide particles are magnetic when exposed to a magnetic field, yet lose their magnetism when the field is removed. And, in an alternating magnetic field, the particles heat up. Researches are developing a process by which the particles are injected into the bloodstream of a patient and guided to the tumor site using a magnet, and then, by applying a very localized alternating magnetic field, the particles are heated to 50 to 60°C to destroy the tumor without harming the surrounding healthy tissue [21].

Scientists elsewhere have been able to demonstrate that semiconductor nanocrystals (quantum dots) coated with homing peptides can target specific types of cancer cells in live mice. As a next step, the group is working to synthesize quantum dots that are functionalized with both homing peptides and cancer treatment drugs in order to target and destroy the cancerous tissue [21].

Meanwhile, while many drug-targeting techniques have focused on the use of two chemically distinct colloid particles – nanoscale liposomes and biodegradable polymers – that carry the drug payload inside a hollow shell; one company has turned its attention to a fundamentally different carrier – colloidal gold nanoparticles. Gold particles are inert, and they are said to bind protein biologics avidly and preferentially on their surface through available thiol groups [21].

Thanks to a "leaky hose effect" that results when blood vessel networks develop hastily to feed a fast growing tumor, nanometer-sized gold particles (typically 35 nm), carrying cancer drugs on their surface, are able to sneak through the porous blood vessel network, bringing the drugs right to the cancerous site.

In its work to date using the cancer drug TNF (tumor necrosis factor, an anticancer protein biologic that has shown promise in killing a variety of solid tumor types) bound to colloidal gold nanoparticles, the company has found that the drug was preferentially sequestered in tumors in mice and dogs at higher rates, compared to testing using unbound TNF. However, initial work also showed that the gold TNF also aggregated preferentially in the livers and spleens (the chief organs of the immune system) of the tested animals. To make the gold "more invisible" to these organs, the developers added a linear form of polyethylene glycol (PEG) to the colloidal gold nanoparticles just like adding brush bristles to the surface. This improvement resulted in "ten times more TNF getting to the tumor than with the non-PEG version, and with no active sequestration in the liver and spleen," say the process developers [21].

Of course, in addition to overcoming potential manufacturing, price-related, and performance hurdles, researchers working in the medical arena must get their nano-technology-based routes through the arduous approval process of the U.S. Food and Drug Administration and comparable regulatory–approval processes elsewhere, and these regulatory hurdles will ultimately dictate which of today's encouraging nanotechnology-related medical breakthroughs will ever be able to fulfill their commercial promise.

17.9   MICROELECTRONICS APPLICATIONS

Discuss microelectronics applications.

SOLUTION

Progressive miniaturization of materials continues to be the hallmark of the electronics and microelectronics industry. Ongoing size reduction and increasing packaging density of microelectronic components has led to an increasing demand for smaller particle sizes of various materials that are used to manufacture these components.

As a result, researchers are developing ways to use nanotechnology to improve the fabrication and functionality of photonic devices that are used in today's broadband optical networks (for both traditional fiber-optic and novel wireless applications). Growing demand for low-cost, high-speed voice and data transmission components, ultrahigh-density memory media, ultrahigh-frequency electronic devices, integrated devices with ultralow energy consumption, and the overall trend toward miniaturization of photonic components are all factors that are driving research activity in this area.

In microelectronics applications, the use of nanoscaled powders of precious metal provides significant advantages during the production of smaller and smaller electronic circuits with decreasing layer thickness, and such an approach can also reduce the overall consumption of precious metals.

17.10   FUTURE ACTIVITES

Discuss future activities in nanotechnology.

SOLUTION

For nanotechnology's most ardent supporters, the scope of the emerging field seems to be limited only by the imaginations of those who would dream at these unprecedented dimensions. However, considerable technological and financial obstacles still need to be reconciled before nanotechnology's full promise can be realized.

For instance, ranking high among the challenges is the ongoing need to develop and perfect reliable techniques to produce (and mass produce) nanoscaled particles that have not just the desirable particle sizes and particle size distributions, but also a minimal number of structural defects and acceptable purity level, as these latter attributes can drastically alter the anticipated behavior of the nanoscaled particles. Experience to date shows that the scaleup issues associated with moving today's promising nanotechnology-related developments from laboratory- and pilot-scale demonstrations to full-scale commercialization can be considerable.

In addition to the inherent challenges associated with designing and scaling up various methodologies to produce nanoscaled materials that have the right particle attributes, additional challenges arise in terms of handling and using these minute particles as functional additives in other matrix materials. For instance, the extremely high surface area of these minute particles creates problems that must be reconciled related to excessive attractive or repulsive surface charges, unwanted nanoparticle agglomeration, problems with dispersion and blending, and so on.

Meanwhile, nanoscaled materials generally command very high prices compared to conventional, macroscopic particles that have essentially the same chemical composition. Such high costs result from the energy that is required to reduce the particle size and from the high research and development and prototyping costs that are incurred during the discovery phase of any novel materials and the related manufacturing processes. However, many industry observers note that the current prices for many nanoscaled materials represent experimental quantities produced at pilot facilities, so the prices are likely to come down, once steady-state, commercial-scale manufacturing conditions are perfected and pursued [8].

Ultimately, premium costs associated with nanoscaled particles, devices, and systems will have to be proven and justified in performance. For instance, if a low-cost, micron-scale powder will suffice in a particular application, the end user is not likely to pay a premium price for its nanoscaled counterpart, whose cost may be orders of magnitude higher [26]. Companies need a strong motivation to justify why they should replace an existing approach with a newer, nanotechnology based approach [27]. Such capacity and cost issues will be key factors that will continue to influence market development and commercial adoption of nanotechnology-based materials and processes into the future.

Beyond just their efforts to produce and use nanometer-sized particles of various materials, some nanotechnology-related scientists and engineers are pursuing far more ambitious – and some would say fantastic or futuristic – applications of this powerful new technological paradigm. For instance, the research community is working toward being able to design and manipulate nanoscaled objects, devices, and systems by the manipulation of individual atoms and molecules [27].

Such forward-looking researchers hope that by using atom-by-atom construction techniques, they will someday be able to create not just substances with remarkable functionality, but also tiny, bacterium-sized devices and machines (thus far dubbed nanorobots), that could be programmed, say, to repair clogged arteries, kill cancer cells, and even fix cellular damage caused by aging. Advanced development is also under way to use nanotechnology to develop advanced sensors to improve the detection of chemical, biological, radiological, and nuclear hazards [28, 29].

Most concede, however, that real, commercial-scale success in this arena is still years, if not decades, away. Nonetheless, despite the fact that the quickly evolving field of nanotechnology integrates the well-established disciplines of chemistry, physics, biology, material science, and all branches of engineering, it also represents a brave new world – a futuristic, imaginative journey that some might say is worthy of a Jules Verne novel.

At a time when most industrialized nations are investing heavily in nanotechnology research and development, and a plethora of potential applications – both practical and fantastic – are being considered, cautious observers note that the potential environmental, health, and safety risks associated with nanotechnology are not being adequately studied and that the complex ethical, legal, and societal implications of this powerful new technological paradigm are not being explored or debated on a large enough scale [30].

While some expectations from nanotechnology may be highly hyped and over estimated in the short-term, many feel that the long-term implications for health care, productivity, and the environment, among others, are underestimated when one considers the depth and breadth of technological breakthroughs recorded to date, and pace at which further research and developments are being undertaken [15].

18 Ventilation

Indoor air pollution is rapidly becoming a major health issue in the United States. Indoor pollutant levels are quite often higher than outdoors, particularly where buildings are tightly constructed to save energy. Since most people spend nearly 90% of their time indoors, exposure to unhealthy concentrations of indoor air pollutants is often inevitable. The degree of risk associated with exposure to indoor pollutants depends on how well buildings are ventilated and the type, mixture, and amounts of pollutants in the building [1].

Alternately, industrial ventilation is the field of applied science concerned with controlling airborne contaminants to produce healthy conditions for workers and a clean environment for the manufacture of products. To claim that industrial ventilation will prevent contaminants from entering the workplace is naive and unachievable. More to the point, and within the realm of achievement, is the goal of controlling contaminant exposure within prescribed limits. To accomplish this goal, one must be able to describe the movement of particle contaminants in quantitative terms that take into account:

1. The spatial and temporal rate at which contaminants are generated;
2. The velocity field of the air in the workplace;
3. The spatial relationship between source, workers, and openings through which air is withdrawn or added;
4. Exposure limits (time-concentration relationships) that define unhealthy conditions [1].

In general, most particulate control/recovery equipment are more economically efficient when handling higher concentrations of contaminants, all else being equal. Therefore, the gas handling system should be designed to concentrate contaminants in the smallest possible volume of air. This is important since, exclusive of the fan, the cost of the control equipment is based principally upon the volume of gas to be handled and not on the quantity of particulate to be remove. The reduction of emissions by process and system control is an important adjunct to any cleaning technology.

Although ventilation does not remove the contaminants (particularly nanoagents) from the workplace, it does provide an opportunity to either recover/control or dilute (into the atmosphere) any problem emissions. Regarding manufacture and

Nanotechnology: Basic Calculations for Engineers and Scientists, by Louis Theodore
Copyright © 2006 John Wiley & Sons, Inc.

production, some European nations and unions either require or recommend that closed operations be employed.

It is important to note that because of the size of nanoparticles, their behavior in a ventilation system will approach that of a gas. Thus, it is safe to conclude that the ventilation procedures and calculations that are currently employed in practice may be applied to nanoparticles.

The reader is referred to the work of Heinsohn [31] for an excellent treatment of ventilation. In addition, it should be noted that several of the Problems to follow were excerpted and/or adapted from publications [32, 33] resulting from NSF sponsored faculty workshops.

18.1 INDOOR AIR QUALITY

Indoor air quality is a relatively recent concern in environmental management. Answer the following four questions

1. Explain why indoor air quality is a concern.
2. Explain what has caused indoor air quality to be of greater concern now than in the past.
3. Describe some of the immediate and long-term health effects of indoor air quality exposure.
4. What are some of the costs associated with indoor air quality problems?

SOLUTION

1. Indoor air quality (IAQ) is a major concern because indoor air pollution may present a greater risk of illness than exposure to outdoor pollutants. People spend nearly 90% of their time indoors. This situation is compounded as sensitive populations, for example, the very young, the very old, and sick people who are potentially more vulnerable to disease, spend many more hours indoors than the average population.

2. Indoor air quality problems have become more serious and of greater concern now than in the past because of a number of developments that are believed to have resulted in increased levels of harmful chemicals in indoor air. Some of those developments are the construction of more tightly sealed buildings to save on energy costs, the reduction of the ventilation rate standards to save still more energy, the increased use of synthetic building materials and synthetics in furniture and carpeting that can release harmful chemicals, and the widespread use of new pesticides, paints, and cleaners.

3. Some of the immediate health effects of indoor air quality problems are irritation of the eyes, nose and throat, headaches, dizziness and fatigue, asthma, pneumonitis, and 'humidifier fever.' Some of the long-term health effects of

indoor air quality problems are respiratory diseases and cancer. These are most often associated with radon, asbestos, and second hand tobacco smoke.

4. The U.S. EPA, in a report to congress in 1989, estimated that the costs of IAQ problems were in the tens of billions of dollar per year. The major types of costs form IAQ problems are direct medical costs, lost productivity due to absence from the job because of illness, decreased efficiency on the job, and damage to materials and equipment.

18.2 INDOOR AIR/AMBIENT AIR COMPARISON

Compare indoor air pollution with ambient air pollution. Show why indoor air quality can be of greater concern than ambient air quality.

SOLUTION

Outdoor air pollution and indoor air pollution share many of the same pollutants, concerns, and problems. Both can have serious negative impacts on the health of the population. Not too many years ago, it was a common practice to advise people with respiratory problems to stay indoors on days when pollution outdoors was particularly bad. The assumption was that the indoor environment provided protection against outdoor pollutants. Recent studies conducted by the U.S. EPA have found, however, that the indoor levels of many pollutants are often two to five times, and occasionally more than 100 times, higher than corresponding outdoor levels. Such high levels of pollutants indoors are of even greater concern than outdoors because most people spend more time indoors than out. As indicated earlier, estimates indicate that most people spend as much as 90% of their time indoors. Indoors is where most people work, attend school, eat, sleep, and even where much recreational activity takes place.

18.3 SOURCES OF CONTAMINANTS IN INDOOR AIR

Discuss in some detail the sources of contaminants in indoor air. The answer should include indoor sources, outdoor sources, structural sources, product sources, and sources due to activities carried out in the indoor space.

SOLUTION

Indoor sources of air contaminants include consumer and commercial products, building or structural materials, and personal activities.

Among the consumer and commercial products that release pollutants into the indoor air are pesticides, adhesives, cosmetics, cleaners, waxes, paints, automotive products, paper products, printed materials, air fresheners, dry cleaned fabrics, and furniture. In addition to the "active" ingredient in all the products mentioned, many products contain so-called inert ingredients that are also contaminants when released into indoor air. Examples would be solvents, propellants, dyes, curing agents, flame retardants, mineral spirits, plasticizers, perfumes, hardeners, resins, binders, stabilizers, and preservatives. Aerosol products produce droplets, which remain in the air long enough to be inhaled. This allows the inhalation of some chemicals that would not be volatile enough to be inhaled otherwise. One consumer product that produces indoor air contaminants and merits special mention is tobacco smoke.

The single most important building or structural source of contaminant is formaldehyde contained in building materials such as plywood, adhesives, insulating materials such as urea formaldehyde foam, floor and wall coverings. Depending on the individual type of construction and maintenance practices, there can be many other building sources of IAQ problems. Damp or wet wood, insulation, walls, and ceilings can be breeding places for allergens and pathogens that can become airborne. Allergens and pathogens can also originate from poorly maintained humidifiers, dehumidifiers, and air conditioners. If the building has openings to the soil, radon can enter the building in those areas where radon occurs. The building's heating plant can be the source of contaminants such as CO and NO_x. Automobile exhaust from attached garages is another source of carbon monoxide and nitrogen oxides. Particulates such as asbestos from crumbling insulation and lead form the sanding of lead-based paints are additional contaminants that can become part of the indoor air.

Personal activities can be sources of indoor air contaminants such as pathogenic viruses and bacteria, and a number of harmful chemicals such as products of human and animal (pet) respiration. Houseplants can release allergenic spores. Pet products, such as flea powder, can be sources of pesticides and pets produce allergenic dander when they lick themselves.

Outdoor sources of indoor air contaminants are widely varied. Polluted outdoor air can enter the indoor space through open windows, doors and ventilation intakes. Most outdoor air is less contaminated than indoor air. In some cases, however, such as with a nearby smokestack, a parking lot, heavy street traffic, or an underground garage, outdoor air can be a significant source of indoor contaminants. Outdoor pesticide applications, barbecue grills, and garbage storage areas can also bring outdoor contaminants into the building if placed close to a window or door, or the intake of a ventilation system. Improper placement of the intake of a ventilation system near a loading dock, parking lots, the exhaust from restrooms, laboratories, manufacturing spaces, and other exhausts of contaminated air is a major source of indoor air pollution. Other outdoor sources of indoor air pollutants are hazardous chemicals entering the structure from the soil. Examples are the aforementioned radon gas, methane and other gases from sanitary landfills, and vapors from leaking underground storage tanks of gasoline, oil, and other chemicals penetrating into basements. Polluted water can give off substantial quantities of harmful chemicals during showering, dishwashing, and similar activities.

18.4 INDUSTRIAL VENTILATION SYSTEM

List and describe the major components of an industrial ventilation system.

SOLUTION

The major components of an industrial ventilation system include the following:

1. Exhaust hood;
2. Ductwork;
3. Contaminant control device;
4. Exhaust fan; and,
5. Exhaust vent or stack.

Several types of hoods are available. One must select the appropriate hood for a specific operation to effectively remove contaminants from a work area and transport them into the ductwork. The ductwork must be sized such that the contaminant is transported without being deposited within the duct; adequate velocity must be maintained in the duct to accomplish this. Selecting a control device that is appropriate for the contaminant removal is important to meet certain pollution control removal efficiency requirements. The exhaust fan is the work horse of the ventilation system. The fan must provide the volumetric flow at the required static pressure, and must be capable of handling contaminated air characteristics such as dustiness, corrosivity, and moisture in the air stream. Properly venting the exhaust out of the building is equally necessary to avoid contaminant recirculation into the air intake or into the building through other openings. Such problems can be minimized by properly locating the vent pipe in relation to the aerodynamic characteristics of the building (see next Chapter for more details). In addition, all or a portion of the cleaned air may be recirculated to the workplace. Primary (outside) air may be added to the workplace and is referred to as makeup air; the temperature and humidity of the makeup air may have to be controlled. It also may be necessary to exhaust a portion of the room air.

A line diagram of a typical industrial ventilation system is provided in Figure 18.1. Note that either the control device or the fan (or both) can be located in the room/workplace.

18.5 DILUTION VENTILATION vs. LOCAL EXHAUST SYSTEMS

Define dilution ventilation and local exhaust systems, and discuss their differences.

SOLUTION

Exposure to contaminants in a workplace can be reduced by proper ventilation. Ventilation can be provided either by dilution ventilation or by a local exhaust system. In dilution ventilation, air is brought into the work area to dilute the contaminant

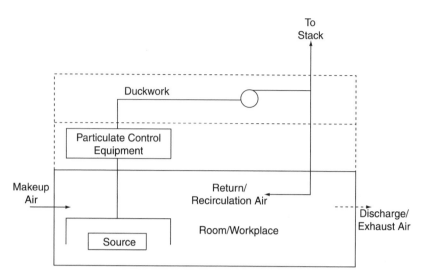

Figure 18.1   Industrial ventilation system components.

sufficiently to minimize its concentration and subsequently reduce worker exposure. In a local exhaust system the contaminant itself is removed from the source through hoods.

A local exhaust is generally preferred over a dilution ventilation system for health hazard control because a local exhaust system removes the contaminants directly from the source, whereas dilution ventilation merely mixes the contaminant with uncontaminated air to reduce the contaminant concentration. Dilution ventilation may be acceptable when the contaminant concentration has a low toxicity, and the rate of contaminant emission is constant and low enough that the quantity of required dilution air is not prohibitively large. However, dilution ventilation generally is not acceptable when the acceptable concentration is less than 100 ppm.

In determining the quantity of dilution air required, one must also consider mixing characteristics of the work area in addition to the quantity (mass or volume) of contaminant to be diluted. Thus, the amount of air required in a dilution ventilation system is much higher than the amount required in a local exhaust system. In addition, if the replacement air requires heating or cooling to maintain an acceptable workplace temperature, then the operating cost of a dilution ventilation system may further exceed the cost of a local exhaust system.

The amount of dilution air required in a dilution ventilation system can be estimated using the following expression:

$$q = K(q_c/C_a)$$

where q = dilution air flowrate; K = dimensionless mixing factor; q_c = flowrate of pure contaminant vapor; and, C_a = acceptable contaminant concentration. For more details, see Problem 18.9.

18.6   VENTILATION DEFINITIONS

Define the following terms commonly used in a ventilation system:

1. Velocity pressure (VP)
2. Static pressure of hood (SP_h)
3. Coefficient of hood entry (C_e)
4. Fan static pressure (FSP)
5. Fan total pressure (FTP)
6. Break horsepower (BHP)
7. Mechanical efficiency (ME)
8. Fan rating table
9. Fan performance curve
10. System performance curve
11. Operating point
12. Threshold limit value (TLV)
13. Lower explosive limit (LEL)

SOLUTION

1. *Velocity pressure (VP).* Velocity pressure is a measure of the kinetic energy of air in motion. Velocity pressure is determined by subtracting the static pressure from the total pressure. In fluid dynamics the velocity head is related to the velocity of using the Bernoulli equation. However, in the ventilation field the velocity head is converted to a convenient unit of inches of water, which is the velocity pressure. When the air is at a density of 0.075 lb/ft^3, the velocity pressure in inches of water can be related to the velocity in ft/min by the following approximate relationship:

$$V = 4005\sqrt{VP}$$

   where $V =$ velocity, ft/min: $VP =$ velocity pressure, in H$_2$O; and $4005 =$ conversion factor.

   VP is usually measured using a pitot tube, which actually measures both the total pressure and the static pressure; the difference is the velocity pressure.

2. *Static pressure of hood (SP_h).* In designing ventilation systems one has to keep track of the pressure losses in the system. A properly selected fan is expected to meet the estimated pressure losses in the system. One such loss in a ventilation system is the static pressure of the hood (SP_h). For a simple hood, SP_h is the sum of the losses due to the energy required to accelerate the room air to the duct velocity and the hood entry loss (h_e), which is

due to the air entering the hood. The acceleration loss is approximately equal to one velocity pressure of the duct (VP_d). The hood entry loss for a simple hood is represented as a function of velocity pressure. Thus, the static pressure of a hood for a simple hood is described as follows.

$$SP_h = VP_d + h_e$$
$$SP_h = VP_d + (F_h)(VP_d)$$

where VP_d = velocity pressure in the duct, in H_2O; and F_h = hood entry loss factor.

For compound hoods additional losses must be included. For example, the SP_h for a typical slot hood (hoods having double entry losses with an aspect ratio of 0.2 or less, commonly used to provide uniform exhaust air flow and an adequate capture velocity over a finite length of contaminant generation) is as follows:

$$SP_h(\text{slot hood}) = 1.78VP_s + F_h(VP_d) + VP_d$$

where VP_s = Velocity pressure of the slot hood, in H_2O.

3. *Coefficient of hood entry (C_e).* The coefficient of entry is the square root of the ratio of VP_d to SP_h. Thus,

$$C_e = \sqrt{\frac{VP_d}{SP_h}}$$

When there is no hood entry loss ($h_e = 0$), the coefficient of entry becomes 1.0 ($SP_h = VP_d + 0$) and the air flow will be maximum. This maximum air flow is the "theoretical flow." In actuality there will always be a hood entry loss and the actual flow (appropriately called the "actual flow") will be lower than the theoretical flow. Hence, the term C_e can also be defined as the ratio of the actual flow to the theoretical flow.

4. *Fan static pressure (FSP).* FSP is the net suction and positive pressure the fan must provide to the ventilation system to transport a given amount of gas flow. The FSP is calculated from the following expression:

$$FSP = SP_{out} - SP_{in} - VP_{in}$$

where FSP = fan static pressure, in H_2O; SP_{out} = static pressure at the outlet of the fan, in H_2O; and SP_{in}, VP_{in} = static pressure and velocity pressure at the inlet of the fan, in H_2O, respectively. (*Note*: SP_{in} carries a negative sign because it represents a suction pressure).

Selection of a fan for a ventilation system is based on the amount of resistance the fan has to overcome for maintaining a desired air flow rate. Fan rating tables published by manufacturers are used to select the fan size.

Fan rating tables are normally provided in inches of water, fan static pressure versus flow rate, or fan total pressure versus flow rate.

5. *Fan total pressure (FTP)*. The fan total pressure is calculated from the following expression:

$$FTP = (SP_{out} + VP_{out}) - (SP_{in} + VP_{in}) = SP_{out} - SP_{in} - VP_{in} + VP_{out}$$
$$= FSP + VP_{out}$$

6. *Break horse power (BHP)*. Break horse power is the amount of energy required to operate a fan assuming no drive losses between the fan and motor. Manufacturers use standardized procedures to develop *BHP* versus *Q* (flow rate) data, and provide them in fan rating tables. The actual amount of power required to run the fan is higher than the *BHP* rating due to drive train losses. The following expression may be used to calculate the *BHP* of the fan:

$$BHP = \frac{Q(FTP)}{6360} = 1.57 \times 10^{-4} Q(FTP)$$

where BHP = break horse power, hp; Q = flow rate, ft^3/min; and FTP = fan total pressure, in H_2O.

7. *Mechanical efficiency (η)*. Mechanical efficiency describes the amount of energy actually required to operate a fan compared to the energy estimated based on no drive losses. The following equation, which is a ratio of the BHP to the actual power required, can be used to estimate the mechanical efficiency of a fan (η).

$$\eta = BHP/W_a$$

where η = efficiency of the fan; and W_a = actual power required, hp.

8. *Fan rating table*. Fan manufacturers test their fans following standardized testing procedures. The variables used in a typical fan rating table are the flow rate, Q, the fan static pressure FSP, or fan total pressure FTP, fan speed in revolutions per minute, RPM, and the BHP. From the fan rating table, one may draw curves relating different variables such as: Q versus FSP, Q versus BHP, and Q versus RPM. It is important to note that unless specified the values in the table are based on an air density of 0.075 lb/ft^3 at mean sea level. If the air density at a site differs from this value due to temperature, altitude, or other factors, then the fan table values must be adjusted accordingly.

9. *Fan performance curve*. A typical fan performance curve relates the flow rate to the fan static pressure at a specific fan rpm. Since fans can run at different speeds, one can draw a performance curve for each speed. The data to draw this performance curve are obtained from fan rating tables.

10. *System performance curve.* A system performance curve describes the relationship between the flow rate and fan static pressure of a ventilation system. Once a system is designed, i.e., once the hood configuration, the duct diameter, the duct length, the contaminant control device, the ventilation duct diameter, and the vent length are fixed, the static pressure that the fan must develop under different air flow rates can be calculated and represented as a system curve. This system curve can be calculated assuming turbulent flow for which the resistance to flow is approximately proportional to the square of the flow rate (or velocity) according to the following equation:

$$FSP = KQ^2$$

where FSP = fan static pressure, in H_2O; K = a constant; and Q = flow rate ft^3/min.

11. *Operating point.* The operating point indicates the actual flow rate and the actual fan static pressure at which the fan will operate when it is connected to the ventilation system. The operating point occurs at the intersection of the fan performance curve and the system performance curve.

12. *Threshold limit value (TLV).* The *TLV* is a guideline concentration of a contaminant used in the control of health hazards in a workplace setting. It represents a time-weighted concentration to which nearly all workers may be exposed to 8 hours/day over extended periods of time without adverse health effects. These values are established by the American Conference of Governmental Industrial Hygienists (ACGIH). See Chapter 25 for additional details.

13. *Lower explosive limit (LEL).* The LEL is the molar or volume concentration of a gas or vapor in air below which the mixture is not flammable or explosive. This is in contrast to the upper explosive limit (UEL), which represents the molar or volume concentration of a gas or vapor in air above which the mixture is not flammable or explosive. The range of gas or vapor concentrations between the UEL and LEL corresponds to its explosive or flammable region at which the mixture will explode or ignite if it comes into contact with an ignition source. See Chapter 26 for additional details.

18.7   AIR EXCHANGE RATE

Discuss in some detail the term "air exchange rate." The answer should explain what is meant by the term "air exchange rate" and then discuss some of the variables that affect it. The discussion should address the following terms: infiltration, exfiltration, wind effects, stack effects, combustion effects, natural ventilation, and forced ventilation.

SOLUTION

The term "air exchange rate" is the rate at which indoor air is replaced with outdoor air. The units of the air exchange rate are "air changes per hour" or "ach." If the volume of air in a building is replaced twice in 1 hour, the air exchange rate would be two. If the volume of air in a building is replaced once in 2 hours, the air exchange rate would be 0.5. The air exchange rate can be calculated by dividing the rate at which outdoor air enters the building in m^3/hr (or ft^3/hr) by the volume of the building in m^3 (or ft^3). If the air exchange rate were 1 ach, it would not mean that every molecule of indoor air would have been replaced at the end of 1 hour. Just which molecules were replaced would depend on a number of factors. Some of those factors are infiltration, exfiltration, wind effects, stack effects, combustion effects, natural ventilation, and forced ventilation.

Infiltration and exfiltration refer to the uncontrolled leakage of air into or out of the building through cracks and other unintended openings in the outer shell of the building. In addition to leakage around windows and doors, infiltration and exfiltration can occur at points such as openings for pipes, wires, and ducts. The rate of infiltration and exfiltration can vary greatly depending on such factors as wind temperature differences between indoors and outdoors, as well as the operation of stacks and exhaust fans.

Wind effects result from wind striking one side of a building causing positive pressure on that side and lower pressure on the opposite side (the leeward side). Air is forced into the building on the windward side and out the leeward side. Some buildings may be somewhat protected from wind effects by terrain, trees, and other buildings.

The tendency of warm air to rise in a room or through the levels of a multilevel building result in what is known as stack effects. In winter, when there is a large temperature difference between indoor and outdoor air, rising warm air escapes through openings at the top of the building and outdoor air is drawn in at the bottom of the building. The effect is usually less pronounced in the summer because of smaller temperature differences and the direction of the flow may be reversed.

Combustion effects often arise from fires in fireplaces, stoves, and heating systems. The combustion uses up indoor air (oxygen), which causes pressure in the building to drop. Outdoor air is then drawn in. This effect can double infiltration rates. Use of outdoor air in a heating system or fireplace substantially reduces this effect.

Natural ventilation is air that is drawn into a building through windows, doors, and other controlled openings. Natural ventilation results from wind striking the building and/or temperature differences between the outdoor air and the indoor air.

Forced ventilation refers to drawing air into a building through fans and ducts. The effectiveness in removing contaminants from indoor air by the use of forced ventilation can vary widely. Fans used to exhaust specific sources of pollutants (such as the kitchen stove) can be very effective. Most forced ventilation systems are used to circulate air-conditioned air. Whole house fans and the forced ducted ventilation systems of large buildings must be carefully balanced by the air supply to prevent backdrafts of contaminants from stacks and heating plants.

18.8   ACCIDENTAL EMISSION

A certain poorly ventilated chemical storage room (10 ft × 20 ft × 8 ft) has a ceiling fan but no air conditioner. The air in the room is at 51°F and 1.0 atm pressure. Inside this room, a 1 lb bottle of iron (III) sulfide (Fe_2S_3) sits next to a bottle of sulfuric acid containing 1 lb H_2SO_4 in water. An earthquake (or perhaps the elbow of a passing technician) sends the bottles on the shelf crashing to the floor where the bottles break, and their contents mix and react to form iron (III) sulfate [$Fe_2(SO_3)$] and hydrogen sulfide (H_2S).

Calculate the maximum H_2S concentration that could be reached in the room assuming rapid complete mixing by the ceiling fan with no addition of outside air (poor ventilation).

SOLUTION

Balance the chemical equation:

amount	Fe_2S_3	+	$3H_2SO_4$	→	$Fe_2(SO_4)_3$	+	$3H_2S$
before	1 lb		1 lb		0		0
reaction	0.0048 lbmol		0.010 lbmol		0		0

The molecular weights of Fe_2S_3 and H_2SO_4 are 208 and 98, respectively.

The terms *limiting reactant* and *excess reactant* refer to the actual number of moles present in relation to the stoichiometric proportion required for the reaction to proceed to completion. See Problem 3.7. From the stoichiometry of the reaction, 3 lbmol of H_2SO_4 are required to react with each lbmol of Fe_2S_3. Therefore, sulfuric acid is the limiting reactant and the iron (III) sulfide is the excess reactant. In other words, 0.0144 lbmol of H_2SO_4 is required to react with 0.0048 lbmol of Fe_2S_3, or 0.030 lbmol of Fe_2S_3 is required to react with 0.010 lbmol of H_2SO_4.

Calculate the moles of H_2S generated n_{H_2S}:

$$n_{H_2S} = (0.010 \text{ lbmol } H_2SO_4)(3H_2S/3H_2SO_4)$$
$$= 0.010 \text{ lbmol}$$

Next, convert the moles to mass:

$$m_{H_2S} = (0.010 \text{ lbmol } H_2S)(34 \text{ lb/lbmol } H_2S)$$
$$= 0.34 \text{ lb}$$

The final H_2S concentration in the room in ppm, C_{H_2S}, can now be calculated. At $32°$ and 1 atm, 1 lbmol of an ideal gas occupies 359 ft^3; at $51°$ 1 lbmol occupies

$$V = 359\left(\frac{460 + 51}{460 + 32}\right) = 373 \text{ ft}^3$$

Therefore,

$$C_{H_2S} = \frac{(0.34 \text{ lb})\left(\dfrac{373 \text{ ft}^3}{\text{lbmol air}}\right)\left(\dfrac{\text{lbmol air}}{29 \text{ lb}}\right)(10^6)}{1600 \text{ ft}^3}$$

$$= 2733 \text{ ppm}$$

This concentration of H_2S far exceeds an acceptable value.

18.9   DILUTION VENTILATION APPLICATION [34]

Estimate the dilution ventilation required in an indoor work area where a toluene-containing adhesive in a nanotechnology process is used at a rate of 3 gal/8 h workday. Assume that the specific gravity of toluene (C_7H_8) is 0.87, that the adhesive contains 4 vol% toluene, and that 100% of the toluene is evaporated into the room air at 20°C. The plant manager has specified that the concentration of toluene must not exceed 80% of its threshold limit value (TLV) of 100 ppm.

As described earlier, the following equation can be used to estimate the dilution air requirement:

$$q = K(q_c/C_a)$$

where q = dilution air flowrate; K = dimensionless mixing factor that accounts for less than complete mixing characteristics of the contaminant in the room, the contaminant toxicity level and the number of potentially exposed workers; usually, the value of K varies form 3 to 10, where 10 is used under poor mixing conditions and when the contaminant is relatively toxic; q_c = volumetric flowrate of pure contaminant vapor, c; C_a = acceptable contaminant concentration in the room, volume or mole fraction (ppm $\times$ 10^{-6}).

SOLUTION

The dilution air can be estimated from (see Problem statement)

$$q = K(V/C_a)$$

Since the TLV for toluene is 100 ppm and C_a is 80% of the TLV,

$$C_a = [0.80(100)] \times 10^{-6} = 80 \times 10^{-6} \quad \text{(volume fraction)}$$

The mass flowrate of toluene is

$$\dot{m}_{\text{tol}} = \left(\frac{3\,\text{gal}_{\text{adhesive}}}{8\,\text{h}}\right)\left(0.4\,\frac{\text{gal}_{\text{toluene}}}{1\,\text{gal}_{\text{adhesive}}}\right)\left[\frac{(0.87)(8.34\,\text{lb})}{1\,\text{gal}_{\text{toluene}}}\right]$$

$$= 1.09\,\text{lb/h}$$

$$= \left(\frac{1.09\,\text{lb}}{1\,\text{h}}\right)\left(\frac{454\,\text{g}}{1\,\text{lb}}\right)\left(\frac{1\,\text{h}}{60\,\text{min}}\right)$$

$$= 8.24\,\text{g/min}$$

Since the molecular weight of toluene is 92,

$$\dot{n}_{\text{tol}} = 8.24/92$$
$$= 0.0896\,\text{gmol/min}$$

The resultant toluene vapor volumetric flowrate q_{tol} is calculated directly form the ideal gas law:

$$q_{\text{tol}} = \frac{(0.0896\,\text{gmol/min})[0.08206\,\text{atm} \cdot \text{L}/(\text{gmol} \cdot \text{K})](293\,\text{K})}{1\,\text{atm}}$$

$$= 2.15\,\text{L/min}$$

Therefore, the required diluent volumetric flowrate is (with $K = 5$)

$$q = \frac{(5)(2.15\,\text{L/min})}{80 \times 10^{-6}}$$

$$= 134\,375\,\text{L/min}$$

$$= \left(134\,375\,\frac{\text{L}}{\text{min}}\right)\left(\frac{1\,\text{ft}^3}{28.36\,\text{L}}\right)$$

$$= 4748\,\text{ft}^3/\text{min}$$

18.10 VINYL CHLORIDE APPLICATION

The vinyl chloride fugitive emission rate in a process was estimated to be 10 g/min by a series of bag tests conducted for the major pieces of connections (i.e., flanges and valves) and pump seals. Determine the flowrate of air (25°C) necessary to maintain a level of 1.0 ppm by dilution ventilation. Correct for incomplete mixing by employing a safety factor of 10. Also consider partially enclosing the process and using local exhaust ventilation. Assume that the process can be carried out in a hood with an opening of 30 in. wide by 25 in. high with a face velocity greater than 100 ft/min to ensure high capture efficiency. What will be the flowrate of air required for local exhaust ventilation? Which ventilation method seems better?

SOLUTION

Convert the mass flowrate of the vinyl chloride (VC) to volumetric flowrate q, in cm^3/min and acfm. First, use the ideal gas law to calculate the density.

$$\rho = \frac{P(MW)}{RT}$$

$$= \frac{(1 \text{ atm})(78 \text{ g/gmol})}{\left(82.06 \dfrac{\text{cm}^3 \cdot \text{atm}}{\text{mol} \cdot \text{K}}\right)(298 \text{ K})}$$

$$= 0.00319 \text{ g/cm}^3$$

$$q = (\text{mass flowrate})/(\text{density})$$

$$= (10 \text{ g/min})/(0.00319 \text{ g/cm}^3)$$

$$= 3135 \text{ cm}^3/\text{min}$$

$$= 0.1107 \text{ acfm}$$

Calculate the air flowrate in acfm, q_{air}, required to meet the 1.0 ppm constraint with the equation

$$q_{air} = (0.1107 \text{ acfm})/10^{-6}$$

$$= 1.107 \times 10^5 \text{ acfm}$$

Apply the safety factor to calculate the actual air flowrate for dilution ventilation:

$$q_{air,dil} = (10)(1.107 \times 10^5 \text{ acfm})$$

$$= 1.107 \times 10^6 \text{ acfm}$$

Now consider the local exhaust ventilation by first calculating the face area of the hood A, in square feet:

$$A = (\text{Height})(\text{Width})$$

$$= (30 \text{ in.})(25 \text{ in.})(\text{ft}^2/144 \text{ in.}^2)$$

$$= 5.21 \text{ ft}^2$$

The air flowrate in acfm $q_{air,exh}$, required for a face velocity of 100 ft/min is then

$$q_{air,exh} = (5.21 \text{ ft}^2)/(100 \text{ ft/min})$$
$$= 521 \text{ acfm}$$

Since the air flowrate for dilution ventilation is approximately 2000 times higher than the local ventilation air flowrate requirement, and considering the high cost of large blowers to handle high air flowrates, the local ventilation method appears to be the better method for this case.

18.11   VENTILATION MODELS [35, 36]

Your consulting firm has received a contract to develop, as part of an emergency preparation plan, a mathematical model describing the concentration of a nano-chemical in a medium-sized ventilated laboratory room. The following infor-mation/data (SI units) is provided:

V = volume of room, m^3

v_0 = volumetric flow rate of ventilation air, m^3/min

c_0 = concentration of the nanochemical in ventilation air, $gmol/m^3$

c = concentration of the nanochemical leaving ventilated room, $gmol/m^3$

c_1 = concentration of the nanochemical initially present in ventilated room, $gmol/m^3$

r = rate of disappearance of the nanochemical in the room due to reaction and/or other effects, $gmol/m^3$-min

As an authority in the field (having taken the Theodore course on Health, Safety and Accident Management), you have been requested to:

1. Obtain the equation describing the concentration in the room as a function of time if there are no "reaction" effects, that is, $r = 0$.
2. Obtain the equation describing the concentration in the room as a function of time if $r = -k$. Note once again that the minus sign is carried since the agent is disappearing.
3. Obtain the equation describing the concentration in the room as a function of time if $r = -kc$. Note once again that the minus sign is carried since the chemical is disappearing.
4. For part 2, discuss the effect on the resultant equation if k is extremely small, that is, $k \to 0$.
5. For part 3, discuss the effect on the resultant equation if k is extremely small, that is, $k \to 0$.

6. For part **1**, qualitatively discuss the effect on the final equation if the volumetric flow rate, v_0 varies sinusoidally.

7. For part **1**, qualitatively discuss the effect on the final equation if the inlet concentration, c_0, varies sinusoidally.

SOLUTION

Use the laboratory room as the control volume. Apply the concentration law for mass to the nanochemical.

$$\left\{\begin{matrix}\text{rate of mass}\\\text{in}\end{matrix}\right\} - \left\{\begin{matrix}\text{rate of mass}\\\text{out}\end{matrix}\right\} + \left\{\begin{matrix}\text{rate of mass}\\\text{generated}\end{matrix}\right\} = \left\{\begin{matrix}\text{rate of mass}\\\text{accumulated}\end{matrix}\right\}$$

Employing the notation specified in the problem statement gives:

$$\{\text{rate of mass in}\} = v_0 c_0$$
$$\{\text{rate of mass out}\} = v_0 c$$
$$\{\text{rate of mass generated}\} = rV$$
$$\{\text{rate of mass accumulated}\} = \frac{dV_c}{dt}$$

Substituting above gives

$$v_0 c_0 - v_0 c + rV = \frac{dV_c}{dt}$$

Since the laboratory room is constant, V may be taken out of the derivative term. This leads to

$$\frac{v_0}{V}(c_0 - c) + r = \frac{dc}{dt}$$

The term V/v_0 represent the average residence time the nanochemicals reside in the room and is usually designated as τ. The above equation may then be rewritten as

$$\frac{dc}{dt} = \frac{c_0 - c}{\tau} + r$$

1. If $r = 0$,

$$\frac{dc}{dt} = \frac{c_0 - c}{\tau}$$

separating variables

$$\frac{dc}{c_0 - c} = \frac{dt}{\tau}$$

$$\int_{c_i}^{c} \frac{dc}{c_0 - c} = \int_0^t \frac{dt}{\tau}$$

$$\ln\left(\frac{c_0 - c}{c_0 - c_i}\right) = \frac{t}{\tau}$$

$$\left(\frac{c_0 - c}{c_0 - c_i}\right) = e^{-\frac{t}{\tau}}$$

$$c = c_0 - (c_0 - c_i)e^{-\frac{t}{\tau}} = c_0 + (c_i - c_0)e^{-\frac{t}{\tau}}$$

2. If $r = -k$

$$\frac{dc}{dt} = \frac{c_0 - c}{\tau} - k = \frac{c_0}{\tau} - k - \frac{c}{\tau} = \left(\frac{c_0 - k\tau}{\tau}\right) - \frac{c}{\tau}$$

$$\frac{dc}{\left[\left(\dfrac{c_0 - k\tau}{\tau}\right) - \dfrac{c}{\tau}\right]} = dt$$

$$\frac{dc}{[(c_0 - k\tau) - c]} = \frac{dt}{\tau}$$

$$\int_{c_i}^{c} \frac{dc}{[(c_0 - k\tau) - c]} = \int_0^t \frac{dt}{\tau}$$

$$-\ln\left[\frac{(c_0 - k\tau) - c}{(c_0 - k\tau) - c_i}\right] = \frac{dt}{\tau}$$

$$\frac{(c_0 - k\tau) - c}{(c_0 - k\tau) - c_i} = e^{-\left(\frac{t}{\tau}\right)}$$

$$c = c_i e^{-\frac{t}{\tau}} + (c_0 - k\tau)\left[1 - e^{-\left(\frac{t}{\tau}\right)}\right]$$

3. If $r = -kc$

$$\frac{dc}{dt} = \frac{c_0 - c}{\tau} - kc = \frac{c_0}{\tau} - \frac{c}{\tau} - kc$$

$$= \frac{c_0}{\tau} - c\left(\frac{k\tau + 1}{\tau}\right)$$

$$\frac{dc}{\left[\dfrac{c_0}{\tau} - \left(\dfrac{1+k\tau}{\tau}\right)c\right]} = dt$$

$$\int_{c_i}^{c} \frac{dc}{[c_0 - (1+k\tau)c]} = \int_{0}^{t} \frac{dt}{\tau}$$

$$-\left(\frac{1}{1+k\tau}\right)\ln\left[\frac{c_0 - (1+k\tau)c}{c_0(1+k\tau)c_i}\right] = \frac{t}{\tau}$$

$$\left(\frac{c_0 - (1+k\tau)c}{c_0 - (1+k\tau)c_i}\right) = e^{-(\frac{t}{\tau})(1+k\tau)}$$

$$c = c_i e^{-(\frac{t}{\tau})(1+k\tau)} + \left(\frac{c_0}{1+k\tau}\right)\left[1 - e^{-(\frac{t}{\tau})(1+k\tau)}\right]$$

4. If $k = 0$; see (2) solution

$$c = c_i e^{-\frac{t}{\tau}} + (c_0 - k\tau)[1 - e^{-\frac{t}{\tau}}]$$

$$= c_i e^{-\frac{t}{\tau}} + c_0 - c_0 e^{-\frac{t}{\tau}}$$

$$= c_0 + (c_i - c_0)e^{-\frac{t}{\tau}}$$

5. If $k = 0$; see (3) solution

$$c = c_i e^{\left(-\frac{t}{\tau}\right)(1+k\tau)} + \left(\frac{c_0}{1+k\tau}\right)\left[1 - e^{\left(-\frac{t}{\tau}\right)(1+k\tau)}\right]$$

$$= c_i e^{-\frac{t}{\tau}} + c_0 - c_0 e^{-\frac{t}{\tau}}$$

$$= c_0 + (c_i - c_0)e^{-\frac{t}{\tau}}$$

6. If v_0 varies and τ varies solving the equation becomes more complex. Variations need to be included in the describing equation

$$\frac{dc}{dt} = \frac{c_0 - c}{\tau} + r; \qquad \tau = \tau(t)$$

This may require numerical solution.

7. If both $c_0 = c_0(t)$ and $v_0 = vd(t)$, the solution in part **6** again applies.

18.12   MINIMUM VENTILATION FLOWRATE

Refer to Problem 18.11.

1. Calculate minimum air ventilation flow rate containing 10 ng/m^3 nanoparticles into the room to assure that the nanoagent concentration does not exceed 35.0 ng/m^3. The nanoagents are appearing in the laboratory at a rate of 250 ng/min. Assume steady-state conditions.
2. Calculate the steady-state concentration in the laboratory: the initial concentration of the nanochemical is 500 ng/m^3. There is no additional source of nanoagents (generated) and the ventilation air is essentially pure, that is, there is no background nanoagent concentration.
3. If the room volume is 142 m^3, the flowrate of the 10 ng/m^3 ventilation air is $12.1 \text{ m}^3/\text{min}$, and nanoparticles are being generated at a steady rate of 30 ng/min, calculate how long it would take for the concentration to reach 20.7 ng/m^3. The initial concentration in the laboratory is 85 ng/m^3.
4. Refer to part 3. How long would it take to achieve a concentration of 13.65 ng/m^3? How long would it take to reach 12.2 ng/m^3?

SOLUTION

1. The applicable model for this case is:

$$v_0(c_0 - c) + rV = V\frac{dc}{dt}$$

Under steady-state, $dc/dt = 0$. Pertinent information includes

$$rV = 250 \text{ ng/min}$$
$$c_0 = 10 \text{ ng/m}^3$$
$$c = 35 \text{ ng/m}^3$$

Substituting gives

$$v_0 = \frac{-rV}{c_0 - c}$$
$$= \frac{rv}{c - c_0}$$
$$= \frac{250}{35 - 10}$$
$$= 10 \text{ m}^3/\text{min} = 353 \text{ ft}^3/\text{min}$$

2. The applicable model is:

$$v_0(c_0 - c) + rV = V\frac{dc}{dt}$$

For steady-state condition, $dc/dt = 0$. In addition, based on the information provided, $r = 0$ and $c_0 = 0$. Therefore, and as expected,

$$c = 0$$

3. First note that

$$\tau = 142/12.1 = 11.73 \text{ min}$$

$$k = r/V = 30/142 = 0.211 \text{ ng/m}^3\text{-min}$$

The applicable describing equation is:

$$c = c_i e^{-\frac{t}{\tau}} + (c_0 + k\tau)[1 - e^{-(\frac{t}{\tau})}]$$

Substituting gives

$$20.7 = 85e^{-\frac{t}{11.73}} + (10 + 2.48)[1 - e^{-\frac{t}{11.73}}]$$

Solving by trial and error gives (approximately)

$$t = 29 \text{ min}$$

4. First calculate the steady-state concentration for this condition. The applicable model is obtained from part **2**, after setting $dc/dt = 0$:

$$\frac{c_0 + k\tau}{\tau} - \frac{c}{\tau} = 0$$

Solving and substituting gives

$$c = c_0 + k\tau$$
$$= 10 + (0.211)(11.73)$$
$$= 12.48 \text{ ng/m}^3$$

Since this is the steady-state concentration, it will take an infinite period of times to read this value.

The steady-state concentration represents the minimum concentration achievable in the laboratory based on the conditions specified. Therefore, the concentration will never read a value of 12.2 ng/m^3.

19 Dispersion Considerations

The environment has always been polluted to some extent through natural phenomena and/or human activities. In recent times, technological advancement, industrial expansion, and urbanization have been the major contributors to the global environment. This problem will become more pronounced as nanoemissions continue to increase.

This Chapter focuses on some of the practical considerations of dispersion in the environment, with particular emphasis on dispersion of particulates by dilution in the atmosphere. The underlying principles are applied and expanded to provide useful "design" equations. To assist readers who would like to refer to the original sources, the notation used by the original investigators is retained [37].

The Chapter concludes with the development of the describing equations for dispersion in water systems. However the bulk of the problems focus on atmospheric dispersion applications.

The effective height of an emission is considered in view of the various equations and correlations currently in use. Atmospheric dispersion equations for continuous sources are reviewed; the effects of multiple sources as well as discharges from line and area sources are also discussed. The atmospheric presentation concludes with what has been referred to in the literature as a "puff" model – namely, an equation that can be used for estimating the effect of discharges from instantaneous (as opposed to continuous) sources [37].

In summary, a preliminary description of any exposure should be obtained which answers the following questions [38,39]:

1. Where, when, and how will the release occur?
2. What is in the immediate vicinity of the release?
3. What is the quantity, physical state, and chemical identity of the released material?
4. What are the concentrations and durations of exposure in the area of the toxicant's release?
5. Will the chemical agent be distributed to a larger area, and, if so, what will be its form (physical and chemical), concentration, and duration of residence throughout the area of distribution? This description should include the concentrations at various locations and times throughout its residence, and it

should include air and waterborne materials as well as those taken up by biological materials such as plants and animals.

Finally it should be noted that the describing dispersion equations almost exclusively apply to gases. The author believes that these same equations can be applied with the same level of confidence to nanoparticles. This conclusion is based on the realization that particles in the nanorange approach the size of gaseous molecules and are therefore subject to similar "dispersion" behavior.

19.1   ATMOSPHERIC DEPOSITION CALCULATION

ETS engineers have been requested to determine the minimum distance downstream from a source emitting a nanosized particle that will be free of particulate deposit. The discharge is located 50 ft above ground level. Assume ambient conditions are at 60°F and 1 atm and neglect meteorological aspects. Additional data are given below.

Particle size range $= 250–5000$ nm
Specific gravity $= 13.1$
Wind Speed $= 1.2$ mi/h

SOLUTION

A particle diameter of 250 nm is used to calculate the minimum distance downstream free of dust since the smallest particle will travel the greatest horizontal distance. To determine the velocity, first calculate the particle density using the specific gravity given and determine the properties of the gas (assume air).

$$\rho_p = (13.1)(62.4)$$
$$= 817.4 \text{ lb/ft}^3$$
$$\rho_a = P(MW)/RT$$
$$= (1)(29)/[(0.73)(60 + 460)]$$
$$= 0.0764 \text{ lb/ft}^3$$

Viscosity of air, μ, at $60°F = 1.22 \times 10^{-5}$ lb/(ft · s)
Since the velocity is in the Stokes' law range (because of particle size), the terminal settling velocity in feet per second is

$$v = \frac{gd_p^2\rho_p C}{18\mu} = \frac{(32.2)[(0.25)/(25\,400)(12)]^2(817)(1.0)}{(18)(1.22 \times 10^{-5})}$$
$$= 8.06 \times 10^{-5} \text{ ft/s}$$

The approximate time for descent is

$$t = h/v$$
$$= 50/8.06 \times 10^{-5}$$
$$= 6.20 \times 10^5 \, \text{s}$$

Thus, the distance traveled, d, is

$$d = tu$$
$$= (6.20 \times 10^5)(1.2/3600)$$
$$= 207 \, \text{miles}$$

The reader should note that this is a conservative calculation for the settling velocity since the Cunningham Correction Factor (C) was not included in the analysis; that is, $C = 1.0$. To include this effect, refer to Problem 11.10. For a 250 nm (or 0.25 μm) particle, C is approximately 1.68. Including this effect leads to

$$v = 13.54 \times 10^{-5} \, \text{ft/s}$$
$$t = 3.69 \times 10^5 \, \text{s}$$
$$d = 1.23 \, \text{miles}$$

19.2 GROUND DEPOSITION OF PARTICLES

A nano plant manufacturing a new detergent explodes one windy day. It disperses 10 000 lb of soap particles (s.g. $= 11.8$) into the atmosphere (70°F, density $= 0.0752$ lb/ft^3). If the wind is blowing 2.0 miles/hr from the west and the particles range in diameter from 10 to 1000 nm, calculate the distances from the plant where the particles will start to deposit and where they will cease to deposit. Assume the particles are blown vertically 40 ft in the air before they start to settle. Also assuming even ground-level distribution through an average 20 ft wide path of settling, calculate the average height of the particles on the ground in the settling area. Assume the bulk density of the particle equal to half the actual density.

SOLUTION

Stokes regime applies because of the size of the particles. Once again, note that the smallest particle will travel the greatest distance while the largest will travel the shortest distance. For the minimum distance, use the smallest particle:

$$d_{\text{p}} = 1000 \, \text{nm} = 1.0 \, \mu\text{m} = 3.28 \times 10^{-6} \, \text{ft}$$

Using Stokes' law with the appropriate Cunningham Correction Factor

$$v = \frac{d_p^2 g \rho_p C}{18\,\mu}; \quad C = 1.16$$

$$= \frac{(3.28 \times 10^{-6})^2 (32.2)(11.8)(62.4)(1.16)}{18(1.18 \times 10^{-5})}$$

$$= 1.393 \times 10^{-3}\ \text{ft/s}$$

The descent time t is

$$t = H/v$$

$$= 40/1.393 \times 10^{-3} = 28.7 \times 10^3\ \text{s}$$

The horizontal distance traveled, L, is

$$L = (28.7 \times 10^3)\left(\frac{2.0}{(60)(60)}\right)(5280)$$

$$= 8.42 \times 10^4\ \text{ft}$$

For the maximum distance, use the smallest particle:

$$d_p = 0.01\ \mu\text{m} = 3.28 \times 10^{-8}\ \text{ft}$$

The velocity v is

$$v = \frac{(32.2)(3.28 \times 10^{-8})^2 (11.8)(62.4)}{(18)(1.18 \times 10^{-5})} C; \quad C = 22.2$$

$$= 2.667 \times 10^{-6}\text{ft/s}$$

The descent time t is

$$t = H/v = 40/2.667 \times 10^{-6}$$

$$= 1.5 \times 10^{-7}\ \text{s}$$

The horizontal distance traveled, L, is

$$L = (1.5 \times 10^7)\left(\frac{2.0}{(60)(60)}\right)(5280)$$

$$= 4.4 \times 10^7\ \text{ft}$$

To calculate the depth D, the volume of particles (actual), V_{act}, is first determined:

$$V_{\text{act}} = (10\ 000)/[(11.8)(62.4)]$$

$$= 13.6\ \text{ft}^3$$

The bulk volume V_B is

$$V_B = 13.6/0.5 = 27.2 \text{ ft}^3$$

The length of the drop area, L_D, is

$$L_D = 4.4 \times 10^7 - 8.4 \times 10^4$$

$$= 4.4 \times 10^7 \text{ ft}$$

Since the width is 20 ft, the deposition are A is

$$A = (4.4 \times 10^7)(20) = 8.8 \times 10^8$$

$$V_B = AD$$

$$27.2 = (8.8 \times 10^8)D$$

Therefore,

$$D = 3.09 \times 10^{-8} \text{ ft}$$

$$= 0.94 \times 10^{-8} \text{ m}$$

$$= 9.4 \text{ nm}$$

The deposition can be, at best, described as a "sprinkling."

19.3 PLUME RISE

If a waste source from a nano plant emits a gas with a buoyancy flux of $40 \text{ m}^4/\text{s}^3$, and the wind averages 5 m/s, find the plume rise at a distance of 600 m downwind from a stack that is 75 m high under unstable atmospheric conditions. Use the equation proposed by Briggs.

SOLUTION

Several plume rise equations are available. Briggs used the following equations to calculate the plume rise:

$$\Delta h = 1.6F^{1/3}u^{-1}x^{2/3} \qquad x < x_f$$

$$= 1.6F^{1/3}u^{-1}x_f^{2/3} \qquad \text{if } x \geq x_f$$

$$x^* = 14F^{5/8} \qquad \text{when } F < 55 \text{ m}^4/\text{s}^3$$

$$= 34F^{2/5} \qquad \text{when } F \geq 55 \text{ m}^4/\text{s}^3$$

$$x_f = 3.5x^*$$

where Δh = plume rise, m; F = buoyancy flux, $m^4/s^3 = 3.7 \times 10^{-5}\dot{Q}_H$; u = wind speed, m/s; x^* = downward distance, m; x_f = distance of transition from first stage of rise to the second stage of rise, m; and, $\dot{Q}_H$ = heat emission rate, kcal/s.

If the term $\dot{Q}_H$ is not available, the term F may be estimated by

$$F = (g/\pi)q(T_s - T)/T_s$$

where g = gravity term, 9.8 m/s^2; q = stack gas volumetric flowrate, m^3/s (actual conditions); and, T_S, T = stack gas and ambient air temperature, K, respectively.

Calculate x_f to determine which plume equation applies since F is less than 55 m^4/s^3,

$$x^* = 14F^{5/8}$$

$$= (14)(40)^{5/8}$$

$$= 140 \text{ m}$$

$$x_f = 3.5x^*$$

$$= (3.5)(140)$$

$$= 491 \text{ m}$$

Since $x > x_F$, the plume rise is therefore

$$\Delta h = 1.6F^{1/3}u^{-1}x_f^{2/3}$$

$$= 1.6(40)^{1/3}(5)^{-1}(491)^{2/3} = 68.2 \text{ m}$$

19.4   PASQUILL–GIFFORD MODEL

A nano plant emits approximately 160 g/s of fine particulates. The effective stack height is 120 m and the wind speed is 2 m/s. At one hour before sunrise, the sky is clear. A dispersion study requires information on the approximate distance of the maximum concentration under these conditions. (*Hint:* calculate the concentrations for the downward distances of 0.1, 1.0, 5, 10, 20, 25, 30, 50, and 70 km.)

SOLUTION

The coordinate system used in making atmospheric dispersion estimates of gaseous pollutants, as suggested by Pasquill and modified by Gifford, is described below. The origin is at ground level or beneath the point of emission, with the x-axis extending horizontally in the direction of the mean wind. The y-axis is in the horizontal plane perpendicular to the x-axis, and the z-axis extends vertically. The plume travels along or parallel to the x-axis (in the mean wind direction). The concentration, C, of gas or aerosol at (x, y, z) from a continuous source with an effective height, H_e, is given by:

$$C(x, y, z, H_e) = (\dot{m}/2\pi\sigma_y\sigma_z u)[e^{-(1/2)(y/\sigma_y)^2}][e^{-(1/2)((z-H_e)/\sigma_z)^2}) + e^{-(1/2)((z+H_e)/\sigma_z)^2}]$$

where H_e = effective height of emission (sum of the physical stack height, H_s and the plume rise Δh), m; u = mean wind speed affecting the plume, m/s; $\dot{m}$ = emission rate of pollutants, g/s; σ_y, σ_z = dispersion coefficients or stability parameters, m; C = concentration of gas, g/m^3; and, x, y, z = coordinates, m.

The assumptions made in the development of the above equation are: (1) the plume spread has a Gaussian (normal) distribution in both the horizontal and vertical planes, with standard deviations of plume concentration distribution in the horizontal and vertical directions of σ_y and σ_z, respectively; (2) uniform emission rate of pollutants $\dot{m}$; (3) total reflection of the plume at ground level ($z = 0$) conditions; and (4) the plume moves downstream (horizontally in the x direction) with mean wind speed, u. Although any consistent set of units may be used, the cgs system is preferred [37].

For concentrations calculated at ground level ($z = 0$), the above equation simplifies to

$$C(x, y, 0, H_e) = (\dot{m}/2\pi\sigma_y\sigma_z u)[e^{-(1/2)(y/\sigma_y)^2}][e^{-(1/2)(H_e/\sigma_z)^2}]$$

If the concentration is to be calculated along the centerline of the plume ($y = 0$), further simplification gives

$$C(x, 0, 0, H_e) = (\dot{m}/\pi\sigma_y\sigma_z u)[e^{-(1/2)(H_e/\sigma_z)^2}]$$

In the case of a ground-level source with no effective plume rise ($H_e = 0$), the equation reduces to

$$C(x, 0, 0, 0) = (\dot{m}/\pi\sigma_y\sigma_z u)$$

It is important to note that the two dispersion coefficients are the product of a long history of field experiments, empirical judgements, and extrapolations of the data from those experiments. There are few knowledgeable practitioners in the dispersion modeling field who would dispute that the coefficients could easily have an inherent uncertainty of $\pm$ 25% [37].

The six applicable stability categories for these coefficients are shown in Table 19.1. Note that A, B, C refer to daytime with unstable conditions; D refers to overcast or neutral conditions at night or during the day; E and F refer to night-time stable conditions and are based on the amount of cloud cover. "Strong" incoming solar radiation corresponds to a solar altitude greater than $60°$ with clear skies (e.g., sunny midday in midsummer); "slight" isolation (rate of radiation from the sun received per unit of Earth's surface) corresponds to a solar

TABLE 19.1   Stability Categories

Surface Wind Speed at 10 m(m/s)	Day			Night	
	Incoming Solar Radiation			Thinly Overcase or > 4/8 Low Cloud	< 3/8 Cloud
	Strong	Moderate	Slight		
2	A	A–B	B	–	–
2–3	A–B	B	C	E	F
3–5	B	B–C	C	D	E
5–6	C	C–D	D	D	D
6	C	D	D	D	D

altitude from 15° to 35° with clear skies (e.g., sunny midday in midwinter). For the A–B, B–C and C–D stability categories, one should use the average of the A and B values, B and C values, and C and D values, respectively. Figures 19.1 and 19.2 provide the variation of σ_y and σ_z with stability categories and distances [37].

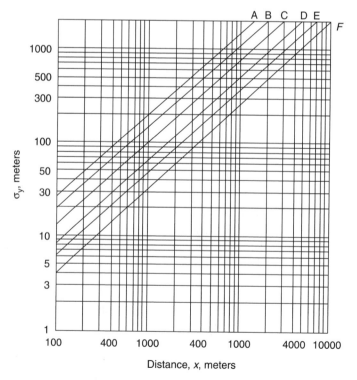

Figure 19.1   Dispersion coefficient, y direction.

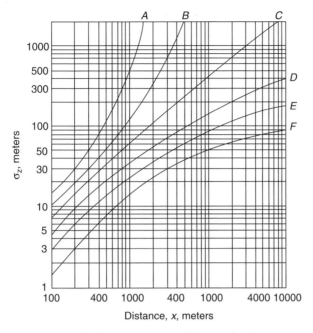

Figure 19.2 Dispersion coefficient, z direction.

Since it is night time, the sky is clear, and the wind speed is 2 m/s. Therefore, the stability category is F (see Table 19.1). Determine the values of σ_y and σ_z for each downward distance for stability category F from Figures 19.1 and 19.2 (see Table 19.2).

Since the maximum ground-level ($z = 0$) concentration along the x-axis is desired and the maximum concentration occurs along the centerline ($y = 0$) of the plume, the applicable equation is

$$C(x, 0, 0, H_e) = (\dot{m}/\pi\sigma_y\sigma_z u)[e^{-1/2(H_e/\sigma_z)^2}]$$

The concentrations of the nanosized particulates at the downward distances are obtained by completing Table 19.3. For example, at 1.0 km,

$$C(1.0, 0, 0, H_e) = [(160)/(\pi)(34)(14)(2)][e^{-1/2(120/14)^2}]$$
$$= 5.39 \times 10^{-18} \text{ g/m}^3$$

From Table 19.3, the maximum concentration is approximately 1.12×10^{-4} g/m^3 and the corresponding location is between 20 and 25 km.

TABLE 19.2   Value of σ_y and σ_z

x (km)	σ_y (m)	σ_z (m)
0.1	4	2.3
1.0	34	14.0
5	148	34.0
10	275	46.5
20	510	60.0
25	610	64.0
30	720	68.0
50	1120	78.0
70	1500	84.0

TABLE 19.3 Concentrations of Particulates

x (km)	C (g/m^3)
0.1	0 (infinitely small)
1.0	5.39×10^{-18}
5	9.95×10^{-6}
10	7.11×10^{-5}
20	1.12×10^{-4}
25	1.12×10^{-4}
30	1.09×10^{-4}
50	8.89×10^{-5}
70	7.43×10^{-5}

19.5   GROUND-LEVEL PARTICLE DEPOSITION

Provide procedures that can be employed to estimate particle deposition at ground level conditions.

SOLUTION

Particulate emissions are subject to downwash settling through the atmosphere due to the action of gravity. For the particles, especially large ones, an additional external force term must be included in the analysis. A rather simple approach is to superimpose the settling velocity of the particle on the initial point of discharge in the following manner:

$$c = \frac{Q}{2u\sigma_y\sigma_z}\exp\left\{-\frac{1}{2}\left[\left(\frac{y}{\sigma_y}\right)^2+\left(\frac{z-H^*}{\sigma_z}\right)^2\right]\right\}$$

where

$$H^* = H - vx/u; \qquad H^* > 0$$

and v is the terminal settling velocity of the particle in question. This effectively "repositions" the particle downstream from the source and eliminates the need of developing and solving a revised equation. If this equation is applied to ground-level ($z = 0$) centerline ($y = 0$) conditions, the following equation results;

$$c = \frac{Q}{2\pi u \sigma_y \sigma_z} \exp\left[-\frac{1}{2}\left(\frac{H^*}{\sigma_z}\right)^2 \right]$$

The rate of deposition of particles per unit area R is then given by the product of the local ground level concentration and the terminal vertical settling velocity of the particle.

$$R = cv$$

$$= \frac{Qv}{2\pi u \sigma_y \sigma_z} \exp\left[-\frac{1}{2}\left(\frac{H^*}{\sigma_z}\right)^2 \right]$$

A consistent set of SI units is g/s-m^2 for R, g/s for Q, m/s for v and u, and m for σ and H^* [37–39].

It is important to note that the deposition rate is a strong function of particle diameter through the term v, which appears twice in the deposition flux equation for R. The above equation must be modified to treat process gas streams discharging particles of a given size distribution. The suggested procedure is somewhat similar to that for calculating overall collection efficiencies for particulate control equipment. (See also Part 2.) For this condition, the overall rate is given by [40]

$$R = \sum_{i=1}^{m} \left(\frac{Q_i v_i}{2\pi u \sigma_y \sigma_z}\right) \exp\left[-\frac{1}{2}\left(\frac{H_i^*}{\sigma_z}\right)^2 \right]$$

Note that the particle size distribution has been divided into n size ranges with $i = $ size range in question; $Q_i = $ discharge rate of particles in size range i; $v_i = $ settling velocity of particles evaluated at the average particle size in i; and, $H_i^* = $ corrected effective height evaluated at the average particle size in range i. This calculational scheme should be considered only as a crude but practical approximation to the actual phenomena. Fundamental treatment of the complete physical process is difficult and has yet to be provided.

Another approach for predicting dust deposition (from stacks) has been developed by Bosanquet et al. [41]; this formulation has been checked experimentally against data for power plants in England. Along the axis of the plume, the deposition rate based on Bosanquet *et al.*'s finding is given by

$$w = \frac{16.4Q_p}{H^2} F\left(\frac{v_t}{u}, \frac{x}{H}\right)$$

where Q_p is expressed in milligrams per second, H in meters, and w in mg/m²-hr. The value of the function F is obtained from a graph with the dimensionless parameter x/H as the abscissa and the dimensionless parameter v_t/u as a series of parametric lines. This graph is available in the literature [37, 39]. The value of the function F varies roughly from 0.5 to 50, with the upper range occurring for large values of v_t and small values of x.

19.6 LINE AND AREA SOURCES

Provide calculational procedures to treat line and area sources.

SOLUTION

The development on atmospheric dispersion in the three previous Problems was limited to emissions from a "point" (e.g., stack) source. Although most dispersion applications involve point sources, in some instances the location of the emission can be more accurately described physically and mathematically by either a line source or an area source.

Line sources have conditionally been applied to roadways and streets along which there are well-defined movements of motor vehicles. For these sources, data are required on the width of the roadway and its center strip, the types and amounts (grams per second per meter) of emissions, the number of lanes, the emissions from each lane, and the height of emissions. In some situations (e.g., a traffic jam at a toll booth, a series of industries located along a river, heavy traffic along a straight stretch of highway) the problem may be modeled as a continuous emitting infinite line source. Concentrations downwind of a continuously emitting infinite line source, when the wind direction is normal to the line, can be calculated from

$$c(x, y, 0; H) = \frac{2q}{\sqrt{2\pi}\sigma_z u} \exp\left[-\frac{1}{2}\left(\frac{H}{\sigma_z}\right)^2\right]$$

Here q is the source strength per unit distance (e.g., g/m-s). Note that the horizontal dispersion parameter σ_y does not appear in this equation, since it is assumed that lateral dispersion from one segment of the line is compensated by dispersion in the opposite direction from adjacent segments. Also, y does not appear, since the concentration at a given x is the same for any value of y. Concentrations from infinite line sources, when the wind is not perpendicular to the line, can also be approximated. If the angle between the wind direction and line source is ϕ one may write

$$c(x, y, 0; H) = \frac{2q}{\sin \phi \sqrt{2} \pi \sigma_z u} \exp\left[-\frac{1}{2}\left(\frac{H}{\sigma_z}\right)^2\right]$$

This equation should not be used where ϕ is less than 45°C. When the continuously emitting line source is reasonably short or "finite," one can account for the edge effects caused by the two ends of the source. If the line source is perpendicular to the wind direction, it is convenient to define the x-axis in the direction of the wind and also passing through the sampling point downwind. The ends of the line source then are at two positions in the crosswind direction, with y_1 less than y_2. The concentration along the x-axis at ground level is then given by

$$c(x, 0, 0; H) = \frac{2q}{\sqrt{2} \pi \sigma_z u}\left\{\exp\left[-\frac{1}{2}\left(\frac{H}{\sigma_z}\right)^2\right]\right\}$$
$$\times \int_{p_1}^{p_2} \frac{1}{\sqrt{2\pi}} \exp\left(-\frac{1}{2}p^2\right) dp$$

where $p_1 = y_1/\sigma_y$ and $p_2 = y_2/\sigma_y$. Once the limits of integration have been established, the value of the integral may be determined from standard table of integrals, as provided in Table 19.4 [37].

To use Table 19.4, which is a form of the error function, consider the following example. Suppose $p_1 = -2.62$ and $p_2 = 1.34$. Go down the z column to 1.3, across the 1.3 row to the column headed by 4, and read 0.9099. Then go down the z column to −2.6, across the −2.6 row to the column headed by 2, and read 0.0044. The integral is

$$\int_{-2.62}^{1.34} \frac{1}{\sqrt{2\pi}} \exp\left(-\frac{1}{2}p^2\right) dp = 0.9099 - 0.0044 = 0.9055$$

Area sources include the multitude of minor sources with individually small emissions that are impractical to consider as separate point or line sources. Area sources are typically treated as a grid network of square areas, with emissions distributed uniformly within each grid square. Area source information requirements include types and amounts of emissions, the physical size of the area over which emissions are prorated, and representative stack height for the area. In dealing with dispersion of pollutions in areas [42] having large numbers of sources (e.g., as in fugitive dust from (coal) piles), there may be too many sources to consider each source individually. Often an approximation can be made by combining all the emissions in a given area and treating this area as a source having an initial horizontal standard deviation, σ_{y_0}. A virtual distance x_y can then be found that will give

TABLE 19.4 Cumulative Areas Under a Standard Normal Distribution

z	0	1	2	3	4	5	6	7	8	9
−3	0.0013	0.0010	0.0007	0.0005	0.0003	0.0002	0.0002	0.0001	0.0001	0.0000
−2.9	0.0019	0.0018	0.0017	0.0017	0.0016	0.0016	0.0015	0.0015	0.0014	0.0014
−2.8	0.0026	0.0025	0.0024	0.0023	0.0023	0.0022	0.0021	0.0021	0.0020	0.0019
−2.7	0.0035	0.0034	0.0033	0.0032	0.0031	0.0030	0.0029	0.0028	0.0027	0.0026
−2.6	0.0047	0.0045	0.0044	0.0043	0.0041	0.0040	0.0039	0.0038	0.0037	0.0036
−2.5	0.0062	0.0060	0.0059	0.0057	0.0055	0.0054	0.0052	0.0051	0.0049	0.0048
−2.4	0.0082	0.0080	0.00478	0.0075	0.0073	0.0071	0.0069	0.0068	0.0066	0.0064
−2.3	0.0107	0.0104	0.0102	0.0099	0.0096	0.0094	0.0091	0.0089	0.0087	0.0084
−2.2	0.0139	0.0136	0.0132	0.0129	0.0126	0.0122	0.0119	0.0116	0.013	0.0110
−2.1	0.0179	0.0174	0.0170	0.0166	0.0162	0.0158	0.0154	0.0150	0.0146	0.0143
−2.0	0.0228	0.0222	0.0217	0.0212	0.0207	0.0202	0.0197	0.0192	0.0188	0.0183
−1.9	0.0287	0.0281	0.0274	0.0268	0.0262	0.0256	0.0250	0.0244	0.0238	0.0233
−1.8	0.0359	0.0352	0.0344	0.0336	0.0329	0.0322	0.0314	0.0307	0.0300	0.0294
−1.7	0.0446	0.0436	0.0427	0.0418	0.0409	0.0401	0.0392	0.0384	0.0375	0.0367
−1.6	0.0548	0.0537	0.0526	0.0516	0.0505	0.0495	0.0485	0.0475	0.0465	0.0455
−1.5	0.0668	0.0655	0.0643	0.0630	0.0618	0.0606	0.0594	0.0582	0.0570	0.0559
−1.4	0.0808	0.0793	0.0778	0.0764	0.0749	0.0735	0.0722	0.0708	0.0694	0.0681
−1.3	0.0968	0.0951	0.0934	0.0918	0.0901	0.0885	0.0869	0.0853	0.0838	0.0823
−1.2	0.1151	0.1131	0.1112	0.1093	0.1075	0.1056	0.1038	0.1020	0.1003	0.0985
−1.1	0.1357	0.1335	0.1314	0.1292	0.1271	0.1251	0.1230	0.1210	0.1190	0.1170
−1.0	0.1587	0.1562	0.1539	0.1515	0.1492	0.1469	0.1446	0.1423	0.1401	0.1379
−0.9	0.1841	0.1814	0.1788	0.1762	0.1736	0.1711	0.1685	0.1660	0.1635	0.1611
−0.8	0.2119	0.2090	0.2061	0.2033	0.2005	0.1977	0.1949	0.1922	0.1894	0.1867
−0.7	0.2420	0.2389	0.2358	0.2327	0.2297	0.2266	0.2236	0.2206	0.2177	0.2148
−0.6	0.2743	0.2709	0.2676	0.2643	0.2611	0.2578	0.2546	0.2514	0.2483	0.2451
−0.5	0.3085	0.3050	0.3015	0.2981	0.2946	0.2912	0.2877	0.2843	0.2810	0.2776
−0.4	0.3446	0.3409	0.3372	0.3336	0.3300	0.3264	0.3228	0.3192	0.3156	0.3121
−0.3	0.3821	0.3783	0.3745	0.3707	0.3669	0.3632	0.3594	0.3557	0.3520	0.3483
−0.2	0.4207	0.4168	0.4129	0.4090	0.4052	0.4013	0.3974	0.3936	0.3897	0.3859
−0.1	0.4602	0.4562	0.4522	0.4483	0.4443	0.4404	0.4364	0.4325	0.4286	0.4247
−0.0	0.5000	0.4960	0.4920	0.4880	0.4840	0.4801	0.4761	0.4721	0.4681	0.4641
0.0	0.5000	0.5040	0.5080	0.5120	0.5160	0.5199	0.5239	0.5279	0.5319	0.5359
0.1	0.5398	0.5438	0.5478	0.5517	0.5557	0.5596	0.5636	0.5675	0.5714	0.5753
0.2	0.5793	0.5832	0.5871	0.5910	0.5948	0.5987	0.6026	0.6064	0.6103	0.6141
0.3	0.6179	0.6217	0.6255	0.6293	0.6331	0.6368	0.6406	0.6443	0.6480	0.6517
0.4	0.6554	0.6591	0.6628	0.6664	0.6700	0.6736	0.6772	0.6808	0.6844	0.6879
0.5	0.6915	0.6950	0.6985	0.7019	0.7054	0.7088	0.7123	0.7157	0.7190	0.7224
0.6	0.7257	0.7291	0.7324	0.7357	0.7389	0.7422	0.7454	0.7486	0.7517	0.7549

(Continued)

TABLE 19.4   *Continued*

z	0	1	2	3	4	5	6	7	8	9
0.7	0.7580	0.7611	0.7642	0.7673	0.7703	0.7734	0.7764	0.7794	0.7823	0.7852
0.8	0.7881	0.7910	0.7939	0.7967	0.7995	0.8023	0.8051	0.808	0.8106	0.8133
0.9	0.8159	0.8186	0.8212	0.8238	0.8264	0.8289	0.8315	0.8340	0.8365	0.8389
1.0	0.8413	0.8438	0.8461	0.8485	0.8508	0.8531	0.8554	0.8577	0.8599	0.8621
1.1	0.8643	0.8665	0.8686	0.8708	0.8729	0.8749	0.8770	0.8790	0.8810	0.8830
1.2	0.8849	0.8869	0.8888	0.8907	0.8925	0.8944	0.8962	0.8980	0.8997	0.9015
1.3	0.9032	0.9049	0.9066	0.9082	0.9099	0.9115	0.9131	0.9147	0.9162	0.9177
1.4	0.9192	0.9207	0.9222	0.9236	0.9251	0.9265	0.9278	0.9292	0.9306	0.9319
1.5	0.9332	0.9345	0.9357	0.9370	0.9382	0.9394	0.9406	0.9418	0.9430	0.9441
1.6	0.9452	0.9463	0.9474	0.9484	0.9495	0.9505	0.9515	0.9525	0.9535	0.9545
1.7	0.9554	0.9564	0.9573	0.9582	0.9591	0.9599	0.9608	0.9616	0.9625	0.9633
1.8	0.9641	0.9648	0.9656	0.9664	0.9671	0.9678	0.9686	0.9693	0.9700	0.9706
1.9	0.9713	0.9719	0.9726	0.9732	0.9738	0.9744	0.9750	0.9756	0.9762	0.9767
2.0	0.9772	0.9778	0.9783	0.9788	0.9793	0.9798	0.9803	0.9808	0.9812	0.9817
2.1	0.9821	0.9826	0.9830	0.9834	0.9838	0.9842	0.9846	0.9850	0.9854	0.9857
2.2	0.9861	0.9864	0.9868	0.9871	0.9874	0.9878	0.9881	0.9884	0.9887	0.9890
2.3	0.9893	0.9896	0.9898	0.9901	0.9904	0.9906	0.9909	0.9911	0.9913	0.9916
2.4	0.9918	0.9920	0.9922	0.9925	0.9927	0.9929	0.9931	0.9932	0.9934	0.9936
2.5	0.9938	0.9940	0.9941	0.9943	0.9945	0.9946	0.9948	0.9949	0.9951	0.9952
2.6	0.9953	0.9955	0.9956	0.9957	0.9959	0.9960	0.9961	0.9962	0.9963	0.9964
2.7	0.9965	0.9966	0.9967	0.9968	0.9969	0.9970	0.9971	0.9972	0.9973	0.9974
2.8	0.9974	0.9975	0.9976	0.9977	0.9977	0.9978	0.9979	0.9979	0.9980	0.9981
2.9	0.9981	0.9982	0.9982	0.9983	0.9984	0.9984	0.9985	0.9985	0.9986	0.9986
3	0.9987	0.9990	0.9993	0.9995	0.9997	0.9998	0.9998	0.9999	0.9999	1.0000

this standard deviation. This is just the distance that will yield the appropriate value for σ_y from Figure 19.1. Values of x_y will vary with stability. The equations for points sources may be used, determining σ_y as a function of $x + x_y$. This procedure effectively treats the area source as a crosswind line source with a normal distribution; this is a fairly good approximation for the distribution across an area source. The initial standard deviation for a square area source can be approximated by $\sigma_{y0} = s/4.3$, where s is the length of a side of the area.

19.7   INSTANTANEOUS "PUFF" MODEL

A 20 m high tank in a plant containing toxic nanoparticles suddenly explodes. The explosion causes an emission of 400 g/s for 3 min. A school is located 200 m west and 50 m south of the plant. If the velocity is 3.5 m/s from the east, how many seconds after the explosion will the concentration reach a maximum in the school? Humans will be adversely affected if the concentration of the nanoparticles is greater than 1.0 μg/L. Is there any impact on the students in the school? Assume that stability category *D* applies (see Problem 19.4 for a description of stability categories).

SOLUTION

A rather significant amount of data and information is available for sources that emit continuously to the atmosphere. Unfortunately, little is available on instantaneous or "puff" sources. Other than computer models that are not suitable for classroom and/or illustrative example calculation, only Turner's *Workbook of Atmospheric Dispersion Estimates* [42], provides a simple equation that may be used for estimation purposes. Cases of instantaneous releases, as from an explosion, or short-term releases on the order of seconds, are also and often of practical concern.

To determine concentrations at any position downwind, one must consider both the time interval after the time of release and diffusion in the downwind direction, as well as lateral and vertical diffusion. Of considerable importance, but very difficult to determine, is the path or trajectory of the puff. This is most important if concentrations are to be determined at specific points. Determining the trajectory is of less importance if knowledge of the magnitude of the concentration for particular downwind distances or travel times is required without the need to know exactly at what points these concentrations occur [37].

An equation that may be used for estimates of concentration downwind from an instantaneous release from height, *H*, is

$$C(x, y, 0, H) = \left[\frac{2m_T}{(2\pi)^{1.5}\sigma_x\sigma_y\sigma_z}\right]\left\{\exp\left[-0.5\left(\frac{x - ut}{\sigma_x}\right)^2\right]\right\}$$
$$\times \left\{\exp\left[-0.5\left(\frac{H}{\sigma_z}\right)^2\right]\right\}\left\{\exp\left[-0.5\left(\frac{y}{\sigma_y}\right)^2\right]\right\}$$

where m_T = total mass of the release; u = wind speed; t = time after the release; x = distance in the x direction; y = distance in the y direction; and, σ_x, σ_y, σ_z = dispersion coefficients in the x, y and z direction, respectively. The dispersion coefficients above are not necessarily those evaluated with respect to the dispersion of a continuous source at a fixed point in space. This equation can be simplified for centerline concentrations and ground-level emissions by setting $y = 0$ and $H = 0$, respectively. The dispersion coefficients in the above equation refer to dispersion statistics following the motion of the expanding puff. The σ_x is the standard deviation of the concentration distribution in the puff in the downwind direction, and t is the time after release. Note that there is essentially no dilution in the downwind direction by wind speed. The speed of the wind mainly serves to give the downwind position of the center of the puff, as shown by examination of the exponential term involving σ_x. In general, one should expect the σ_x value to be about the same as σ_y.

Unless another model is available for treating instantaneous sources, it is recommended that the above equation be employed. The use of appropriate values of σ for this equation is not clear cut. As a first approximation the reader may consider employing the values provided in Problem 19.4.

For the problem at hand, first determine the total amount of toxic agent released, m_T.

$$m_T = (400 \text{ g/s})(3 \text{ min})(60 \text{ s/min})$$
$$= 72\ 000 \text{ g}$$

The values of the three standard deviations are obtained from Figures 19.1 and 19.2, and Problem 19.4.

$\sigma_x = \sigma_y = 16$ m at downwind distance $= 200$ m and stability category D
$\sigma_z = 8.5$ m at downwind distance $= 200$ m and stability category D

The time at which the maximum concentration will occur at the school is

$$t = (\text{downward distance})/(\text{wind speed})$$
$$= (200 \text{ m})/(3.5 \text{ m/s})$$
$$= 57.1 \text{ s}$$

The maximum concentration at the school can now be calculated using the equation given above.

$$C(x, y, 0, H) = \left[\frac{2m_T}{(2\pi)^{1.5}\sigma_x\sigma_y\sigma_z}\right]\left\{\exp\left[-0.5\left(\frac{x - ut}{\sigma_x}\right)^2\right]\right\}$$
$$\left\{\exp\left[-0.5\left(\frac{H}{\sigma_z}\right)^2\right]\right\}\left\{\exp\left[-0.5\left(\frac{y}{\sigma_y}\right)^2\right]\right\}$$

Note that for the maximum concentration, $x = ut$ and the first exponential term at the maximum concentration becomes 1.0 (unity).

Substituting yields

$$C(200, 50, 0, 20) = \left[\frac{2(72\ 000)}{(2\pi)^{1.5}(16)(16)(8.5)}\right][1]\left\{\exp\left[-0.5\left(\frac{20}{8.5}\right)^2\right]\right\}$$
$$\times \left\{\exp\left[-0.5\left(\frac{50}{16}\right)^2\right]\right\}$$
$$= 2.00 \times 10^{-3} \text{ g/m}$$

The concentration can also be expressed in units of micrograms/liter:

$$C = (2.00 \times 10^{-3} \text{g/m}^3)(1 \times 10^6 \,\mu\text{g/g})/(1000 \, L/\text{m}^3) = 2.00 \,\mu\text{g/L}$$

Since the calculated maximum concentration is 2.00 μg/L, which is greater than 1.0 μg/L, there is an impact on students in the school.

19.8 INSTANTANEOUS "PUFF" SOURCES [35]

The U.S. EPA considers it a health hazard if a particular nanoagent has a ground-level concentration (GLC) greater than 1×10^{-9} g/m^3 (1.0 ng/m^3). An explosion in a chemical plant releases 1000 kg of this agent as a "puff" from ground-level conditions. A
residential community is located 4000 m downwind and 1000 m crosswind from the source of the explosion. If the wind speed is 6 m/s and the stability category is "unstable," how much time do the residents have to evacuate the town?

Suggested values of σ_y and σ_z for quasi-instantaneous sources are given in Table 19.5.

SOLUTION

From Table 19.5, determine the values of the dispersion coefficients, σ_y and σ_z.

$$\sigma_y = 300 \text{ m}$$
$$\sigma_z = 220 \text{ m}$$

Also, assign a value to the dispersion coefficient, σ_x

$$\sigma_x = \sigma_y = 300 \text{ m}$$

Since the explosion occurs at ground level, $H = 0$:

$$C(x, y, 0, H) = \left[\frac{2m_T}{(2\pi)^{1.5} \sigma_x \sigma_y \sigma_z} \right] \left\{ \exp\left[-0.5 \left(\frac{x - ut}{\sigma_x} \right)^2 \right] \right\}$$

$$\times \left\{ \exp\left[-0.5 \left(\frac{H}{\sigma_z} \right)^2 \right] \right\} \left\{ \exp\left[-0.5 \left(\frac{y}{\sigma_y} \right)^2 \right] \right\}$$

TABLE 19.5   Values of σ_y and σ_z

	$x = 100$ m		$x = 4$ km	
	σ_y	σ_z	σ_y	σ_z
Unstable	10.	15.	300	220
Neutral	4.	3.8	120	50
Very stable	1.3	0.75	35	7

Set the left-hand side (LHS) of the instantaneous puff equation to the maximum GLC allowed and solve for the time, t:

$$1 \times 10^{-9} = \left[\frac{2(1000)(1000)}{(2\pi)^{1.5}(300)(300)(220)}\right]\left\{\exp\left[-0.5\left(\frac{4000-6t}{300}\right)^2\right]\right\}[1]$$

$$\times \left\{\exp\left[-0.5\left(\frac{1000}{300}\right)^2\right]\right\}$$

$$t = 442 \text{ s}$$

$$= 7.4 \text{ min}$$

19.9 U.S. EPA DISPERSION MODELS

Describe the U.S. EPA's computer dispersion models.

SOLUTION

Models are available, individually, from the National Technical Information Service (NTIS) on microcomputer disks. Regulatory and Air Quality models are also available as a package. Most of the disks must be transported (uploaded) to IBM 3090 machines; however, some models do have PC executable codes. Other air quality models are available from the Office of Research and Development (ORD). Flynn and Theodore [39] provide summary descriptions of over two dozen NTIS and ORD models.

More recently, the U.S. EPA formulated the concept of an integrated modeling and analysis framework. The Models-3/Community Multiscale Air Quality (CMAQ) modeling system is a multilayered system of components that help build, evaluate, and apply models to air quality-related problems. Models-3/CMAQ contain an air quality modeling tool for simulating urban-to-regional-scale pollution problems relating to topospheric ozone, acid/nutrient deposition, and visibility. Although fine particulate matter ($PM_{2.5}$) is included, the analysis is yet to address nanoparticles. The system also facilitates the integration of scientific and technology advancements of other federal agencies and research institutions, thereby allowing a more unified and comprehensive approach to air quality modeling.

As with all models, any application of an air quality model may have deficiencies that cause estimated concentrations to be in error. When practical they should be compared with observed air quality data and their validity determined in order to obtain a measure of confidence in the estimates.

The model validation process should consist of a series of analytical steps:

1. Comparing estimated concentrations with measured air quality data;
2. Determining the cause of discrepancies;
3. Correcting and improving data bases;

4. Modifying the model (if necessary) in a manner that provides a better mathematical representation of physical reality; and

5. Documenting, for others, the accuracy of the estimates.

19.10   DISPERSION IN WATER SYSTEMS AND SOILS [37, 39]

Discuss dispersion in water systems and soils.

SOLUTION

Enormous amounts of waste dumped into water systems are degrading water quality and causing increased human health problems. In assessing this pollution, there are two distinct problem areas. The first, and worst, is in marine estuaries and associated coastal waters. As fewer and fewer alternatives remain for land disposal, wastes are finding their way more often into water. The second area consists of the oceans themselves, although it is believed that currently not much of a problem exists, because relatively little waste is dumped directly into the oceans.

Municipal sewage and agricultural runoff are major sources of water pollution; yet industry gets the brunt of the blame for toxic pollution in marine waters. Nevertheless, in the aggregate, industrial discharges represent the largest source of toxic pollutants entering the marine environment. The major reason for this is that marine disposal, if available, is much cheaper for industry than other ways of disposing of wastes.

In addition to the normal, everyday emissions into water systems is the ever-present threat of a discharge resulting from an accident, an emergency, or a combination of these. Added to this list are nanoparticles. The dispersion and ultimate fate of such emissions is a major concern to the environmental engineer. It is for this reason that the Problems on dispersion applications in water systems has been included. Much material is available in one of the classic works in this field by Thomann and Mueller [43].

In general, the role of the water quality engineer and scientist is to analyze water quality problems by dividing each case into its principal components. These are:

1. *Inputs*: the discharge into the environment of residue from human and natural activities.

2. *Reactions and physical transport*: the chemical and biological transformation, and water movement, that result in different levels of water quality at different locations in time in an aquatic ecosystem.

3. *Output*: the resulting concentration of a substance such as dissolved oxygen or nutrients at a particular location in a river or stream and during a particular time of the year or day.

For example, the simplest physical model that can be used to describe a stream or river is a tubular flow reactor [44]. The describing equations then become one dimensional in the Cartesian (rectangular) coordinate system in the direction of the moving water. Most streams and rivers are subjected to sources or sinks of a pollutant which are distributed along the length of the stream. An example of an external source is runoff from agricultural areas, whereas oxygen-demanding material distributed over the bottom of the stream exemplifies an in-stream, or internal source. The concentration in a stream or river due to multiple point and/or distributed sources is the linear summation of the responses due to the individual sources plus the response due to any upstream boundary conditions. For streams with multiple sources in which flows and velocities vary, but are constant for a given length (reach), solutions are available although this situation is sometimes treated analytically as a one-dimensional transport equation in spherical coordinates [43, 44].

A rigorous treatment of dispersion in soils is beyond the scope of this book. However, some qualitative discussion is warranted because of the potential and existing problems already described. Two main problems arise because dispersion in soil (or land) is anisotropic (i.e., it varies with direction) and the permeability is not only a variable but also an unknown.

The variation in permeability with direction reflects the differences in path length through which a fluid element moves and the forces it experiences in moving through the porous media in a given direction. The anisotropy of a porous medium (like a soil) is undoubtedly related to the circumstances under which the bed was formed. A porous medium may be expected to be anisotropic if the elementary particles are asymmetric and have, on average, a particular orientation. Even beds of symmetric particles can be anisotropic if they are present in certain regular patterns. In naturally occurring soils, however, the packing pattern is sufficiently random to preclude the occurrence of anisotropy in beds made up of symmetric particles. The size distribution of the particles making up the bed can also affect the magnitude of the permeability, but ordinarily this does not contribute to the anisotropy.

19.11   CANAL CONCENTRATION PROFILE [39, 44]

A large concentration of toxic waste has been dumped into a stationary body of water in a long shallow straight canal. The dumped waste, which may be assumed to be at a concentration of 0.13 lbmol/ft^3, is initially located uniformly through a 20 ft section of the canal. Calculate the concentration of the waste in the canal as a function of position and time if the "effective" diffusivity for the waste-water may be assumed to be 5.0 ft^2/h.

SOLUTION

Based on the Problem statement, one may assume that the waste is initially bounded by two parallel planes $2a$ units of length apart at a concentration of c_{A_0} equal to

0.13 lbmol/ft^3. This region is located at the midpoint of the canal, which is assumed to be infinite in length. The actual process may be best described by a one-dimensional, unsteady-state mass transfer process. The describing equation in Cartesian (rectangular) coordinates with y representing the longitudinal direction of the canal is

$$\frac{\partial c_A}{\partial t} = D_{AB}\frac{\partial^2 c_A}{\partial y^2}$$

The solution to this equation is

$$C_A(y,\,t) = \int_{-\infty}^{+\infty}\frac{c_A(y')}{\sqrt{4\pi Dt}}\exp\left[-\frac{(y-y')^2}{4Dt}\right]dy'$$

The boundary condition may now be substituted between $-\infty$ to $-a$, $-a$ to $+a$, and $+a$ to $+\infty$.

$$C_A(y,\,t) = 0 + \int_{-a}^{+a}\frac{c_{A_0}}{\sqrt{4\pi Dt}}\exp\left[-\frac{(y-y')^2}{4Dt}\right]dy' + 0$$

Let

$$\zeta = \frac{y-y'}{\sqrt{4Dt}}$$

so that

$$y' = y - \sqrt{4Dt}\;\;\zeta$$

and

$$dy' = -\sqrt{4Dt}\;\;d\zeta$$

When

$y' = -\infty,\; \zeta = +\infty$
$y' = -a,\; \zeta = (a+y)/\sqrt{4Dt}$
$y' = +\infty,\; \zeta = -\infty$
$y' = +a,\; \zeta = (-a+y)/\sqrt{4Dt}$

Changing variables yields

$$c_A(y, t) = \frac{c_{A_0}}{\sqrt{4\pi Dt}} \int_{(y+a)/\sqrt{4Dt}}^{(y-a)/\sqrt{4Dt}} e^{-\zeta^2}(-\sqrt{4Dt})\, d\zeta$$

$$= \frac{c_{A_0}}{\sqrt{\pi}} \left[\int_0^{(y+a)/\sqrt{4Dt}} e^{-\zeta^2} d\zeta - \int_0^{(y-a)/\sqrt{4Dt}} e^{-\zeta^2} d\zeta \right]$$

The first integral I is given by

$$I = \frac{\sqrt{\pi}}{2} \, \text{erf}\left(\frac{y+a}{\sqrt{4Dt}}\right)$$

Integral II is

$$II = \frac{\sqrt{\pi}}{2} \, \text{erf}\left(\frac{y-a}{\sqrt{4Dt}}\right)$$

Therefore

$$C_A(y, t) = \frac{c_{A_0}}{\sqrt{4\pi}} \left[\frac{\sqrt{\pi}}{2} \, \text{erf}\left(\frac{y+a}{\sqrt{4Dt}}\right) - \frac{\sqrt{\pi}}{2} \, \text{erf}\left(\frac{y-a}{\sqrt{4Dt}}\right) \right]$$

$$= \frac{c_{A_0}}{2} \left[\text{erf}\left(\frac{y+a}{\sqrt{4Dt}}\right) - \text{erf}\left(\frac{y-a}{\sqrt{4Dt}}\right) \right]$$

One need only substitude for c_{A_0}, D_{AB}, and a (10 ft) to obtain c_A as a function of position and time. This is left as an exercise for the reader. Tabulated values of the error function erf, which is also referred to as the probability integral, are available in the literature [39, 44].

19.12   ACCIDENTAL/EMERGENCY DISCHARGE INTO A LAKE/RESERVOIR [37, 39]

A near core meltdown at a nuclear power plant brought about the implementation of an emergency response procedure. Part of the response plan resulted in the steady discharge of a radioactive nanoparticulate effluent into a nearby reservoir. The mass flowrate of discharge was 120 000 lb/h with a radioactive waste concentration of 10^6 picocuries/L (pCi/L) over an 11 h period. (One gram of radium undergoes 3.7×10^{10} nuclear disintegrations in 1s. This number of disintegrations is known as a curie, Ci, which is the unit used to measure nuclear activity). The reservoir volume and the net throughput volumetric flowrate are approximately (annual average) 3.6×10^8 ft^3 and 200 ft^3/s, respectively. If the waste decays in a first order manner with a decay constant of 0.23 (h)$^{-1}$, determine the following:

1. The equilibrium concentration associated with the steady waste discharge.
2. The maximum concentration.
3. The time after termination of the waste discharge for the concentration to reach an acceptable level of 10 pCi/L.

SOLUTION

Thomann and Mueller [43] have provided simple, easy-to-use equations that can be employed to describe the concentrations of species in different bodies of water for (a large number of) various conditions. For the case of a steady continuous discharge of a pollutant species into a lake or reservoir undergoing an irreversible first-order reaction, they have shown that the concentration of the pollutant can be described by the following equation.

$$C = \frac{(\dot{m})\left\{1 - \exp\left[-\left(\frac{1}{t_d} + k\right)t\right]\right\}}{q(1 + kt_d)}$$

where C = concentration of pollutant at time t; $\dot{m}$ = mass flowrate of pollutant discharge; q = net volume flowrate through lake or reservoir; t_d = lake retention time = V/q; V = lake or reservoir volume; and k = reaction velocity constant, (time)$^{-1}$. The concentration at a time θ flowing termination of the discharge may be calculated using the following equation:

$$C = C_o\left\{1 - \exp\left[1 - \left(\frac{1}{t_d} + k\right)\theta\right]\right\}$$

where C_o is the concentration that would be achieved if the pollution discharge continued indefinitely.

The describing equation for the equilibrium concentration is obtained using the equation provided above:

$$C = \frac{(\dot{m})\left\{1 - \exp\left[-\left(\frac{1}{t_d} + k\right)t\right]\right\}}{q(1 + kt_d)}$$

Setting $t = \infty$ leads to

$$C_{eq} = \frac{\dot{m}}{q(1 + kt_d)}$$

Values may now be substituted into this equation. Assume that the effluent has the properties of water. Also, note that $t_d = V/q$:

$$\dot{m} = \left(120\,000\ \frac{\text{lb}}{\text{h}}\right)\left(\frac{1}{62.4\ \frac{\text{lb}}{\text{ft}^3}}\right)\left(\frac{\text{L}}{0.0353\ \text{ft}^3}\right)\left(\frac{10^6\ \text{pCi}}{\text{L}}\right)$$

$$= 5.45 \times 10^{10}\ \text{pCi/h}$$

$$q = (200\ \text{ft}^3/\text{s})(3600\ \text{s/h})$$

$$= 7.2 \times 10^5\ \text{ft}^3/\text{h}$$

$$t_d = (3.6 \times 10^8\ \text{ft}^3)/(7.2 \times 10^5\ \text{ft}^3/\text{h})$$

$$= 500\ \text{h}$$

$$k = 0.23\ \text{h}^{-1}$$

Substituting the values of $\dot{m}$, q, t_d, and k into the equation yields

$$C_{eq} = \frac{\dot{m}}{q(1 + kt_d)}$$

$$= \frac{5.45 \times 10^{10}\ \dfrac{\text{pCi}}{\text{h}}}{\left(7.2 \times 10^5\ \dfrac{\text{ft}^3}{\text{h}}\right)\left[1 + \left(0.23\dfrac{1}{\text{h}}\right)(500\ \text{h})\right]}$$

$$= 652\ \text{pCi/ft}^3$$

$$= 652\ \text{pCi/ft}^3\ (0.0353\ \text{ft}^3/\text{L})$$

$$= 23.02\ \text{pCi/L}$$

The maximum concentration, C_{max}, is achieved when the discharge is stopped, that is, when $t = 11$ h. This is obtained from the following equation:

$$C_{max} = \frac{(\dot{m})\left\{1 - \exp\left[-\left(\dfrac{1}{t_d} + k\right)t\right]\right\}}{q(1 + kt_d)}$$

Substituting yields

$$C_{max} = C_{eq}\left\{1 - \exp\left[-\left(\frac{1}{t_d + k}\right)t\right]\right\}$$

$$= 23.02 \frac{pCi}{L}\left\{1 - \exp\left[-\left(\frac{1}{500\ h} + 0.23\frac{1}{h}\right)(11\ h)\right]\right\}$$

$$= 21.23\ pCi/L$$

The time, θ, after termination of the waste discharge, that the concentration will reach an acceptable level is given by

$$C = C_{max}\left\{1 - \exp\left[-\left(\frac{t}{t_d} + k\right)t\right]\right\}$$

$$10\ \frac{pCi}{L} = 21.23\ \frac{pCi}{L}\left\{1 - \exp\left[-\left(\frac{1}{500\ h} + 0.23\ \frac{1}{h}\right)\theta\right]\right\}$$

$$\frac{10}{21.23} = 1 - \exp(-0.232\theta)$$

$$0.471 = \exp(-0.232\theta)$$

$$\ln(0.471) = -0.232\theta$$

$$\theta = -0.753/-0.232$$

$$= 3.25\ h$$

20 Ethics

Ethics is defined as that branch of philosophy dealing with the rules of right conduct. When defining environmental ethics, right conduct with respect to the environment must be evaluated. The American Consulting Engineers Council for Professional and Ethical Conduct Guidelines state that "consulting engineers shall hold paramount to the safety, health and welfare of the public in the performance of their professional duties. Consulting engineers shall at all times recognize that their primary obligation is to protect the safety, health, property and welfare of the public. If their professional judgment is overruled under circumstances where the safety, health, property or welfare of the public are endangered, they shall notify their client and such other authority as may be appropriate. Consulting engineers shall approve only engineering work which, to the best of their knowledge and belief, is safe for public health, property and welfare and in conformity with accepted standards."

Ethics responsibilities are discussed further in the Problems in this Chapter. For now, it is important to realize that, the above statement of ethics can be extended to apply to all individuals. Whether one's particular area of responsibility lies with the government, industry, or family, or whether the individual is an engineer, attorney, politician, or scientist, actions should aim to promote the general health and well-being of the community in which one works and lives and the surrounding environment.

The last seven Problems presented in this Chapter have been primarily drawn (with permission) from the work of Wilcox and Theodore [45]. The authors used case studies to discuss ethical issues; both believe the case study method is one of the best ways to engage students and professionals in the discussion of moral challenges facing the engineering and scientific communities. Each of the problems/case studies contains a problem statement and data, and a solution that is highlighted by questions for discussion that may suggest an accompanying "answer."

20.1 DETERMINATION OF ETHICAL VALUES

How do individuals and society determine ethical values?

Nanotechnology: Basic Calculations for Engineers and Scientists, by Louis Theodore
Copyright © 2006 John Wiley & Sons, Inc.

SOLUTION

Without a doubt, culture has a strong bearing on the formation of ethical standards by individuals and by a society. The beliefs, knowledge, and traditions that make up a culture provide society with a world view that frames the range of acceptable behavior for any given individual and for the society as a whole. It is essential to recognize, however, that the source of ethical standards must be external to the individual, whether grounded in religious faith, social tradition, or professional or governmental policy. An individual who claims for him or herself the right to establish the ethical rules by which they live can just as easily disestablish those rules should they subsequently prove inconvenient to observe.

20.2 DO'S AND DON'TS

Provide a list of ethical do's and don'ts.

SOLUTION

As one moves into professional life, the engineer and scientist learn a long list of ethical do's and don'ts. Some examples are:

1. Never falsify data;
2. Never offer or accept a bribe;
3. Judge colleagues, employees, employers, clients, and others on their professional merits and not on any other basis such as race, gender, religion, age, national origin, or disability;
4. Never draw misleading conclusions from the data you have collected;
5. Avoid conflicts of interest;
6. Accept work only in those areas in which you are properly trained;
7. Obey all laws, rules and regulations;
8. Never break confidentiality;
9. Always do the best possible work of which you are capable;
10. Statements to the public must always be truthful and objective;
11. The health and safety of the public must always the highest priority; and
12. Always act to protect the environment.

20.3 CODES OF ETHICS

Provide three codes of ethics that have been prepared by professional societies.

SOLUTION

Most professionals belong to one or more professional organizations, which commonly adopt a code of ethics for their members. Each code is designed to provide members with an ethical foundation from which to work. Three such codes are provided below.

1. The American Society of Civil Engineers code:

   Engineers shall hold paramount the safety, health and welfare of the public in the performance of their professional duties. [This is known as CANON 1]

   Engineers who have knowledge or reason to believe that another person or firm may be in violation of any of the provisions of CANON 1 shall present such information to the proper authority in writing and shall cooperate with the proper authority in furnishing such further information or assistance as may be required.

2. The National Society of Professional Engineers code:

   Engineers shall hold paramount the safety, health and welfare of the public in the performance of their professional duties [II. RULES OF PRACTICE].

   Engineers shall at all times recognize that their primary obligation is to protect the safety, health, property and welfare of the public. If their professional judgment is overruled under circumstances where the safety, health, property or welfare of the public are endangered, they shall notify their employer or client and such authority as may be appropriate.

3. IEEE code:

   We, the members of the IEEE ... commit ourselves to the highest ethical and professional conduct and agree:

   1. to accept responsibility in making engineering decisions consistent with the safety, health and welfare of the public, and to disclose promptly factors that might endanger the public or the environment.

It is clear that these codes place the ethic of protecting the public and the environment over the ethic of faithful service to the client or employer.

Variations in circumstances can produce an infinite number of dilemmas arising from conflicts and difficult choices. One needs to approach each case by reducing the problem to its essentials and then applying the rules and principles that have been agreed upon as professionals to the root cause of the ethical question. That may not make action any easier, but should make the ethical decision a clearer one to choose.

20.4 THE HEAVY METAL DILEMMA

Fact Pattern

A university that has a program in environmental engineering has been given a grant to do research in the area of nanosized heavy metals in lake sediments. The overseer of this research is a professor who is well respected not only at the university but also in the field. After this professor secures the grant, he puts to work various graduate and undergraduate students to carry out the research in the lab. The experiment involves examining the binding capacity of metals in sediments; more specifically, the ability of certain metals to form metal sulfides, thereby rendering the heavy metals harmless to any life in the sediment.

Paul, one of the ten student members of the research team, is selected to set up this experiment, which he does as follows. A long plastic cylinder is used as housing. The bottom of this cylinder is sealed, with the top left open. Sediment collected from the bottom of the lake is placed in the cylinder, and then water is placed on top of the sediment, to simulate lake conditions. A key parameter of this experiment is that the water be kept at a pH of 7, that being the normal pH of lakes. Over time, the water is sampled and analyzed for metals. The only maintenance on this experiment is that the overlying water must be refilled periodically because the volume decreases due to evaporation. This water is to be replaced with a special buffer made up in the lab and kept in two-liter flasks. Paul is to perform this experiment for four different cylinders so as to generate a reasonable amount of data from the experiment.

For about two months, Paul tends to the experiment. Occasionally he takes samples from the overlying water, and very often he has to refill the cylinders with buffer solution. There are some instances, however, when Paul cannot be in the lab to tend to his experiment; either he has to attend class or has to take a day off from work. Under these circumstances, one of the other members of the research team takes over Paul's duties.

One day when Paul is in the lab analyzing his samples of the overlying water from the sediment cylinders, he notices something peculiar. His samples read very low for the first two months of sampling, and then the concentration of metal shoots up suddenly. This happens on the same day of sampling for each of the four cylinders. The data seem to suggest someone tampering with his experiment; the concentrations he finds break all the rules of kinetics. Paul asks the other members of the research team what they think, but none of them knows what to make of the results. The only thing Paul can do is go to the professor and try to explain what went wrong with the experiment.

Paul does just that. In the meantime, a conversation takes place between Melinda and Jonas, two of the other lab researchers.

"It's a tough break for Paul if that experiment went bad," Jonas says to Melinda.

"Yeah, it really is," Melinda agrees.

"It would be a shame if someone really did tamper with his sediment cylinders. He put so much work into that experiment," Jonas continues.

"That *would* be terrible."

"Especially if it was one of his coworkers."

"Yeah, but I highly doubt anyone in here would be capable of something like that," Melinda responds.

"I'm not so sure about that," Jonas answers quickly. "Did you happen to see the jar containing the buffer solution that Paul uses to refill his cores?"

"Yeah, sure. I made up the buffer."

"Well, did you happen to see the label on the jar?" Jonas asks as he moves to the bench where the very jar sits.

"*Of course*! I made up the buffer and put the label on there *myself*,"says Melinda.

"Why don't you look at the jar more closely?" Jonas replies.

As Jonas says this, Melinda looks at the jar and sees the source of the problem. Her face goes slightly pale as she answers him quickly, "Listen, I didn't realize that there happened to be two labels on the jar, one on either side. When I made the buffer solution, I removed one of the labels, and replaced it with a label that said buffer. I didn't know that there was a label on the other side of the flask that said that the flask contained 3 percent nitric acid. I also did not look into the jar to see if there was any liquid. How was I supposed to know? It's not my fault."

"Whether or not it's your fault is debatable," Jonas responds, "but what is definite is the fact that you now know what happened and you don't seem inclined to tell anyone. Take some responsibility for your actions. When Paul added the buffer with acid to the cylinders, the pH must have dropped down to around 1. At that pH, all of the metal sulfides were dissolved, freeing up all of the metal. That accounts for his results being the way they are."

Just as Jonas says this, Paul comes back into the lab. "Did you get hold of him?" asks Jonas about Paul's attempt to see the professor.

"No," replies Paul, "but I'm going to go back in fifteen minutes and try to talk to him again."

Questions for Discussion

1. What are the facts in this case?
2. Is the confusion between the buffer and the acid Melinda's fault?
3. Should Melinda tell the professor what happened so that he doesn't blame Paul?
4. Is Melinda unethical in her decision to refrain from telling Paul what happened?
5. If Melinda does not tell the professor, should Jonas tell the professor?

20.5 LET THEM WORRY ABOUT IT

Fact Pattern

John works at a wastewater treatment plant in Anytown, New York. As an entry-level engineer at the plant, he monitors the operating conditions of the chlorine contact basins that are presently in operation. The different parameters that John is to monitor include the flow of wastewater into the chlorine contact basins, the coliform bacteria count of the wastewater entering and leaving the tanks, and the chlorine residual of the effluent flow. The purpose of this monitoring is to ensure that the coliform bacteria count is within an acceptable level, one that conforms to the Environmental Protection Agency's maximum contaminant level (MCL). Coliform bacteria is not harmful to humans, but this type of bacteria is an indicator of the presence of other types that *will* cause harm. Thus far, John has not recorded any values for coliform bacteria that have exceeded the acceptable level, but, as an enterprising engineer, John discovers something else of interest.

Taking all of the data for the concentration of coliform bacteria from the contact basins from the past six months, John plots concentration versus time. Upon doing this, he notices a slight upward trend in the data, one that would never be noticeable when looking at the concentration of bacteria from day to day. The plant still is meeting the maximum contaminant level for coliform bacteria, but if this upward trend were to continue, in a few months the MCL would be reached. As soon as John is sure that his analysis of the coliform counts for the past few months is correct, his initial thought is to take his results to William, his supervisor. The first opportunity that John has to mention his findings to William is while his boss is alone in his office.

"William, do you have a minute?" asks John.

"Only if it's about the plans for the flow model you are working on," replies William. "The big guys upstairs want to know what kind of progress we are making."

"It's coming along well, but I have to do a bit more research on it. I wanted to talk to you about something else, though; it's about the coliform levels in the contact basin effluent."

"What about them?" asks William. "Don't tell me we've got a reading that is not in compliance! I mean, we didn't have a plant failure just now, did we?"

"No, nothing like that. It's just that I've noticed something about the data. The coliform bacteria levels have been slowly increasing over the past few months."

"How bad?"

"Well, if the increase continues at this rate, we would be out of compliance in a couple of months."

At this point, William sits back in his chair and does not say anything. He thinks hard about what John has just said. John then asks, "What are we going to do?"

"What do you mean, 'What are we going to do?'" responds William hesitantly. "We aren't going to do anything."

"Why not? I mean, I understand that we are in compliance now, but in two months or so . . ."

"John, let me explain something to you about the way things work around here. The people upstairs – most specifically, my boss, Mr. Doe – have been hitting us with work nonstop for quite a while now. I can guarantee that if we show him the results you have compiled, he will ask me to assign someone to look into this problem, and he will still expect all of the other work to get done without increasing the number of personnel. We are not out of compliance, so we don't have to tell any one anything."

"William, I can look into it and still handle my other work," replies John. "Let me look into it."

"I can't let you do that, I need you to finish up your part of this flow model. It's more important right now," William answers.

"What happens when we start to approach coliform levels out of compliance?" asks John.

"We let them worry about it upstairs, and we tell them that we could have done something about it if we weren't so overworked. It will be their fault, not ours."

Questions for Discussion

1. What are the facts in this case?
2. Is the decision by William *not* to address the problem an unethical one?
3. Is William justified in his decision?
4. Should John approach someone else about his findings? Should he approach someone of equal or greater authority than William?
5. Should John look into this potential problem further if it does not interfere with his regular work?

20.6 IT'S IN THE AIR

Fact Pattern

When Jim hears about the Montreal Protocol calling for the elimination of certain refrigerants damaging the atmosphere, he's worried. "I've spent my whole life working on refrigeration systems. I know these refrigerants inside and out. I don't think they're polluting the environment. What am I going to do if they eliminate chlorofluorocarbons?"

"Jim, get a grip," says Bill, his partner in J & B Automotive. "We'll just have to get prepared for the future. Let's get one of the new refrigerant manufacturers to come in and talk with us."

"You're right, Bill. Maybe I am overreacting. Perhaps this will be the best thing for us and our business," Jim replies.

When the CFC salesman arrives, he extolls the virtues of the new refrigerants. "They're user-friendly. They're less toxic." The salesman continues, "Did you know that when you introduce this refrigerant into older cars, you'll have to perform a major overhaul, since the current O-ring seal material is not compatible with the new refrigerants?"

"Really?" says Bill. "Wow, I can see us making a lot of money on this."

"Yeah," says the salesman. "And, as the supply of CFC refrigerant starts to dwindle, everyone will have to switch to the new refrigerants because the cost of the older refrigerants will skyrocket."

Following the sales pitch, each of the mechanics is convinced. Jim rushes out to the local automotive supply house and buys 20 canisters of the CFC-based refrigerant. When he returns, Bill is furious.

"What are you doing?"

"I don't want any of my customers having to pay exorbitant prices for refrigerant," answers Jim, "and besides, it's better that this stuff is in the hands of responsible people like us, rather than in the hands of some do-it-yourselfer. They would probably release half the stuff into the atmosphere before they charged their AC system,"

They both chuckled. "You're right about that," says Bill, "but it just seems like we're doing something illegal. And we're not going to make any money on this stuff now."

Questions for Discussion

1. What are the facts in this case?
2. What are the choices available to J & B Automotive concerning the use of refrigerants?
3. What are, the risks associated with each of the choices?
4. What are the benefits to J & B Automotive with each of the choices?
5. What are the risks and benefits to the customers of J & B Automotive?
6. Are each of the choices ethically equivalent?

20.7   CHEAP AT WHAT PRICE

Fact Pattern

Zurich is one of the gardeners on an estate. The head groundskeeper requests that Zurich fertilize the grounds. When Zurich opens the fertilizer, he notices that it is a much cheaper brand than the one normally used; he also notices that there is a warning on the bag about potential harm to animals and advice to keep small animals off lawns for two days after fertilizing them. Zurich asks the head grounds-keeper if the fertilizer was bought accidentally, but he is told that the owner of the estate purchased it to save money. Zurich then asks the *owner* of the estate if he should return this fertilizer and purchase the usual brand because of the warning he has read on the bag of the new kind,

"No, you shouldn't return it. I purchased this fertilizer because it's cheaper," states owner Dan.

"But, what about the warning for the animals?" asks Zurich.

"What animals? I don't have any pets. Now quit bothering me and get back to work," barks Dan. Zurich still has qualms about laying the fertilizer, fearing a neigh-bor's pet may get ill. Yet, since there is a chain-link fence surrounding the property, Zurich believes the fertilizer won't cause any harm. That night, though, there is a heavy storm, and the fertilizer is spread to the neighbor's property by the wind and rain. The next day the neighbor's child becomes ill after playing in the grass. When Zurich hears that news, he fears that the child has become ill from the fertilizer. He informs Dan of his concern.

Dan has already spoken with the child's parents, who were informed by their doctor that the child would be fine in a matter of days, although the doctor isn't exactly sure what has caused the illness. Dan, therefore, does not see the need to tell the neighbors that the fertilizer may have been the cause of the child's stomach-ache. In reality Dan is concerned that the neighbors might sue him if they believe the fertilizer is to blame. Dan thus warns Zurich against telling the neighbors about the possibility of the fertilizer having caused the harm.

"What's the sense in telling them about the fertilizer? The child is fine now, and no permanent harm was done. Who's to know if it was the fertilizer that caused the illness anyway? Therefore, Zurich, I don't want you to mention anything about the fertilizer to our neighbors," Dan states authoritatively.

Zurich knows that the child will regain full health, yet he feels it deceitful not to inform the neighbors of the possibility that the fertilizer is the cause of the child's illness. He knows that he might be in trouble with Dan if he tells the neighbors what happened. Yet Zurich feels guilty and decides he should tell the neighbors just in case the fertilizer might have some unknown side effect or cause long-term damage. He promptly informs the neighbors of the possibility that the fertilizer has caused the child's illness. The neighbors then receive word from their doctor that the fertilizer *did* cause the child's illness, and they decide to sue the owner of the estate. Upon learning that he is being sued, the owner of the estate proceeds to fire Zurich.

Questions for Discussion

1. What are the facts in this case?
2. What ethical dilemma has occurred?
3. In what way is Dan, the owner of the estate, to blame for the child's illness?
4. Is Zurich responsible for the child's illness in any way?
5. If Zurich had been able to determine that the fertilizer would have no side effects or do no permanent damage to the child, should he have told the neighbors?
6. Is the owner of the estate justified in firing Zurich?
7. Are there other actions Zurich could have taken at any time after his first contact with the new fertilizer that might have had better results for the child and himself?

20.8 SAFETY COMES FIRST

Fact Pattern

After several men have been injured in cogeneration facility, the plant manager decides to establish a safety committee which consists of the operations manager and maintenance manager as well as three technicians. Their ultimate goal is to find and eliminate all the safety hazards that exist in this facility. As an incentive, the plant manager decides to reward this committee monetarily whenever a given year passes with the number of injuries being lower than the previous year.

In the course of several years, the committee has eliminated many of the safety hazards in the facility; concomitantly, the cost of workers' compensation has decreased considerably, and the number of injured people per year has reached an all-time low. Both the plant manager and coworkers are very happy with the results generated by this committee.

Although the committee has been solely responsible for the increase in safety awareness at the facility the plant manager receives all the recognition and praise without acknowledging them; he also has not provided any of the promised monetary incentive to them.

The safety committee has become very angry and resentful. "Let's forget about using the allotted safety budget for safety; let's take some of the money for ourselves. We deserve it. And anyway, we were promised some form of monetary incentive!" shouts one of the members.

"Wait! We can't do that. Our coworkers depend on us to use those funds to help eliminate the safety hazards that still exist and to provide safety awareness courses," explains the operations manager.

Questions for Discussion

1. What are the facts in this case?
2. Do you think that the operations manager's comment will affect anyone in the committee?
3. What dilemma is the committee facing?
4. What are the two things that the plant manager failed to give the safety committee?
5. How do you think most of the members in the committee feel about what is happening?
6. What do you think this committee should do?

20.9 INTELLECTUAL PROPERTY

Fact Pattern

Mike has been working for a large defense company for five years and has been thinking of looking for another, higher-paying job. The department he works for makes radios for the military; he writes the software that controls Digital Signal Processing (DSP) chips in the units. Mike also works on data encryption and searches for better ways of making transmissions more secure so enemy forces cannot decode the signals.

Just recently Mike has thought of a better way of encrypting transmissions. His idea has a wide variety of applications in the telecommunications field and in cellular phones. Mike knows that according to his employment contract, any inventions he comes up with while working for the company are the intellectual property of the company. Mike would receive only a dollar from the company for his patent (due to legal formalities), and the company could stand to make millions.

Mike knows that he can leave the company in six months when his employment contract expires, and he will be able to patent his idea and market it to Bell Atlantic, Sprint, or NYNEX. But Mike knows this will be unethical and unlawful. He knows that the company may sue him, saying the invention was discovered while he was still in their employ and therefore belongs to them. Mike does not know what to do.

Question for Discussion
1. What are the facts in this case?
2. What is the ethical dilemma?
3. Does Mike deserve more than one dollar for his invention?
4. What do you think Mike should do?

20.10   THERE'S NO SUCH THING AS A FREE SEMINAR

Fact Pattern

Bryan is the director of the radiology department of a major metro-area hospital. Through some trade publications, Bryan learns of a new type of imaging system involving nanopartides being developed that combines a computerized axial tomography (CAT) scan with a virtual-reality setup; this combination would allow a physician to "fly" through a patient's colon, for example, and look for any abnormalities.

Several competing manufacturers, including New England Imaging (NEI), are developing these systems. The aging imaging equipment at Bryan's hospital is due to be replaced in the near future, and Bryan contacts NEI to request additional information on their new system.

Within several days an NEI representative contacts Bryan and tells him that due to his professional status, he can attend an upcoming two-day seminar on the "virtual CAT scan" in a posh midtown hotel and his registration fee will be waived.

Bryan is no fool; he realizes that this "educational" seminar will most likely be a big sales pitch for NEI's system. However, it would be nice to get away from the often hectic pace of the hospital for a couple of days. Besides, all new medical equipment that is to be purchased by the hospital first must be evaluated by the clinical engineering department; this process involves researching systems from other manufacturers as well as getting feedback from the hospital staff members who use the equipment. Bryan will have *some* input when the final decision is made, but he feels that he is too professional to be "schmoozed" into advocating a system that may turn out to be problematic.

Questions for Discussion
1. What are the facts in this case?
2. What is the ethical issue?
3. Should Bryan attend the seminar?

References: Part 3

1. L. Theodore and R. Kunz, *Nanotechnology: Environmental Implications and Solutions.* Hoboken NJ: John Wiley & Sons, 2005

2. A. Mnyusiwalla, A. Daar and P. Singer, "Mint the Gap: Science and Ethics in Nanotechnology", *Nanotechnology*, 14, R9–R14 (2003).

3. M. Roco, "Nanoscale Science and Engineering: Unifying and Transforming Tools", *AIChE Journal*, 50(5), 890–897 (2004).

4. L. Theodore and M. Blenner, *Unit Operations for the Practicing Engineer.* Hoboken, NJ: John Wiley & Sons, 2006 (in press).

5. J. Santoleri, J. Reynolds and L. Theodore, *Introduction to Hazardous Waste Incineration*, 2nd Edition. Hoboken NJ: John Wiley & Sons, 2000.

6. D. Evans, *Thermodynamic Limits of Nanomachines.* Australian Academy of Science, 2 May, 2003 Available at http://www.science.org.au/sats2003/evans.htm > .

7. M. Wilson, K. Kannangara, G. Smith, M. Simmons, and B. Raguse, *Nanotechnolgy Basic Science and Emerging Technologies*, Boca Raton, FL, Chapman & Hall/CRC, 2002, reprinted 2004.

8. U. Fink R. Davenport, S. Bell and Y. Ishikawa, *Nanoscale Chemicals and Materials – An Overview on Technology, Products and Applications*, Specialty Chemicals Update Program. Menlo Park, CA: SRI Consulting, December, 2002.

9. D. Hairston, "Nano-primed", *Chemical Engineering*, 39–41 (1999).

10. R. Rittinger, *Lehrbuch der Aufbereitungskunde*, Berlin: Ernst and Korn: 1867.

11. W. Walher, W. Lewis, W. McAdams and E. Gilliland, *Principles of Chemical Engineering*, 3rd Edition, New York City: McGraw-Hill, 1937.

12. R. Kick, *Das Gasetz der propertionalen Widerstomale and seine Anwendung*, Leipzig, 1885.

13. A. Bond, *Trans, Am. Inst. Min. Matall. Pet. Eng.*, 193, 484 (1952).

14. R. Perry and D. Green, *Perry's Chemical Engineers Handbook*, 7th Edition. New York City: McGraw-Hill, 1997.

15. M. C Roco, "Nanoscale Science and Engineering: Unifying and Transforming Tools", *AIChE Journal*, 50(5), 890–897 (2004).

16. S. A. Shelley, "Carbon Nanotube: A Small-Scale Wonder". *Chemical Engineering*, 27–29, (2003).

17. A. Hirsch, "Nanotubes: Small Tubes with Great Potential", remarks made at a BASF Aktiengessellschaft event entitled "Journalists and Scientists in Dialogue: Nanotechnology in Chemistry—Experience meets Vision," October 28–29, 2002, Mannheim, Germany.

18. S. Bramer, "The Carbon Nanotube Industry", presented at The Nanoparticles 2002 Conference, October 29, 2002, New York, Business Communications Co. (Norwalk, CT).

19. G. Ondrey, T. Kamiya, and D. Hairston, "Buckyballs Stretch Out in a Nanotube", *Chemical Engineering*, 41–43 (2002).

20. A. Ghosh, "Nanotechnology Will Initiate a Quantum Leap in Manufacturing Efficiency", ARC Insights #2002-49M, October 23, 2002, Dedham, MA; ARC Advisory Group.

21. M. N. Rittner, "World Market Overview for Nanoparticulate Materials", presented at the Nanoparticles 2002 Conference, October 29, 2002, New York, NY, Business Comunications Co. (Norwalk, Conn.).

22. C. Crabb and G. Parkinson, "The Nanosphere: A Brave New World", *Chemical Engineering*, 27–31 (2002).

23. M. Roco and R. Tomelini, eds., "Nanotechnology: Revolutionary Opportunities and Societal Implications", Summary of Proceedings of the 3rd Joint European Commission—National Science Foundation Workshop on Nanotechnology, Lecce, Italy, January 21–February 1, 2002.

24. P. Fairley, "The Start of Something Big", *Chemical Week*, December 12 2001, pp. 23–26.

25. S. A. Shelley, with G. Ondrey, "Nanotechnology – The Sky's the Limit", *Chemical Engineering*, 23–27 (2002).

26. A. Wood and A. Scott, "Nanomaterials", *Chemical Week*, October 16, 2002, pp. 17–21.

27. R. Bailey, "The Smaller the Better", *Reason*, December 1, 2003.

28. G. H. Reynolds, "The Science of the Small", *Legal Affairs*, July 1, 2003.

29. K. E. Drexler, "Machine-Phase Nanotechnology", *Scientific American*, September 16, 2001.

30. A. Mnyusiwalla, A. S. Daar, and P. A. Singer, "Mind the Gap: Science and Ethics in Nanotechnology", *Nanotechnology*, 14, R9–R14 (2003).

31. R. Heinsohn, *Industrial Ventilation: Engineering Principle*, Hoboken, NJ: John Wiley & Sons, 1991.

32. R. Dupont, T. Baxter and L. Theodore, *Environmental Management: Problems and Solutions*, Boca Raton, FL: CRC-Lewis Publishers, 1998.

33. K. Ganesan, L. Theodore and R. Dupont, *Air Toxins: Problems and Solutions*, New York City: Gordon and Breach, 1996.

34. L. Theodore, Personal notes–final exam problem, Accident & Emergency Management, 2000.

35. J. Reynolds, J. Jeris and L. Theodore, *Handbook of Chemical and Environmental Engineering Calculations*, Hoboken, NJ: John Wiley & Sons, 2002.

36. L. Theodore, *Chemical Reaction Kinetics*, A Theodore Tutorial, East Welleston, NY, 1992.

37. L. Theodore, J. Reynolds and F. Taylor, *Accident & Emergency Management*, Hoboken, NJ: John Wiley & Sons, 1989.

38. L. Theodore, personal notes, 2000.

39. A. Flynn and L. Theodore, *Health, Safety, and Accident Management in the Chemical Process Industries*, New York City: Marcel Dekker, 2002.

40. L. Theodore, personal notes, 1979.

41. C. Bosanquet, W. Carey and E. Halton, "Dust Deposition from Chimney Stacks", *Proc. Inst. Mech. Engineers*, 162, 355 (1950).

42. B. Turner, *Workbook of Atmospheric Dispersion Estimates*, USEPA Publication No. AP-26, Research Triangle Park, NC, 1970.

43. R. Thomann and J. Mueller, *Principles of Surface Water Quality and Control*, New York: Harper and Row, 1987.

44. L. Theodore, *Transport Phenomena for Engineers*, Scranton, PA: Intext, 1971.

45. J. Wilcox and L. Theodore, *Engineering and Environmental Ethics*, Hoboken, NJ: John Wiley & Sons, 1998.

PART 4
Environmental Concerns

In the last three and a half decades, people have become aware of a wide range of environmental issues. All sources of air, land, and water pollution are under constant public scrutiny. Increasing numbers of professionals are being confronted with problems related to environmental control. Because some of these issues are relatively new concerns, e.g., emissions from nanoprocesses, individuals must develop a proficiency and an improved understanding of technical and scientific, as well as regulatory issues regarding pollution prevention and remediation in order to cope with these challenges. Any technology can have various and imposing effects on the environment and society. Nanotechnology is no exception, and the results will be determined by the extent to which the technical community manages this technology.

This period has been filled with environmental tragedy as well as with a heightened environmental awareness. The oil spills of the Exxon Valdez in 1989 and in the Gulf War of 1991 showed how delicate oceans and their ecosystems truly are. The disclosures of Love Canal in 1978 and Times Beach in 1979 made the entire nation aware of the dangers of hazardous chemical waste. The discovery of acquired immunodeficiency syndrome virus (AIDS) and the beach washups of 1985 brought the issue of medical waste disposal to the forefront of public consciousness. A nuclear accident placed the spotlight on Chernobyl, and this region is still seeing the effects of that event. More recently, concerns with nanotechnology have justifiably emerged; this is an area that will have to be addressed in the future.

This last Part is divided into seven Chapters covering seven subjects:

1. Regulations
2. Toxicology
3. Noncarcinogens
4. Carcinogens
5. Health Risk Assessment
6. Hazard Risk Assessment
7. Epidemiology

The reader is referred to the literature [1–6] for additional problems and solutions. Several numerical examples related to Toxicology (Chapter 22) can also be found in the Noncarcinogen, Carcinogen, and Health Risk Assessment Chapters that appear in this Part.

It should be noted that nanoenvironmental concerns are starting to be taken seriously around the globe. There are a variety of studies going on into the health and environmental impacts of many applications of nanotechnology. It is in everyone's interest to ensure that any new compound is fully understood and the long-term implications studied before it is commercialized. Class action suits in the United States against both tobacco companies and engineering companies, coupled with a new era of corporate responsibility, have ensured that most companies are well aware of this need. Now that potential risks that may have been overlooked are becoming widely known, these companies are more inclined to be proactive than they have been with risks in the past [7].

21 Environmental Regulations

Many environmental concerns are addressed by existing health and safety legislation. Most countries require a health and safety assessment for any new chemical before it can be marketed. Further, the European Union (EU) recently introducing the world's most stringent labeling system. Prior experience with materials such as PCBs and asbestos, and a variety of unintended effects of drugs such as thalidomide, mean that both companies and governments have an incentive to keep a close watch on potential negative health and environmental effects [8].

It is very difficult to predict future nanoregulations. In the past, regulations have been both a moving target and confusing. What can be said for certain is that there will be regulations, and the probability is high that they will be contradictory and confusing. Past and current regulations provide a measure of what can be expected. And, it is for this reason that this Chapter is included in this book.

Environmental regulations are not simply a collection of laws on environmental topics. They are an organized system of statutes, regulations, and guidelines that minimize, prevent, and punish the consequences of damage to the environment. This system requires each individual – whether an engineer, scientist, attorney, or consumer – to be familiar with its concepts and case-specific interpretations. Specifically, environmental regulations deal with the problems of human activities and the environment, and the uncertainties the law associates with them.

With the onset of the Industrial Revolution, environmental law has increased in popularity due to an interest in public health and safety and the environment. Companies are concerned with exposing themselves to future potentially disastrous liabilities by polluting the environment where their operations are located. Businesses are now taking a proactive approach to environmental compliance. The recent popularity of environmental issues with public and private interest groups has brought about changes in legislation and subsequent advances in technology.

The U.S. EPA's authority is increasingly broadening. It is the largest administrative agency in the federal government and accounts for nearly one-seventh of the federal budget. The agency has nearly 20,000 employees and churns out more pages of regulation than any other administrative agency. The U.S. EPA's comprehensive environmental programs encompass regulations for air pollution, hazardous waste management, solid waste management, drinking water standards, emergency

Nanotechnology: Basic Calculations for Engineers and Scientists, by Louis Theodore
Copyright © 2006 John Wiley & Sons, Inc.

management, and permitting requirements for discharges to the air, land, or waters of the United States and its territories and districts.

Environmental management is accomplished principally in response to requirements in environmental laws and regulations. However, federal, state, and local regulations remain the greatest overall influence on the practice of environmental science, engineering, and sociology. Because of this, environmental regulations become the framework for environmental management, where environmental management has as its objective the protection of human health and well being, and the protection (preservation and conservation) of life forms and their habitats. Finally, understanding specific environmental management objectives in each law is critical to effective environmental management practice.

21.1   THE REGULATORY SYSTEM

Provide an overview of the regulatory system.

SOLUTION

Over the past three plus decades, environmental regulation has become a system in which laws, regulations, and guidelines have become interrelated. Requirements and procedures developed under previously existing laws may be referenced to in more recent laws and regulations. The history and development of the regulatory system has led to laws that focus principally on only one environmental medium, that is, air, water, or land. Some environmental managers feel that more needs to be done to manage all of the media simultaneously, and not treated in a compartmentalized manner. Hopefully, the environmental regulatory system will evolve into a truly integrated, multimedia management framework.

Federal laws are the product of Congress. Regulations written to implement the law are promulgated by the Executive Branch of government, but until judicial decisions are made regarding the interpretations of the regulations, there may be uncertainty about what regulations mean in real situations. Until recently, environmental protection groups were most frequently the plaintiffs in cases brought to court seeking interpretation of the law. Today, industry has become more active in this role. Forum shopping, the process of finding a court that is more likely to be sympathetic to the plaintiffs' point of view, continues to be an important tool in this area of environmental regulation. Many environmental cases have been heard by the Circuit Court of the District of Columbia.

Enforcement approaches for environmental regulations are environmental management oriented, in that they seek to remedy environmental harm, not simply a specific infraction of a given regulation. All laws in a legal system may be used in enforcement to prevent damage or threats of damage to the environment or human health and safety. Tax laws (e.g., tax incentives) and business regulatory laws (e.g., product claims, liability disclosures) are examples of laws not directly focused on environmental protection, but that may also be used to encourage compliance and discourage noncompliance with environmental regulations.

Common law also plays an important role in environmental management. Common law is the set of rules and principles relating to the government and security of persons and property. Common law authority is derived from the usages and customs that are recognized and then enforced by the courts. In general, no infraction of the law is necessary when establishing a common law suit. Legal precedent often forms the foundation for common law court actions. A common law "civil wrong" (e.g., environmental pollution) that is brought to court is called a tort. Environmental torts may arise because of nuisance, trespass, or negligence (see Problem 14.4 for additional details).

Laws tend to be general and contain uncertainties relative to the implementation of principles and concepts they contain. Regulations derived from laws may be more specific, but are also frequently too broad to allow clear translation into environmental technology practice. Permits may be used in the environmental regulation industry to bridge this gap and provide specific technical requirements imposed on a facility by the regulatory agencies for the discharge of pollutants or on other activities carried out by the facility that may impact the environment.

Most major federal environmental laws provide for citizen law suits. This empowers individuals to seek compliance or monetary penalties when these laws are violated and regulatory agencies do not take enforcement action against the violator.

21.2 AIR QUALITY ISSUES

Describe the general subject of air quality issues from a regulatory point of view.

SOLUTION

The atmosphere has always been polluted to some extent through natural phenomena and/or human activities. In recent times, technological advancement, industrial expansion, and urbanization have been the major contributors to global air pollution. Air pollution adversely affects human health and property; the American Cancer Society claims it is a matter of life and breath. An EPA (1994) report on national air quality and emission trends showed that emissions of nitrogen oxides (NO_x) increased 690%, emissions of volatile organic compounds (VOCs) increased by 260%, and sulfur dioxide emissions increased by 210% between 1900 and 1970. The Clean Air Act (CAA) has produced substantial improvements in ambient air quality since its passage in 1970.

The CAA established two levels of National Ambient Air Quality Standards (NAAQS) – primary and secondary standards. Primary air quality standards are promulgated to protect public health, including the health of "sensitive" populations such as asthmatics, children, and the elderly. Secondary air quality standards are set to protect public welfare, including protection against decreased visibility, and damage to animals, crops, vegetation, and buildings.

The U.S. EPA has established NAAQS levels for six principal pollutants: carbon monoxide (CO), lead (Pb), nitrogen dioxide (NO_2), ozone (O_3), particulate matter whose aerodynamic diameter is less than or equal to 10 μm (PM-10) – recently revised to include 2.5 μm (PM-2.5), and sulfur dioxide (SO_2). PM-10 and SO_2 are regulated by primary standards for both short-term (24 h or less) and long-term (annual average) averaging times. Short-term standards are established to protect the public from adverse health effects associated with acute exposure to air pollution, while long-term standards are established to protect the public from chronic exposures to pollution. Besides these six pollutants, "air toxic" compounds or hazardous air pollutants (HAPs) are also regulated following amendments made to the CAA in 1990. These HAPs are chemicals known to be acutely toxic or that are known or suspected of causing cancer or other serious health effects (e.g., reproductive effects) or ecosystems damage. Examples of HAPs include dioxin, benzene, arsenic, beryllium, mercury, and vinyl chloride. The 1990 CAA Amendments list 189 compounds as HAPs [9].

The major sources of the six criteria pollutants for which NAAQSs exist are transportation fuel combustion, industrial processes, nontransportation fuel combustion, and natural sources such as wildfires. Unlike other criteria pollutants, ozone is not emitted directly to the air, but is created in the lower atmosphere when sunlight reacts with NO_x and VOCs. The air toxic compounds are emitted from a wide variety of sources including stationary and mobile point sources, and area sources.

The U.S. EPA routinely monitors concentrations of the criteria pollutants near ground level at various locations across the United States. If monitoring indicates that the concentration of a pollutant exceeds the NAAQS in any area of the country, that area is labeled a nonattainment area for the pollutant, meaning that the area is not meeting the ambient standard. Conversely, any area in which the concentration of a criteria pollutant is below the NAAQS is labeled an attainment area, indicating that the NAAQS is being met.

The attainment/nonattainment designation is made on a pollutant-by-pollutant basis. Therefore, the air quality in an area of the country may be designated attainment for some pollutants and nonattainment for other pollutants at the same time. For example, many cities are designated nonattainment for ozone, but are in attainment for the other criteria pollutants.

The major federal rules that govern air quality in attainment areas are known as the Prevention of Significant Deterioration, or PSD, provisions. These rules are designed to ensure that air quality in "clean" areas (that is, attainment areas) will not degrade, but will remain clean, even as new sources of pollutants are constructed. The PSD program applies to new major sources and major modifications to existing major sources. Note that the term "major" is an air quality term that has different definitions in different parts of the Clean Air Act. In the PSD program, a source is "major" if it is one of 28 listed categories of sources with the potential to emit 100 tons or more per year of any air pollutant regulated by the Clean Air Act, or if it is any other type of source with the potential to emit 250 tons per year or more in one year.

The set of rules and regulations that applies to new or modified emissions sources in nonattainment areas, the areas of the country where NAAQS are not being met, are known as Nonattainment Area New Source Review (NA-NSR). Restrictions on emissions and control technology requirements under NA-NSR provisions are more stringent than under PSD, because the goal of the NA-NSR rules is to improve the air quality until the NAAQS are met.

The New Source Performance Standards (NSPS) were established under the Amendments of 1970, relatively early in the history of air quality regulation, in recognition of the fact that newly constructed sources should be able to operate more "cleanly" than existing, older sources. The NSPS establish the minimum level of control of certain pollutants that specific categories of industrial sources constructed since 1971 must achieve. The emissions limits under NSPS are based on the best technological system of continuous emission reduction available, taking into account costs and other factors of applying the technology.

As outlined previously, the foundation for many of the regulatory programs that result from the Clean Air Act are the NAAQS. However, the U.S. EPA has been able to establish NAAQS for only the six criteria pollutants. With few exceptions until the Amendments of 1990, the U.S. EPA had not been able to address the hundreds of individual toxic or hazardous substances routinely released into the atmosphere by industrial, commercial, and mobile sources.

Prior to the Amendments of 1990, the U.S. EPA was mandated to regulate air-borne toxic pollutants under the National Emission Standards for Hazardous Air Pollutants (NESHAP) program. The goal of the original NESHAP program was to restrict concentrations of identified hazardous pollutants to levels that would prevent adverse health effects "with an ample margin of safety." The U.S. EPA identified a few pollutants (arsenic, asbestos, benzene, beryllium, mercury, radio-nuclides, radon, and vinyl chloride) suspected to be carcinogens for regulation under the program. The U.S. EPA took the position on carcinogens during the development of NESHAP regulations that there is essentially no safe level of exposure to these potentially cancer-causing compounds, i.e., there is no threshold value.

21.3 PARTICULATE LOADING

A hazardous waste incinerator is burning an aqueous slurry of carbon nanotubes with the production of a small amount of fly ash. The waste is 70% water by mass and is burned with 0% excess air (EA). The flue gas contains 0.25 grains (gr) of particulates in each 7 ft^3 (actual) at 620°F. For regulation purposes, calculate the particulate concentration in the flue gas in gr/acf, in gr/scf, and in gr/dscf.

SOLUTION

Calculate the particulate concentration in the flue gas per actual cubic foot:

$$0.25 \, \text{gr}/7.0 \, \text{acf} = 0.0357 \, \text{gr/acf}$$

Convert actual cubic feet to standard cubic feet:

$$60°F = 520°R; \qquad 620°F = 1080°R$$
$$7.0\,\text{acf}(520/1080) = 3.37\,\text{scf}$$

Calculate particulate concentration in standard cubic feet:

$$0.25\,\text{gr}/3.37\,\text{scf} = 0.074\,\text{gr/scf}$$

To determine the volume fraction of water in the flue gas, employ the reaction equation for the combustion of the waste:

$$C + H_2O + O_2 \rightarrow CO_2 + H_2O$$

The number of moles of each component becomes (100 lb basis)

C: $30\,\text{lb}/(12\,\text{lb/lbmol}) = 2.5\,\text{lbmol}$
H_2O: $70\,\text{lb}/(18\,\text{lb/lbmol}) = 3.89\,\text{lbmol}$

Calculate the moles of CO_2 and H_2O in the flue gas:

$$CO_2 = 2.5\,\text{lbmol}$$
$$H_2O = 3.89\,\text{lbmol}$$

The stoichiometric (0% excess air) requirement for this solid waste can now be determined:

O_2:   2.5 lbmol
N_2: $(79/21)(2.5\,\text{lbmol}) = 9.4\,\text{lbmol}$
Air: $9.4 + 2.5 = 11.9$

The oxygen and nitrogen in the flue gas are therefore

$$O_2 = 0.0\,\text{lbmol}$$
$$N_2 = 9.4\,\text{lbmol}$$

Dividing the moles of H_2O by the total moles yields the mole fraction of H_2O in the flue gas:

$$H_2O \text{ fraction} = \text{lbmol } H_2O/\text{lbmol } (CO_2 + H_2O + N_2)$$
$$= 3.89/(2.5 + 3.89 + 9.4)$$
$$= 0.25 \text{ (by mole or volume)}$$

The molar quantity of dry gas is obtained by subtracting the moles of H_2O from the total moles to yield

$$\text{Moles dry gas} = (2.5 + 3.89 + 9.4) - 3.8 = 12.0 \text{ lbmol}$$

The dry volume in dry standard cubic feet (dscf) is obtained by subtracting the volume of H_2O from the total volume to yield

$$\text{Dry Volume} = 3.37(1 - 0.25) = 2.53 \text{ dscf}$$

The particulate concentration on a dscf basis becomes

$$0.25 \text{ gr}/2.53 \text{ dscf} = 0.099 \text{ dr/dscf}$$

21.4   CLEAN AIR ACT ACRONYMS

The Clean Air Act and the people who work with it use a large number of acronyms with dizzying frequency. Knowledge of these acronyms is an absolute necessity for communicating with professionals in the environmental management field.

Provide the meaning of the following acronyms.

1. *MACT*. Explain how MACT applies to attainment area, nonattainment areas, major stationary sources, hazardous air pollutants (Section 112 of Title III), consideration of cost, new sources, existing sources, and the issuing of permits.

2. *RACT*. Explain how RACT is used in an environmental management system. How does RACT apply to existing major stationary sources, National Ambient Air Quality Standards, process modifications, considerations of cost, social impacts, and alternate approaches to meeting Clean Air Act requirements.

3. *CTG*. Explain how CTGs apply to existing non-attainment area major stationary sources.

4. *NSPS*. Discuss how NSPSs are used in an environmental management system. Explain how NSPSs apply to attainment areas, non-attainment

areas, major new stationary sources, modified stationary sources, specific pollutants, energy needs, consideration of costs, work practices, alternate ways of reducing regulated emissions, and the issuing of permits.

5. *LAER*. Describe under what circumstances LAER might be a part of an environmental management system. Discuss how LAER applies to attainment areas, nonattainment areas, major new stationary sources, modified stationary sources, specific pollutants, specific sources, energy needs, consideration of costs, work practices, alternate ways of reducing regulated emissions, and the issuing of permits.

6. *PSD*. Discuss how PSD might be incorporated into an environmental management system. Explain how PSD applies to attainment areas, new stationary sources, the modification of stationary sources, specific sources, specific pollutants, hazardous air pollutants (Section 112 of Title III), specific categories of sources, quantities of pollutant(s) emitted, consideration of costs, considerations of energy use, economic considerations, and the issuing of permits. Indicate why the application of PSDs is controversial.

SOLUTION

Answers can be obtained from a copy of the Clean Air Act or Title 40 of the Code of Federal Regulations (40CFR).

1. *MACT – Maximum Achievable Control Technology*. MACT must be applied in both attainment and nonattainment areas to categories of major stationary sources of any of the 189 hazardous air pollutants (HAPs) listed in Section 112 of Title III of the Clean Air Act of 1990. These HAPs are also known as air toxics. Cost may be considered in the development of MACTs. MACT applies to both new and existing sources but is more stringent for new sources. For new sources, MACT must "...not be less stringent then the emission control that is achieved in practice by the best controlled similar source..." For existing sources, MACT must be set at not less than the level of control achieved by the 12% of sources in the same category with the best record of performance. If a category has fewer than 30 sources, then the best five are used. MACT could be part of a permit.

2. *RACT – Reasonably Available Control Technology*. RACT applies to categories of existing, nonattainment area, major stationary sources of pollutants for which the following National Ambient Air Quality Standards (NAAQS) have been set: the two ozone precursors, volatile organic compounds (VOCs) and nitrogen oxides (NO_x); carbon monoxide (CO); particulate matter smaller than 10 μm (PM-10); sulfur oxides (SO_x); and lead. RACT means devices or system process modifications, and so on, that are reasonably available. RACT development takes into account the need to meet the NAAQS, cost, social and environmental impact, alternate ways of meeting the NAAQS, and other factors.

3. *CTGs – Control Technique Guidelines.* CTGs are documents published by the U.S. EPA detailing what constitutes RACT for each category of existing, nonattainment areas, major stationary sources of VOCs, NO_x, CO, and/or PM-10. EPA has published about 30 CTGs thus far; most of them deal with VOCs. Another dozen CTGs are currently being worked on by the EPA.

4. *NSPS – New Source Performance Standards.* NSPSs are emission standards for categories of major new or significantly modified stationary sources of emissions of VOCs, NO_x, CO, PM-10, acid mist, SO_x, fluorides, H_2S in acid gas, lead, and total reduced sulfur. Development of NSPSs utilizes the "emission reduction achievable" by the best technology that has been "adequately demonstrated." Environmental impact and energy needs are also considered. Cost is considered in a limited way.

    NSPSs are sometimes stated as process or work practices, operational standards, design standards, or other alternate ways of producing low emissions. NSPSs also serve as the minimum standards used in establishing permits for the construction of specific major new stationary emission sources. NSPSs are periodically revised to take advantage of new technology to achieve still lower emissions. NSPSs apply in both attainment and nonattainment areas. In attainment areas, the performance standard is known as the Best Available Control Technology (BACT). In nonattainment areas, the performance standard is known as the Lowest Achievable Emission Rate (LAER).

5. *LAER – Lowest Achievable Emission Rate.* LAER applies to emission standards in permits for construction of new or significantly modified major stationary specific sources of criteria pollutants in a nonattainment area. The criteria pollutants are CO, NO_x, SO_2, lead, total particulate matter, PM-10, and ozone (and its precursor, VOCs). LAERs are applied to specific sources but are established by category of source. LAER is either the lowest rate that any source in that category or a similar category has achieved, or the lowest emission rate in any State Implementation Plan (SIP). Whichever is lower, LAER must be at least as low as any NSPS for that source category. Cost is only considered if it is so high as to render use of the LAER standard not feasible for that specific source.

6. *PSD – Prevention of Significant Deterioration.* To prevent areas that are in compliance with air quality standards from deteriorating in quality as the result of the construction of new sources or major modifications to existing sources of pollutants, permits are required before such construction or modification can take place. These permits apply to new specific stationary sources that could emit 100 tons per year (T/yr) or more of any regulated pollutant (except the 189 HAPs) and belong to one of the 28 categories specifically listed by the U.S. EPA or could emit 250 T/yr or more of a pollutant. Permits are also required if a major modification to an existing plant results in an increase in emissions of any of the following pollutants in the amount given in parenthesis: CO (100 T/yr), NO_x (40 T/yr), SO_2 (40 T/yr), total particulate matter (25 T/yr), PM-10 (15 T/yr), ozone (40 T/yr of VOCs), lead

(0.6 T/yr), asbestos (0.007 T/yr), beryllium (0.0004 T/yr), mercury (0.1 T/yr), vinyl chloride (1 T/yr), fluorides (3 T/yr), sulfuric acid mist (7 T/yr), H_2S (10 T/yr), and reduced sulfur compounds including HS^- (10 T/yr).

The permits require the use of the Best Available Control Technology (BACT). The development of the BACT for each specific source permit takes into account many factors on a case-by-case basis. Some of these factors are: technology, cost, energy use, and the need to limit how much any one plant can be allowed to deteriorate the air quality so that other plants needed for economic development in the area can be built. The BACT is determined by the state within rules issued by the U.S. EPA. The development of BACTs has been and still is the subject of controversy and litigation.

21.5   WATER POLLUTION CONTROL

Discuss some of the key Water Pollution Control Acts.

SOLUTION

Congress put the framework together for water pollution control by enacting the Federal Water Pollution Control Act (FWPCA), the Marine Protection, Research and Sanctuaries Act (MPRSA), the Safe Drinking Water Act (SDWA), and the Oil Pollution Control Act (OPA). Each statute provides a variety of tools that can be used to meet the challenges and complexities of reducing water pollution in the nation. These are discussed below.

21.5.1   Federal Water Pollution Control Act

In 1977, Congress renamed the FWPCA the Clean Water Act (CWA) and substantially revamped and revised the control of toxic water pollutants. The CWA has two basic components: a statement of goals and objectives and a system of regulatory mechanisms calculated to achieve these goals and mechanisms. The objective of the program as outlined in Section 101 of the CWA is to "restore and maintain the chemical, physical and biological integrity of the nation's waters." To achieve this, the CWA provides water quality protection for fish, shellfish, and wildlife for recreational use, and eliminates the discharge of pollutants into the waters of the United States.

21.5.2   Safe Drinking Water Act

The Safe Drinking Water Act (SDWA) was originally passed in 1974 to ensure that public water supplies are maintained at high quality by setting national standards for levels of contaminants in drinking water by regulating underground injection wells, and by protecting sole source aquifers.

The SDWA requires the U.S. EPA to establish maximum contamination level goals (MCLGs) and national primary drinking water regulations (NPDWR) for contaminants that, in the judgment of the U.S. EPA administrator, may cause any adverse effect on the health of persons and that are known or anticipated to occur in public water systems. The NPDWRs are to include maximum contamination levels (MCLs) and "criteria and procedures to assure a supply of drinking water that dependably complies" with such MCLs. If it is not feasible to ascertain the level of a contaminant in drinking water, the NPDWRs may require the use of a treatment technique instead of an MCL. The U.S. EPA was mandated to establish MCLGs and promulgate NPDWRs for 83 contaminants in public water systems. (The SDWA was amended in 1986 by establishing a list of 83 contaminants for which U.S. EPA was to develop MCLGs and NPDWRs.)

21.5.3 Oil Pollution Control Act

The Oil Pollution Control Act (OPA) of 1990 was enacted to expand prevention and preparedness activities, improve response capabilities, ensure that shippers and oil companies pay the costs of spills that do occur, and establish and expand research and development programs. This was primarily in response to the *Exxon Valdez* oil spill in Prince William Sound in 1989.

The OPA established a new Oil Spill Liability Trust Fund, administered by the U.S. Coast Guard. This fund replaces the fund established under the CWA and other oil pollution funds. The new act mandates prompt and adequate compensation for those harmed by oil spills and an effective and consistent system of assigning liability. The Act also strengthens requirements for the proper handling, storage, and transportation of oil, and for the full and prompt response in the event discharges occur.

21.6 WATER QUALITY ISSUES

Discuss the major water quality issues.

SOLUTION

Since ancient times, water has been considered the most suitable medium to clean, disperse, transport, and dispose of waste from human activities. The use of this resource for culinary, agricultural, industrial, and recreational needs puts an increasingly stringent demand on its management for multiple purposes. With the advent of industrialization and increasing population, the range of requirements for water has expanded as has the demand for higher quality water. Over time, water requirements have emerged for drinking and personal hygiene, fisheries, agriculture (irrigation and livestock supply), navigation for transport of goods, industrial production, cooling for power plants, hydropower generation, and recreational activities such as swimming or fishing. Fortunately, the largest demands for water quantity, such as for agricultural irrigation and industrial cooling, require the least stringent level of water quality. Drinking water supplies and specialized industrial manufacturers exert the most sophisticated demands on water quality, but their quantitative needs are relatively moderate.

Water pollution refers to any change in natural waters that may impair further use of the waters, caused by the introduction of organic or inorganic substances or by a change in temperature of the water. Wastewaters emanate from the following primary sources: municipal wastewater, industrial wastewater, agricultural runoff, and stormwater and urban runoff.

In light of the complexity of factors determining water quality, and the large number of variables involved in describing the status of water bodies in quantitative terms, it is difficult to develop a universally applicable standard that can define the baseline chemical or biological quality for water. Rather, the description of water quality is made either through quantitative measurements such as physiochemical determinations (particulate matter, biological solids, and so on) and biochemical/biological tests (BOD_5 measurement, toxicity tests, and so on), or through semiqualitative and qualitative descriptors such as biotic indices, visual aspects, species inventories, odors, and so on, BOD (biological oxygen demand) is discussed later.

The basic objective of the field of water quality management is to determine the environmental controls that must be instituted to achieve a specific environmental quality objective. A multidisciplinary approach is usually needed when a receiving water must meet competing agriculture, municipal, recreational, and industrial requirements. In many cases, a cost–benefit ratio must be established between the benefit derived from a specified water quality and the cost of achieving that quality.

Many wastewater treatment technologies have been developed for different circumstances and requirements. The most common system practiced in urban areas is the use of wastewater treatment plants based upon a combination of physical, biological, and chemical treatment steps. These are generally divided into five consecutive treatment stages:

1. *Preliminary treatment* – screening of large material.
2. *Primary treatment* – removal of settleable solids, which are separated as sludge for disposal.
3. *Secondary (biological) treatment* – biodegradable organic waste is decomposed by microorganisms.
4. *Tertiary (advanced) treatment* – further physical/chemical treatment of secondary effluents, for example, through chemical coagulation and/or filtration, for the removal of nonbiodegradable residual contaminants.
5. *Sludge treatment* – dewatering, stabilization, and ultimate disposal of sludge generated in the previous treatment steps.

In the United States, municipal treatment plants are required to treat to secondary treatment levels. The use of tertiary treatment is not common but has been increasingly based on more stringent stream standards. Increased emphasis has recently been placed upon the removal of secondary pollutants such as nutrients and organics and upon water reuse for industrial and agricultural purposes. This emphasis has generated research, both fundamental and applied, which has improved both the design and operation of wastewater treatment facilities.

In order to attain a water quality management goal, activities must be centered upon the assignment of allowable discharges to a water body so that a designated water use and water quality standards are met using the basic principles of cost-benefit engineering analysis of the effect of waste load inputs on water quality. The analysis must also include economic impacts, which in turn, must also recognize the sociopolitical constraints that are operating in the overall problem context.

Water quality standards are usually based on one of two primary criteria: stream standards or effluent standards. Stream standards are based upon receiving-water quality concentrations determined from threshold values of specific pollutants that are required to be maintained in the receiving water to maintain a specific beneficial use of the water into which the waste is being discharged. Effluent standards establish the concentration of pollutants that can be discharged (the maximum concentration of a pollutant, mg/L or the maximum load, lb/d) to a receiving water or upon the degree of treatment required for a given type of wastewater discharge, no matter where in the United States the waste originates. These effluent limitations are related to the characteristics of the discharger, not the receiving stream.

21.7 CLEAN WATER ACT AND PWPs

Describe the interrelationship between the Clean Water Act (CWA) and priority water pollutants (PWPs). With reference to the PWPs, provide the chemical formula and molecular weight for the following five pollutants: anthracene, 2-chloronapthalene, 4,4-DDT, ethylbenzene, and vinyl chloride.

SOLUTION

The Clean Water Act addresses a large number of issues related to water pollution management. Control of industrial wastewaters is primarily the responsibility of the National Pollutant Discharge Elimination System (NPDES), originally established by Public Law 92-500. Any municipality or industry that discharges wastewater in the United States must obtain a discharge permit under the regulations set forth by the NPDES. Under this system, there are three classes of pollutants (conventional pollutants, priority pollutants, and nonconventional/nonpriority pollutants). Conventional pollutants are substances such as biochemical oxygen demand (BOD), suspended solids (SS), pH, oil and grease, and coliforms. Priority pollutants were so designated on a list of 129 substances originally set forth in a consent decree between the U.S. EPA and several environmental organizations. This list was incorporated into the 1977 amendments and has since been reduced to 126 substances. Most of the substances on this list are organics, but it does include most of the heavy metals. These substances are generally considered to be toxic. However, the toxicity is not absolute; it primarily depends on the concentration. In recent years, pollution prevention programs have been implemented to reduce their use in industrial processes. The literature [9] provides extensive details on these priority (water) pollutants. The third class of pollutants could include any pollutant not in the first two categories. Examples of substances that are presently regulated in the third category are nitrogen, phosphorous, and sodium.

The chemical formula (and molecular weight) for each pollutant is provided below [9].

Anthracene: $C_{14}H_{10}$ (178.2)
2-Chloronapthalene: $C_{10}H_7Cl$ (162.6)
4,4-DDT: $C_{14}H_9Cl_5$ (354.5)
Ethylbenzene: C_8H_{10} (106.18)
Vinyl chloride: $CH_2:CHCl$ (62.5)

21.8   WASTEWATER COMPOSITION

Define the following regulatory terms used as water quality descriptors and briefly explain their differences related to the methods used for their determination.

1. BOD
2. COD
3. TOC
4. Solids analysis

SOLUTION

1. The biochemical oxygen demand (BOD) test is used to determine the amount of biodegradable contents in a sample of wastewater. As discussed in the introductory section, the BOD test measures the amount of oxygen consumed by living organisms (mainly bacteria) as they metabolize the organic matter present in a waste. The test simulates conditions as close as possible to those that occur naturally [10]. As is true for any bioassay, the success of a BOD test depends on the control of such environmental and nutritional factors as pH and osmotic conditions, essential nutrients, constant temperature, and population or organisms representative of natural conditions.

   Although it theoretically takes an infinite amount of time for all the oxidizable material in a sample of water to be consumed, it has been empirically determined that a period of 20 days is required for near completion. Because the 20-day waiting period is too long to wait in most cases, a 5-day incubation period (the time in which 70 to 80% of available material is usually oxidized) has been adopted as standard procedure. This shorter test is referred to as BOD5. The BOD5 test also serves the purpose of avoiding the contribution of nitrifying bacteria to the overall DO (dissolved oxygen) measurement, since these bacteria do not become large enough to make a significant oxygen demand until about 8 to 10 days from the start of the BOD test [11].

2. While the BOD test is the only test presently available that gives a measure of the amount of biologically oxidizable organic material present in a body of water, it does not provide an accurate picture of the total amount of overall oxidizable material and thus the total oxygen demand. For such measurements, the *chemical oxygen demand* (COD) *test* is used. In this test, a very strong oxidizing chemical, usually potassium dichromate, is added to samples of different dilution. To ensure full oxidation of the various compounds found in the samples, a strong acid and a chemical catalyst are added. One of the products of this oxidation–reduction reaction is the chromate ion, which gives a very sharp color change that can be easily detected in a spectrophotometer. Thus, the greater the absorbance measured (inverse of transmittance), the greater the amount of oxidation that has taken place; also, the more oxidizable material originally present, the greater the oxygen demand of that material. To find the corresponding concentration of DO in the sample, the absorbance measured is related to an absorbance–concentration graph of a standard whose oxygen concentration is known.

3. A third method for measuring the organic matter present in a wastewater is the total organic carbon (TOC) test. The test is performed by placing a sample into a high-temperature furnace or chemically oxidizing environment. The organic carbon is oxidized to carbon dioxide. The carbon dioxide that is produced can then be measured. While the TOC test does not directly measure the concentration of organic compounds, it does not provide a direct measurement of the rate of reaction or the degree of biodegradability. For this reason, the TOC test has been accepted as a monitoring technique but has not been utilized in the establishment of treatment regulations.

    The BOD, COD, and TOC tests provide estimates of the general organic content of a wastewater. However, because the particular composition of the organics remains unknown, these tests do not reflect the response of the wastewater to various types of biological treatment technologies. It is therefore necessary to separate the wastewater into its specific components.

4. The solids analysis focuses on the quantitative investigation and measurement of the specific content of solid materials present in a wastewater. The total solids are the materials that remain after the water from the solution has evaporated. For any solution, the total solids consist of suspended and dissolved particles. Dissolved particles are classified according to size as either soluble or colloidal.

    When a solution is poured through a filter (usually 0.45 μm in size), those solids that are too large to pass through the filter openings will be retained.

These solids are called suspended solids. The filter containing the suspended solids is heated to remove the water and to produce the total suspended solids (TSS). The TSS are further heated in a muffle furnace at 550°C. At such high temperatures, the volatile suspended solids (VSS), mostly organics, will be vaporized, leaving behind the remaining fixed suspended solids. Those solids that are small enough to pass through the filter are called dissolved (soluble) solids. By heating the solution to remove the water, the total dissolved solids are obtained. As with the suspended solids, these solids are further heated at extreme temperatures to remove the volatile dissolved solids (VDS), leaving the fixed dissolved solids.

The assumption is made that all volatile solids (VSS and VDS) are organic compounds. A distinction is made as to whether such solids may or may not be adsorbed by a specific treatment medium. The soluble organics that are nonsorbable are further separated into degradable and nondegradable constituents.

21.9   SOLID WASTE MANAGEMENT ISSUES

Discuss solid waste issues from a regulatory perspective.

SOLUTION

Disposal of the solid waste produced by society, including industries and homes, has drawn a great deal of public attention. The term solid waste includes all of the heterogeneous mass of throw-aways from urban areas, as well as the more homogeneous accumulation of agricultural, industrial, and mineral wastes. In urban communities the accumulation of solid waste is a direct and primary consequence of life. This accumulation adds up to approximately 4 lb of solid waste per person every day in the United States.

When the population was smaller, the available landfill space was seemingly unlimited, the waste materials predominantly natural (as opposed to synthetic), and the understanding of possible harmful effects minimal, public concern regarding solid waste management was usually limited to local issues of dependable collection and mitigation of nuisance factors such as truck traffic and odor. However, times have changed. The population has increased, the use of synthetic materials is widespread, and many regions in the United States have only limited landfill space remaining. Advances have been made in the understanding of the chemical and physical properties of waste materials and the changes they undergo in the natural environment. Public acceptance for recycling has increased and the tolerance for risks has decreased.

The need for a nationally based solid waste management system became apparent in the 1970s. Properly dealing with a large amount of complex waste is what solid waste management is all about. In 1976, the Resource Conservation and Recovery Act (RCRA) was enacted to deal with the problems of solid and hazardous waste collection, transport, treatment, storage, and disposal in the United States. The goals of RCRA are:

1. To protect human health and the environment;
2. To reduce waste and conserve energy and natural resources; and,
3. To reduce or eliminate the generation of hazardous waste as expeditiously as possible.

Subtitle D of RCRA applies to municipal solid waste facilities and defines municipal solid waste as follows:

1. Garbage: milk cartons, coffee grounds, etc.
2. Refuse: metal scraps, wallboard, etc.
3. Sludge: from a waste treatment plant or a pollution control facility.
4. Other discarded materials, including solids, semi-solids, liquids, or contained gaseous materials resulting from industrial, commercial, mining, agricultural, and domestic sources, e.g., boiler slag, fly ash, etc., that is not otherwise considered hazardous,

Subtitle C applies to hazardous waste management, and requires all generators, transporters, storers, treaters, and disposers of hazardous waste to comply with a complex set of administrative requirements for record keeping, conformance to manifest systems, development of plans and training for securing facility permits, and to provide financial assurances to cover closure costs for their facilities.

The predominant method for municipal and hazardous waste disposal in the United States remains landfilling, despite the emphasis that RCRA has on waste minimization and recycling. Current U.S. EPA initiatives on pollution prevention have encouraged the analysis of alternatives to landfilling of solid waste. Recently, these alternative and combined methods for solid waste management have received considerable attention, achieved some success and simultaneously introduced other concerns. For example, incineration of solid waste greatly reduces the need for landfill space and, if so designed, can recover some of the energy value of the waste being treated. However, incineration can produce harmful air quality effects and generates an ash that concentrates heavy metals that may require secure landfill disposal. Composting of organic materials is another alternative to landfilling that reduces volume and produces a useful soil container. It, however, requires significant space and operator attention, and can generate undesirable odor problems.

Additional information on solid waste, particularly as it applies to regulations, is available in the literature [12–15].

21.10 HAZARDOUS WASTE INCINERATOR

By federal law, hazardous waste incinerators must meet a minimum destruction and removal requirement, i.e., that the principal organic hazardous constituents

(POHCs) of the waste feed be incinerated with a minimum destruction and removal efficiency (DRE) of 99.99% ("four nines"). The destruction and removal efficiency is the fraction of the inlet mass flow rate of a particular chemical that is destroyed and removed in the incinerator or, equivalently,

$$DRE = 1 - \frac{\text{mass flow of chemical out}}{\text{mass flow of chemical in}}$$

A feed stream to an incinerator contains 245 g/hr of a nanoorganic. Calculate the maximum outlet flow rate of the organic from the incinerator by law.

SOLUTION

Substituting data given in the problem statement into the DRE equation provides the means of determining an acceptable mass flow rate of the nanoorganic out of the incinerator as follows:

$$DRE = 1 - \frac{\text{mass flow of chemical out}}{\text{mass flow of chemical in}} = 0.9999$$

$$= 1 - \frac{\text{mass flow of chemical out}}{245 \text{ g/hr}} = 0.9999$$

$$\text{Mass flow of chemical out} = (1 - 0.9999)(245 \text{ g/hr})$$
$$= (0.0001)(245 \text{ g/hr}) = 0.245 \text{ g/hr}$$

21.11 NANOTECHNOLOGY ENVIRONMENTAL REGULATIONS OVERVIEW

Provide an overview of nanotechnology environmental regulations.

SOLUTION

Completely new legislation and regulatory rulemaking may be necessary for environmental control of nanotechnology. However, in the meantime, one may speculate on how the existing regulatory framework might be applied to the nanotechnology area as this emerging field develops over the next several years. One experienced Washington DC attorney has done just that, as summarized below [16–18]. The reader is encouraged to consult the cited references as well as the text of the laws that are mentioned and the applicable regulations derived from them.

Commercial applications of nanotechnology are likely to be regulated under TSCA, which authorizes the U.S. EPA to review and establish limits on the manufacture, processing, distribution, use, and/or disposal of new materials that EPA determines to pose "an unreasonable risk of injury to human health or the environment." The term chemical is defined broadly by TSCA. Unless qualifying for an exemption under the law (a statutory exemption requiring no further approval by EPA), low-volume production, low environmental releases along with low volume, or plans for limited test marketing, a prospective manufacturer is subject to the full-blown procedure. This requires submittal of said notice, along with toxicity and other data to EPA at least 90 days before commencing production of the chemical substance.

Approval then involves recordkeeping, reporting, and other requirements under the statute. Requirements will differ, depending on whether EPA determined that a particular application constitutes a "significant new use" or a "new chemical substance." The EPA can impose limits on production, including an outright ban when it is deemed necessary for adequate protection against "an unreasonable risk of injury to health or the environment." The EPA may revisit a chemical's status under TSCA and change the degree or type of regulation when new health/environmental data warrant. It was expected to be issuing several new TSCA test rules in 2004 [19]. If the experience with genetically engineered organisms is any indication, there will be a push for EPA to update regulations in the future to reflect changes, advances, and trends in nanotechnology [20].

Workplace exposure to a chemical substance and the potential for pulmonary toxicity is subject to regulation by the OSHA, including the requirement that potential hazards be disclosed on a MSDS (Material Safety Data Sheet). (An interesting question arises as to whether carbon nanotubes, chemically carbon but with different properties because of their small size and structure, are indeed to be considered the same as or different from carbon black for MSDS purposes.) Both governmental and private agencies can be expected to develop the requisite threshold limit values (TLVs) for workplace exposure. (See Problem 22.4 for additional information of TLVs). Also, the U.S. EPA may once again utilize TSCA to assert its own jurisdiction, appropriate or not, to minimize exposure in the workplace. Furthermore, NIOSH (National Institute for Occupational Safety and Health) was anticipated to provide workplace guidance for nanomanufactures and their employees in 2005; this has not occurred as of the writing of this book. This was almost definitely wishful thinking given the past performance of similar bureaucratic agencies, e.g., the U.S. EPA. Adding to NIOSH's dilemma is the breadth of the nano field and the lack of applicable toxicology (see Chapter 22) and epidermiology (see Chapter 27) data.

Another likely source of regulation would fall under the provisions of the Clean Air Act (CAA) for particulate matter less than 2.5 μm ($PM_{2.5}$). Additionally, an installation manufacturing nanomaterials may ultimately become subject as a "major source" to the CAA's Section 112 governing hazardous air pollutants (HAP).

Wastes from a commercial-scale nanotechnology facility would be classified under RCRA, provided that it meets the criteria for RCRA waste. RCRA requirements could be triggered by a listed manufacturing process or the act's specified hazardous waste characteristics. The type and extent of regulation would depend on how much hazardous waste is generated and whether the wastes generated are treated, stored, or disposed of on site.

21.12   NANOTECHNOLOGY OPPONENTS

Briefly describe the regulatory process that opponents of nanotechnology could pursue.

SOLUTION

Opponents of nanotechnology may be able to use the National Environmental Policy Act (NEPA) to impede nanotechnology research funded by the U.S. government. A "major Federal action significantly affecting the quality of the human environment" is subject to the environmental impact provision under NEPA. (Various states also have environmental impact assessment requirements that could delay or put a stop to construction of nanotechnology facilities.) Time will tell whether this path will be followed.

22 Toxicology

Toxicology is the science of poisons. It has also been defined as the study of chemical or physical agents that produce adverse responses in biological systems. Together with other scientific disciplines (such as epidemiology, the study of the cause and distribution of disease in human populations, and risk assessment), toxicology can be used to determine the relationship between an agent of interest and a group of people or a community [5, 6].

Of interest to an engineer and scientist are the regulatory and environmental applications of the discipline. The former is of use in interpreting the setting of standards for allowable exposure levels of a given contaminant or agent in an ambient or occupational environment; the latter is of use in estimating the persistence and movement of an agent in a given environment. Both applications can be of direct use to risk assessment activities, and both regulatory toxicology and environmental toxicology closely involve other branches of the discipline. The relationship is particularly close for the regulatory toxicologist who depends largely on the products of descriptive toxicology when making decisions on the risk posed by a specific agent [5, 6].

In a very real sense, the science of toxicology will be significantly impacted by nanotechnology as discussed in the Introduction to Chapter 7. Unique properties cannot be obtained for particles in the nanosize range since properties vary with particle size. This element also applies to toxicological properties. In effect, a particle of one size could be carcinogenic while a particle (of the same material) of another size would not be carcinogenic. Alternately, two different sized nanoparticles of the same substance could have different threshold limit values (TLVs), a topic discussed in Problems 22.4 and 22.6. This problem has yet to be resolved by the toxicologist.

In conclusion, the purpose of the toxicity assessment is to weigh available evidence regarding the potential for particular contaminants to cause adverse effects in exposed individuals and to provide, where possible, an estimate and the increased likelihood and/or severity of adverse effects. Note, as indicated in the introduction to this Part 4, some overlap exists between this Chapter and Chapters 23 to 25.

22.1 THE SCIENCE OF TOXICOLOGY

Discuss the science of toxicology [3, 5, 6].

Nanotechnology: Basic Calculations for Engineers and Scientists, by Louis Theodore
Copyright © 2006 John Wiley & Sons, Inc.

SOLUTION

Toxicology is the science dealing with the effects, conditions, and detection of toxic substances or poisons. Six primary factors affect human response to toxic substances or poisons. These are detailed below:

1. *The chemical itself* – some chemicals produce immediate and dramatic biological effects, whereas others produce no observable effects or produce delayed effects.
2. *The type of contact* – certain chemicals appear harmless after one type of contact (e.g., skin), but may have serious effects when contacted in another way (e.g., lungs).
3. *The amount (dose) of a chemical* – The dose of a chemical exposure depends upon how much of the substance is physically contacted.
4. *Individual sensitivity* – humans vary in their response to chemical substance exposure. Some types of responses that different persons may experience at a certain dose are serious illness, mild symptoms, or no noticeable effect. Different responses may also occur in the same person at different exposures.
5. *Interaction with other chemicals* – toxic chemicals in combination can produce different biological responses than the responses observed when exposure is to one chemical alone.
6. *Duration of exposure* – some chemicals produce symptoms only after one exposure (acute), some only after exposure over a long period of time (chronic), and some may produce effects from both kinds of exposure.

22.2   TOXICOLOGY CLASSIFICATIONS

Prepare a table outlining the different types of toxicology and their corresponding intended applications [21].

SOLUTION

There are many different types of toxicology. Seven of these classes are provided in Table 22.1.

22.3   ROUTES OF EXPOSURE

Briefly discuss the various routes by which a chemical can enter the body [3].

TABLE 22.1   Types of Toxicology

Type	Purpose
Clinical toxicology	To determine the effects of chemical poisoning and the treatment of poisoned people
Descriptive toxicology	To test the toxicology of chemicals
Environmental toxicology	To determine the environmental fate of chemicals and their ecological and health effects
Forensic toxicology	To answer medicolegal questions about health effects
Industrial toxicology	To determine health effects of occupational exposures
Mechanistic toxicology	To describe the biochemical mechanisms that cause health effects
Regulatory toxicology	To assess the risk involved in marketing chemicals and products, and establish their subsequent regulation by government agencies

SOLUTION

To protect the body from hazardous chemicals, one must know the route of entry into the body; these include inhalation, ingestion, and absorption (skin). All chemical forms may be inhaled. After a chemical is inhaled into the mouth, it may be ingested, absorbed into the bloodstream, or remain in the lungs. Various types of personal protective equipment (PPE) such as dust masks and respirators prevent hazardous chemicals from entering the body through inhalation. Ingestion of chemicals can also be prevented by observing basic housekeeping rules, such as maintaining separate areas for eating and chemical use or storage, washing hands before handling food products, and removing gloves when handling food products. Wearing gloves and protective clothing prevents hazardous chemicals from entering the body through skin absorption.

After a chemical has entered the body, the body must break it down or metabolize it, the body may excrete it, or the chemical may remain deposited in the body.

The route of entry of a chemical is often determined by the physical form of the chemical. Physical chemical forms and the routes of entry are listed in Table 22.2.

22.4   THRESHOLD LIMIT VALUE (TLV)

Define TLV and describe the three TLV categories.

TABLE 22.2   Chemical Forms and Routes of Entry

Chemical Form	Principle Danger
Solids and fumes	Inhalation, ingestion, and skin absorption
Dusts and gases	Inhalation into lungs
Liquids, vapors, and mists	Inhalation of vapors and skin absorption

SOLUTION

The threshold limit value (TLV) is a term used to express the maximum airborne concentration of a material to which anyone can be exposed to every day without experiencing adverse effects. The TLV can be categorized in three ways:

1. TLV-TWA is defined as the *time-weighted average* concentration for a normal 8-hour workday and a 40-hour work week, to which nearly all workers may be repeatedly exposed, day after day, without any long-term adverse effect.
2. TLV-STEL is defined as the concentration to which workers can be exposed for a short period of time without suffering from irritation, chronic tissue damage, or mucosis of sufficient degree to increase the likelihood of accidental injury. The *short term exposure limit* (STEL) is a 15-minute time-weighted average exposure, which should *not* be exceeded at any time during a workday.
3. Another concentration that is frequently referred to is the concentration that is *immediately dangerous to life or health* (IDLH). This concentration is the level at or below which a person exposed for 30 minutes will not lose control and will be able to put on protective equipment or take other protective action. (See the next Problem for further details on IDLH.) TLV-C is defined as the threshold limit concentration ceiling value that should *not* be exceeded even for an instant. The American Conference of Governmental Industrial Hygienists (ACGIH) and the Occupational Safety and Health Administration (OSHA) state that a ceiling value should not be exceeded even instantaneously. The National Institute for Occupational Safety and Health (NIOSH) also uses ceiling values. However, its ceiling values are similar to a STEL.

Values of TWA are available in the literature [22]. For example, the TWAs for sulfur dioxide and xylene are 2 ppm (5 mg/m^3) and 100 ppm (435 mg/m^3), respectively. The corresponding STELs are 5 ppm (10 mg/m^3) and 150 ppm (655 mg/m^3).

Most toxicity studies are performed by using test animals. Humans obviously cannot be exposed to lethal concentrations of toxic materials to determine toxicity. The reader should note that there are some differences in chemical tolerance levels for humans and animals. The differences include metabolism and other factors. Thus, toxicity tests are not always easy to interpret. However, the results of animal toxicity tests are used to guide the selection of acceptable exposure limits for humans. This topic will receive additional attention in Chapters 23 to 25.

22.5   TOXICOLOGY TERMINOLOGY

Define the following toxicology terms:

1. Permissible exposure limit (PEL)
2. Immediately dangerous to life and health (IDLH)

3. Lethal dose (LD)
4. Effective dose (ED)
5. Toxic dose (TD)
6. Lethal concentration (LC)

SOLUTION

Permissible exposure limits (PELs) are extracted from the TLVs and other standards including standards for benzene and 13 carcinogens. Since OSHA is a regulatory agency, its PELs are legally enforceable standards and apply to all private industries and federal agencies. See the next Problem for additional details.

The IDLH (immediately dangerous to life and health) is the maximum concentration of a substance to which a human can be exposed for 30 min without experiencing irreversible health effects (see previous Problem).

Dosages of a chemical can be described as a lethal dose (LD), effective dose (ED), or toxic dose (TD). The LD50 or LD_{50} is a common parameter used in toxicology. It represents the dose at which 50% of a test population would die when exposed to a chemical at that dose. Similarly, the lethal concentration (LC) is the concentration of a substance in air that will cause death.

22.6 TLV VERSUS PEL

Explain the difference between TLV and PEL.

SOLUTION

The ACGIH developed the TLVs in the 1960s and used them as guidelines for exposures through inhalation in industry. OSHA developed the PELs; they were extracted from the TLVs and other standards in 1971. However, since OSHA is a regulatory agency, their PELs are legally enforceable standards and apply to all private industries and Federal agencies.

One should note that for each individual chemical, the TLV is generally equal to the PEL, i.e., the PEL is synonymous in most applications with the TLV-TWA.

22.7 TOXICITY FACTORS

Define and compare the following pairs of parameters used in toxicology [1, 3].

1. NOEL and NOAEL
2. LOEL and LOAEL
3. ADI and RfD

SOLUTION

The parameters NOEL and NOAEL, LOEL and LOEAL, and ADI and RfD are used to establish thresholds.

The NOEL (*no observed effect level*) is the highest dose of the toxic substance that will not cause an effect. The NOAEL (*no observed adverse effect level*) is the highest dose of the toxic substance that will not cause an adverse effect.

The LOEL (*lowest observed effect level*) is the lowest dose of the toxic substance tested that shows effects. The LOAEL (*lowest observed adverse effect level*) is the lowest dose of the toxic substance tested that shows adverse effects. The LOEL and LOAEL give no indication of individual variation in susceptibility.

The ADI (*acceptable daily intake*) is the level of daily intake of a particular substance that will not produce an adverse effect. The RfD (*reference dose*) is an estimate of the daily exposure level for the human population. The RfD development follows a stricter procedure than that followed for the ADI. The ADI approach is used extensively by the U.S. Environmental Protection Agency (EPA).

When using data for LOELs, LOAELs, NOELs, or NOAELs, it is important to be aware of their limitations. Statistical uncertainty exists in the determination of these parameters due to the limited number of animals used in the studies to determine the values. In addition, any toxic effect might be used for the NOAEL and LOAEL so long as it is the most sensitive toxic effect and considered likely to occur in humans.

22.8 OSHA AND NIOSH

The *Occupational Safety and Health Act* (OSHA) enforces basic duties that must be carried out by employers. Discuss these basic duties. Also, state the major roles of the *National Institutes of Safety and Health* (NIOSH) and the *Occupational Safety and Health Administration* (OSHA) [1, 3].

SOLUTION

Employers are bound by OSHA to provide each employee with a working environment free of recognized hazards that cause or have the potential to cause physical harm or death. Employers must have proper instrumentation for the evaluation of test data provided by an expert in the area of toxicology and industrial hygiene. This instrumentation must be obtained because the presence of health hazards cannot be evaluated by visual inspection. This data collection effort provides the employer with substantial evidence to disprove invalid complaints by employees alleging a hazardous working situation. This law also gives employers the right to take full disciplinary action against those employees who violate safe working practices in the workplace.

NIOSH recommends standards for industrial exposure that OSHA uses in its regulations. OSHA has the power to enforce all safety and health regulations and standards recommended by NIOSH.

22.9 TOXICOLOGY DETERMINATION

Calculating toxicological effects attributable to an environmental contaminant often begins with the following simple equation:

Health risk = (Human exposure)(Potency of contaminant)

Identify several factors that may lead to an "uncertain" calculation of the actual risk to human health of a particular chemical [3].

SOLUTION

Uncertainties associated with determining the "human exposure" term include the following:

1. Limited knowledge of source characteristics, e.g., how much of a contaminant is released and for how long the release has occurred.
2. Difficulty in describing and calculating the "fate and transport" of the contaminant as it travels from the release point in the environment to the exposed population, i.e., the receptor (sometimes referred to as pathway analysis).
3. Mobility of the exposed individual or population, thereby constantly changing the individual's exposure.

Uncertainties associated with determining the "potency of contaminant" term include the following:

1. Potency factors are often based on animal toxicity studies and then applied to humans.
2. Variable effects on exposed humans due to differences in age, sex, health condition, and so on.
3. Extrapolation from measured high-dose effects to determine low-dose effects.

The "uncertainty" topic is reviewed again in Chapters 23 and 24.

22.10 IDLH AND LETHAL LEVEL

As described earlier, the *immediately dangerous to life and health* (IDLH) level is the maximum concentration of a substance to which one can be exposed for 30 min without irreversible health effects or death. A lethal level is the concentration at which death is almost certain to occur.

Carbon dioxide is not normally considered to be a threat to human health. It is exhaled by humans and is found in the atmosphere at about 3000 parts per million (ppm). However, at high concentrations it can be a hazard and may cause headaches, dizziness, increased heart rate, asphyxiation, convulsions, or coma.

Two large bottles of flammable solvent were ignited by an undetermined ignition source after being knocked over and broken by a janitor while cleaning a 10 ft × 10 ft × 10 ft nanoresearch laboratory. The laboratory ventilator was shut off and the fire was fought with a 10 lb CO_2 fire extinguisher. As the burning solvent had covered much of the floor area, the fire extinguisher was completely emptied in extinguishing the fire.

The IDLH level for CO_2 set by NIOSH is 50 000 ppm. At that level, vomiting, dizziness, disoriented, and breathing difficulties occur after a 30 min exposure. At a 10% level (100 000 ppm), death can occur after a few minutes even if the oxygen in the atmosphere would otherwise support life.

Calculate the concentration of CO_2 in the room after the fire extinguisher is emptied. Does it exceed the IDLH value? Assume that the gas mixture in the room is uniformly mixed, that the temperature in the room is 30°C (warmed by the fire and above the normal room temperature of 20°C), and that the ambient pressure is 1 atm [3, 6].

SOLUTION

First, calculate the number of moles of CO_2, n_{CO_2}, discharged by the fire extinguisher:

$$n_{CO_2} = (10 \text{ lb } CO_2)(454 \text{ g/lb})/(44 \text{ g/gmol } CO_2)$$
$$= 103 \text{ gmol of } CO_2$$

The volume of the room, V, is

$$V = (10 \text{ ft})(10 \text{ ft})(10 \text{ ft})(28.3 \text{ L/ft}^3)$$
$$= 28\,300 \text{ liters}$$

The ideal gas law is used to calculate the total number of moles of gas in the room,

$$n = \frac{PV}{RT}$$
$$= \frac{(1 \text{ atm})(28\,300 \text{ L})}{\left(0.08206 \frac{\text{atm} \cdot \text{L}}{\text{gmol} \cdot \text{K}}\right)(303 \text{ K})}$$
$$= 1138 \text{ gmol gas}$$

The concentration or mole fraction of CO_2 in the room, y_{CO_2}, may now be calculated:

$$y_{CO_2} = (\text{gmol } CO_2)/(\text{gmol gas})$$
$$= (103 \text{ gmol } CO_2)/(1138 \text{ gmol gas})$$
$$= 0.0905$$
$$= 9.05\%$$

The IDLH level is 5.0% and the lethal level is 10.0%. Therefore, the level in the room of 9.05% does exceed the IDLH level for CO_2. It is also dangerously close to the lethal level. The person extinguishing the fire is in great danger and should take appropriate safety measures.

If a dangerous level is present, consideration must be given to using protective equipment such as a respirator. Respirators protect the individual from harmful materials in the air. Air purifying respirators will clean the air but will not protect users against an oxygen-deficient atmosphere. Thus, air-purifying respirators are not used in IDLH applications. The only respirators that are recommended for fighting fires are self-contained breathing apparatuses with full facepieces. Recommendations for the selection of the proper respirator are based on the most restrictive of the occupational exposure limits.

22.11 CHEMICAL EXPOSURE

New York City obtains the majority of its drinking water from several reservoirs in the Catskill region of New York State. At an abandoned site approximately 0.25 miles from one such reservoir, several dozen leaking barrels of two nanoderived chemicals have been discovered in a buried trench.

Given the following limited information for two of these chemicals, discuss each compound in relative terms of its potential hazards to: (1) remediation workers involved in sampling and cleanup of the site, and (2) ground and surface water supplies.

Assume the barrels have corroded and have been leaking slowly over several months. The site consists of porous soil, is at a slightly higher elevation than the reservoir, and is situated over a groundwater source.

Data for the two chemicals are provided in Table 22.3.

SOLUTION

Refer to Table 22.3. Both chemicals are present in liquid form in the trench. Both would also be expected to be present in the vapor state, with LT being the most volatile due to its high Henry's law constant. The principle dangers to workers, therefore, would be from inhalation of vapors and skin absorption of contacted liquids.

TABLE 22.3 Data for Problem 22.11

Chemical	Form	Water Solubility (mg/L)	Henry's Law Constant [atm-m^3/(gmol)]	Density (g/mL)
Nano-LT	Liquid	6900	0.003	1.32
Nano-MKT	Liquid	1.1	2.40	0.91

The higher a compound's water solubility, the more easily the compound will disperse within a water source. The LT would be expected to spread rapidly upon reaching groundwater and would be most likely to find its way into the reservoir. The MKT has a low water solubility. This compound would not travel large distances in stagnant water.

The compound with a relatively low Henry's law constant (LT) has the lowest potential volatility. It would not evaporate readily and would therefore present a higher risk to the ground and surface waters. The MKT has the highest relative volatility and would present the least risk to ground and surface water.

LT's density is greater than that of water, that is, greater than 1.0 g/mL. As such, this compound would be expected to sink to the bottom of the ground and surface water bodies. The MKT has a density less than that of water and would be expected to float on the ground and surface water. Chemicals sinking to and settling on the bottom of a water body generally present a lesser risk to humans and animal life unless disturbed. Chemicals that float on water surfaces are more likely to contact human and animal life because they can be carried large distances by winds and currents.

22.12   THRESHOLD LIMIT VALUES

The dynamic seal on a control valve suddenly starts leaking inert nitrogen containing toxic nanoparticles at a concentration of 0.02 g/cm^3 at $400 \text{ cm}^3/\text{h}$ into a $12 \text{ ft} \times 12 \text{ ft} \times 8 \text{ ft}$ high room. The air in the room is uniformly mixed by a ceiling fan. The background particle concentration is 1.0 ng/cm^3. The air temperature and pressure are $77°F$ and 1 atm, respectively. Calculate the number of minutes after the leak starts that a person sleeping on the job would be at risk of being exposed to this toxic nanoparticle with respect to the STEL. Also calculate the average exposure over an 8 h period relative to the TLV.

The exposure limits have been recently estimated as flows:

1. Short-term exposure limit (STEL) $= 200 \text{ ng/cm}^3$
2. Threshold limit value (TLV) $= 50 \text{ ng/cm}^3$

SOLUTION

1. The particle concentration in the room, C (ng/cm^3), as a function of time, t (min), is calculated as follows:

$$C(\text{ng/cm}^3) = 1.0 + \frac{\left(400 \dfrac{\text{cm}^3}{\text{h}}\right)\left(\dfrac{\text{h}}{60 \text{ min}}\right)\left(\dfrac{\text{m}^3}{10^6 \text{ cm}^3}\right)\left(\dfrac{0.02 \text{ g}}{\text{cm}^3}\right)\left(\dfrac{10^9 \text{ ng}}{\text{g}}\right)(t)}{(12 \text{ ft})(12 \text{ ft})(8 \text{ ft})\left(\dfrac{0.02832 \text{ m}^3}{\text{ft}^3}\right)}$$

$$= 1.0 + 4.087t; \qquad t \equiv \text{minutes}$$

Note that t is the time from the start of the leak, and 1.0 ng/cm^3 is the background concentration in the room.

The time, t, to reach the STEL of 200 ng/cm^3 is obtained by using the mean value theorem since a time-weighted average is required.

$$\int_{t_1}^{t_2} C \, dt = C_{ave}(t_2 - t_1)$$

where C_{ave} = average concentration or STEL; $(t_2 - t_1)$ = range of the time-weighted value, in this case 15 minutes.

Integrating the equation yields the following solution:

$$\int_{t_1}^{t_2} (1.0 + 4.087)dt = (200)(15)$$

$$\left(1.0t + \frac{4.087t^2}{2}\right)\Big|_{t_1}^{t_2} = 3000$$

$$(t_2 - t_1) + (2.044)(t_2^2 - t_1^2) = 3000$$
$$(15) + (2.044)(t_2 - t_1)(t_2 + t_1) = 3000$$
$$(15) + (2.044)(15)(t_2 + t_1) = 3000$$

Noting that the STEL applies over a 15-minute interval, one may substitute $t_2 = 15 + t_1$

$$15 + (30.66)(15 + 2t_1) = 3000$$
$$t_1 = 41.18 \, \text{min}$$

The total time to reach the STEL of 200 ng/cm^3 on a TWA basis is $15 + t_1$, or $t_2 = 56.18$ min. Thus, the STEL is achieved over the 15-minute time interval between 41.2 and 56.2 minutes.

2. For an 8 hour (480 min) period, the average concentration is given by

$$C_A = \frac{1.0 + (1.0 + 4.087t)}{2}$$

$$= 1.5 + 2.044t$$
$$= 1.5 + (2.044)(480)$$
$$= 983 \, \text{ng/cm}^3$$

The TLV is exceeded by nearly a factor of 20. Alternately, the time at which the TLV is exceeded is given by:

$$50 = 1.0 + 2.044t$$
$$t = 24 \, \text{min}$$

23 Noncarcinogens

There is significant interest in health risk calculations/estimations of both noncarcinogens and carcinogens; this interest will almost certainly become magnified in the future with certain nanochemicals. The present approach to these determinations are examined in this and the next Chapter. The reader should note that much of this material is an extension of that presented in Chapter 22.

Webster describes a carcinogen as "any substance that produces cancer." In a very general sense, a noncarcinogen may be defined as a substance that is not a carcinogen, yet harmful to one's health. The term carcinogen is most commonly applied to certain chemical agents introduced into the environment by human activity. Researchers and regulatory personnel label a substance a carcinogen if it causes a statistically significant increase in abnormal cell growth in previously unexposed organisms. The modes of cancer initiation are still not fully understood and the question of usefulness of extrapolating laboratory tests on animals to assess human health risks is particularly complex.

Many of the nanochemicals to be developed in the future will undoubtedly be classified as either carcinogens or noncarcinogens. The analysis that will be applied to these new chemicals will almost certainly be based on existing procedures. This Chapter provides a dozen worked problems for noncarcinogenic chemicals; the next Chapter will treat carcinogenic agents.

The quantitative approach used to describe the potential for noncarcinogenic toxicity to occur in an individual is not expressed as the probability of an individual suffering an adverse effect. (The U.S. EPA does not at the present time use a probabilistic approach to estimate the potential for noncarcinogenic health effects.) Rather the potential for noncarcinogenic effects is evaluated by comparing an exposure level over a specified period of time (e.g., a lifetime) with a reference dose derived for a similar exposure period. This ratio of exposure to toxicity is called a hazard quotient and is described in more detail in Problem 23.1. The noncancer hazard quotient assumes that there is a level of exposure below which it is unlikely for even sensitive populations to experience adverse health effects.

23.1 HAZARD QUOTIENT

Describe the hazard quotient for noncarcinogens.

Nanotechnology: Basic Calculations for Engineers and Scientists, by Louis Theodore
Copyright © 2006 John Wiley & Sons, Inc.

SOLUTION

The noncancer hazard quotient assumes that there is a level of exposure (i.e., RfD) below which it is unlikely for even sensitive populations to experience adverse health effects.

$$\text{Noncancer Hazard Quotient} = \text{EL/RfD}$$

where EL = exposure level (or intake); RfD = reference dose; and EL and RfD are expressed in the same units.

If the EL exceeds this threshold (i.e., EL/RfD > 1.0), there may be concern for potential noncancer effects. As a rule, the greater the value of EL/RfD above unity, the greater the level of concern. However, one should not interpret ratios of EL/RfD as statistical probabilities; a ratio of 0.001 does not mean that there is a one in one thousand chance of the effect occurring. Further, it is important to emphasize that the level of concern does not increase linearly as the RfD is approached or exceeded, because RfDs do not have equal accuracy or precision and are not based on the same severity of toxic effects.

23.2 REFERENCE DOSE

Describe and illustrate the process of correlating a reference dose (RfD) with a schematic of a dose–response curve. Label both axes and the critical points on the curve [3].

SOLUTION

A reference dose (RfD) estimates the lifetime dose that does not pose a significant risk to the human population. This estimate may have an uncertainty of one order of magnitude or more. The RfD is determined by dividing the no observed adverse effect level (NOAEL) dose of a substance by the product of the uncertainty and modifying factors as shown in the following equation

$$\text{RfD} = \frac{\text{NOAEL}}{(\text{UF})(\text{MF})}$$

The uncertainty factor (UF) is usually represented as a multiple of 10 to account for variation in the exposed population (to protect sensitive subpopulations), uncertainties in extrapolating from animals to humans, uncertainties resulting from the use or subchronic data instead of data obtained from chronic studies, and uncertainties resulting from the use of the lowest observable adverse effect level (LOAEL) instead of the NOAEL. The modifying factor (MF) reflects qualitative professional judgment of additional uncertainties in the data. These effects are discussed in more detail in Problem 23.10.

The schematic of the dose–response curve shown in Figure 23.1 illustrates that the value of the reference dose is less than the value of the NOAEL by a safety factor.

A word of interpretation is in order for the dose–response relationship provided in Figure 23.1. From a dimensional analysis point of view, the dose – with unit of mg/kg-day – can be viewed as the intake rate in mg/day per kg mass of the receptor. The response may be viewed as a dimensionless probability term that could represent the number of individuals adversely affected by the dose relative to a given population, e.g., five in a million or 5×10^{-6}. Also note that the slope of the line has the units of kg-day/mg. This slope will be revisited in the next Chapter [6].

23.3 CONCEPT OF THRESHOLD

Discuss the concept of threshold from a noncarcinogenic viewpoint.

SOLUTION

For many noncarcinogenic effects, protective mechanisms are believed to exist that must be overcome before the adverse effect is manifested. For example, where a large number of cells perform the same or similar function, the cell population may have to be significantly depleted before the effect is seen. As a result, a

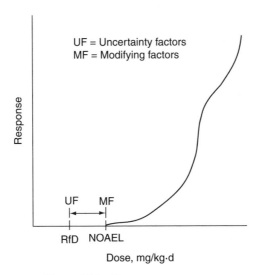

Figure 23.1 Dose–response curve.

range of exposures exists from zero to some finite value that can be tolerated by the organism with essentially no chance of experiencing adverse effects. In developing a toxicity value for evaluating noncarcinogenic effects (i.e., the aforementioned RfD), the approach is to identify the upper bound of this tolerance range (i.e., the maximum subthreshold level). Because significant variability exists in humans, attempts are made to identify a subthreshold level of protection for sensitive individuals in the population. For most chemicals, this level can only be estimated; therefore, the RfD incorporates uncertainty factors (discussed above) indicating the degree or extrapolation used to derive the estimated value. RfD values often contain a statement expressing the overall confidence that the evaluators have in the RfD (high, medium, or low). The RfD is generally considered to have an uncertainty spanning an order of
magnitude or more.

23.4 EXPOSURE DURATION CLASSIFICATION

Discuss the three exposure durations employed in analyzing adverse health effects.

SOLUTION

Three exposure durations that are used in considering the possibility of adverse non-carcinogenic health effects are chronic, subchronic, and shorter-term exposures.

1. *Chronic* exposures for humans range in duration from seven years to a lifetime; such long exposures are almost always of concern at superfund sites (e.g., inhabitants of nearby residencies, year-round users of specified drinking water sources).
2. *Subchronic* human exposures typically range in duration from two weeks to seven years.
3. *Short-term* exposures for less than two weeks are occasionally of concern. Also, if chemicals known to be developmental toxicants are present, short-term exposures of only a day or two can be of concern.

As described earlier, a chronic RfD is defined as an estimate (with an uncertainty spanning perhaps an order of magnitude or greater) of a daily exposure level for the human population, including sensitive subpopulations, that is likely to be without an appreciable risk of deleterious effects during a lifetime. Chronic RfDs are specifically developed to be protective for long-term exposure to a compound. Also, chronic RfDs are often used to evaluate the potential noncarcinogenic effects associated with exposure periods between seven years (approximately 10% of a human lifetime) and a lifetime.

Regarding subchronic exposure, the U.S. EPA has begun developing subchronic RfDs (RfD_{ss}), which are useful for characterizing potential noncarcinogenic effects associated with shorter-term exposures, and developmental $RfD_{dt}s$, which are useful for specifically assessing potential developmental effects resulting from exposure to a compound. These short-term exposures can result when a particular activity is performed for a limited number of years or when a chemical with a short half-life degrades to negligible concentrations within several months. Developmental RfDs are also used to evaluate the potential effects on a developing organism following a single exposure event.

23.5 RISK FOR MULTIPLE AGENTS: CHRONIC EXPOSURE

An individual is exposed to two nanochemicals NC1 and NC2. The exposure levels for NC1 and NC2 are approximately 0.4 and 0.6 mg/kg-day, respectively. The RfDs for NC1 and NC2 have been recently estimated to be 1.0 and 2.0 mg/kg-day, respectively. Based on this information, determine the hazard index, HI, for the scenario.

SOLUTION

To assess overall potential for noncarcinogenic effects posed by more than one agent, an approach has been developed based on EPA's Guidelines for Health Risk Assessment of Chemical Mixtures. This approach assumes that simultaneous subthreshold exposures to several chemicals could result in an adverse health effect. It also assumes that the magnitude of the adverse effect will be proportional to the sum of the ratios of the subthreshold exposures to acceptable exposures. The noncancer HI is equal to the sum of the hazard quotient for each agent as described below. The describing agent takes the form:

$$\text{Hazard Index} = \sum_{i=1}^{N}(EL_i)(RfD_i) = EL_1/RfD_1 + EL_2/RfD_2$$

$$+ \cdots + EL_i/RfD_i + \cdots + EL_N/RfD_N$$

where EL_i = exposure level (or intake) for the ith toxicant RfD_i = reference dose for the ith toxicant. It should be noted that EL and RfD are expressed in the same units (usually mg/kg-day) and represent the same exposure period (i.e., chronic, subchronic, or shorter term).

When the hazard index exceeds unity, there may be concern for potential health effects. While any single chemical with an exposure level greater than the toxicity value will cause the hazard index to exceed unity, the reader should also note that for multiple exposures, the hazard index can also exceed unity even if no single chemical exposure exceeds its RfD. It is also important to calculate the hazard index separately for chronic, subchronic and short-term exposure periods. These later two periods are discussed in the next Problem.

Applying the above equation to the two agent exposure scenario leads to:

$$\begin{aligned} \text{HI} &= (EL_1/RfD_1) + (EL_2/RfD_2) \\ &= (0.4/1) + (0.6/2) \\ &= 0.4 + 0.3 = 0.7 \end{aligned}$$

One can conclude there is no adverse health effect.

23.6   RISK FOR MULTIPLE AGENTS: SUBCHRONIC EXPOSURE

A population has experienced exposure to three compounds, one of which is a nano-chemical, for seven weeks. The exposure intake of agents A1, A2, and A3 are 0.4, 0.6, and 0.1 mg/kg-day, respectively. The RfDs for A1, A2, and A3, are 1.5, 2.5, and 0.2 mg/kg-day, respectively. Calculate the hazard index for this subchronic exposure.

SOLUTION

For each subchronic exposure (i.e., two weeks to seven year exposure), one can cal-culate a separate subchronic hazard index from the ratio of subchronic daily exposure level or daily intake (SDI) to the subchronic reference dose (RfDs). The subchronic HI is then calculated from:

$$\text{Subchronic HI} = \sum_{i=1}^{N} (SDI_i)(RfDs_i) = SDI_1/RfD_1 + SDI_2/RfD_2$$

$$+ \cdots + SDI_i/RfD_i + \cdots + EL_N/RfDs_N$$

where SDI_i = subchronic daily intake for the ith toxicant in mg/kg-day and RfD_i = subchronic reference dose for the ith toxicant in mg/kg-day.

The same procedure above may be applied for simultaneous shorter-term (see Problem 23.5) exposures to several chemicals. There are several limitations to this approach and the earlier approaches that should be noted. As mentioned earlier, the level of concern does not increase linearly as the reference dose is approached or exceeded because the RfDs do not have equal accuracy or precision and are not based on the same severity of effects. Moreover, hazard quotients may be combined for substances with RfDs based on critical effects of varying toxicological significance. Also, RfDs of varying levels of confidence that include different uncertainty adjustments and modifying factors may be combined.

Another limitation with the HI approach is that the assumption of dose additivity is more properly applied to compounds that induce the same effect by the same mechanism of action. Consequently, application of the hazard index equation to a number of compounds that are not expected to induce the same effect by the same mechanism could overestimate the potential for problems. This possibility is generally not of concern if only one or two substances are responsible for driving the HI above unity.

Applying the above equation to the three chemicals leads to:

$$\text{Subchronic HI} = \sum_{i=1}^{N} (SDI_1/RfD_1) + (SDI_2/RfD_2) + (SDI_3/RfD_3)$$

$$= (0.4/1.5) + (0.6/2.5) + (0.1/0.2)$$

$$= 0.267 + 0.24 + 0.5 = 1.007$$

Since the resultant index is close to unity, there should be some concern.

23.7 MULTIPLE EXPOSURE PATHWAYS

In some situations, an individual might be exposed to a substance or combination of substances through several pathways. For example, an individual might be exposed to substance(s) from a hazardous waste site by consuming contaminated drinking water from a well, eating contaminated fish caught near the site, and through inhalation of dust originating from the site. The total exposure to various chemicals will equal the sum of the exposures by all pathways. However, one should not automatically sum risks from all exposure pathways evaluated for a site.

There are two steps required to determine whether risks or hazard indices for two or more pathways should be combined for a single exposed individual or group of individuals. The first is to identify reasonable exposure pathway combinations. The second is to examine whether it is likely that the same individuals would consistently face the "reasonable maximum exposure" (RME) by more than one pathway. Additional details are available in the literature [5].

To assess the overall potential for noncarcinogenic effects posed by several exposure pathways, the total hazard index for each exposure duration (i.e., chronic, subchronic, and shorter-term) should be calculated separately. The equation to employ is provided below:

Total Exposure Hazard Index $=$ Hazard Index (exposure pathway$_1$)

$+$ Hazard Index (exposure pathway$_2$) $+ \cdots$

$+$ Hazard Index (exposure pathway$_i$) $+ \cdots$

$+$ Hazard Index (exposure pathway)$_N$

The calculating scheme is similar to that provided in the two previous Problems. Also, note that the total exposure hazard index should be calculated separately for chronic, subchronic, and shorter-term exposure periods. As before, if the total hazard index for an exposed individual or group of individuals exceeds unity, there may be concern for noncarcinogenic health effects.

23.8   MCL AND RFD

The drinking water maximum contaminant level (MCL) set by the U.S. EPA for atrazine is 0.003 mg/L and its reference dose (RfD) is 3.5 mg/kg-day. How many liters of water containing atrazine at its MCL would a person have to drink each day to exceed the RfD for this triazine herbicide [3]?

SOLUTION

As with most of these calculations, it is assumed that those exposed can be represented by a 70 kg individual. The volume rate, q, of drinking water at the MCL to reach the RfD for atrazine is (using the standard RfD equation)

$$q = (3.5 \text{ mg/kg-day})(70 \text{ kg})/(0.003 \text{ mg/L})$$
$$= 81\,700 \text{ L/day}$$

This is a surprisingly high volume of water on a daily basis. However, based on the MCL and RfD, there appears to be no problem. This large volume indicates that there may be uncertainty (i.e., the product of the uncertainty factors is large) in estimating a reference dose for atrazine. The subject of uncertainty factors is addressed in the next two Problems.

23.9   UNCERTAINLY AND MODIFYING FACTORS

How are uncertainty and modifying factors included in the development of an RfD?

SOLUTION

As noted earlier, the RfD is derived from the NOAEL (or LOAEL) for the critical toxic effect by consistent application of uncertainty factors (UFs) and a modifying factor (MF). The uncertainty factors generally consist of multiples of 10 (although values less than 10 are sometimes used), with each factor representing a specific area of uncertainty inherent in the extrapolation from the available data. The bases for application of different uncertainty factors are explained below.

1. A UF of 10 is used to account for variation in the general population and is intended to protect sensitive subpopulations (e.g., the elderly and children).
2. A UF of 10 is used when extrapolating from animals to humans. This factor is intended to account for the interspecies variability between humans and other mammals.
3. A UF of 10 is used when a NOAEL derived from a subchronic instead of a chronic study is used as the basis for a chronic RfD.
4. A UF of 10 is used when a LOAEL is used instead of a NOAEL. This factor is intended to account for the uncertainty associated with extrapolating from LOAELs to NOAELs.
5. In addition to the UFs listed above, a modifying factor (MF) is also applied. An MF ranging from >0 to 10 is included to reflect a qualitative professional assessment of additional uncertainties in the critical study and in the entire data base for the chemical not explicitly addressed be the preceding uncertainty factors. The default value for the MF is 1.0 [5].

To calculate the RfD, the appropriate NOAEL (or the LOAEL if a suitable NOAEL is not available) is divided by the product of all of the applicable uncertainty factors and the modifying factor. The describing equation is provided below.

$$RfD = \frac{NOAEL \text{ or } LOAEL}{(UF_1 \cdot UF_2 \cdots MF)}$$

23.10 CALCULATING AN RFD FROM NOAEL

The NOAEL for a new nanochemical (NCLT) was recently estimated to be 72 mg/kg-day. Uncertainty factors of 10 need to be applied to items (1) and (2), as presented in the previous Problem. In addition, a modifying factor of 5 is also to be employed. Using the equation provided in Problem 23.9, estimate the RfD for NCLT.

SOLUTION

For this application

$$RfD = NOAEL/(UF_1)(UF_2)(MF)$$

Substitution yields

$$RfD = 72/(10)(10)(5)$$
$$= 0.144 \text{ mg/kg-day}$$

23.11   METAL PLATING FACILITY APPLICATION

Owing to contamination from a metal plating facility, the water from a nearby community water supply well was shown to contain cyanide, nickel, and chromium(III) at concentrations of 200, 950, and 10 200 mg/L, respectively. If the daily water intake is assumed to be 0.2 L, and the body weight of an adult is 70 kg, do these noncarcinogenic chemicals pose a health hazard? The RfDs for the three chemicals are 0.02, 0.02, and 1 mg/kg-day, respectively.

SOLUTION

The exposure level, or intake, or dose, E, for cyanide is:

$$E = \frac{\left(0.2\frac{mg}{L}\right)\left(0.2\frac{L}{day}\right)}{70\,kg}$$

$$= 5.71 \times 10^{-4}\frac{mg}{kg \cdot day}$$

The Hazard Index for cyanide is therefore:

$$HI\,(cyanide) = \frac{5.71 \times 10^{-4}}{0.02} = 0.03$$

Similar calculations employing the same procedure leads to:

$$HI\,(nickel) = 0.14$$
$$HI\,(chromium(III)) = 0.029$$

The individual ratios are not only well below 1 but also the total. Thus, this indicates an acceptable hazard. The reader is encouraged to resolve the problem assuming a daily water intake of 2 L.

23.12   NONCARCINOGEN CALCULATION PROCEDURE

Provide a procedure for estimating risk for noncarcinogens.

SOLUTION

The preliminary assessment of noncarcinogenic risk, as recommended by the U.S. EPA, is typically calculated in four major steps [23]:

1. Identify discrete exposure conditions:
   (a) Exposure route
   (b) Frequency
   (c) Duration
   (d) Administered dose
2. Derive appropriate reference doses for each discrete set of conditions.
3. Evaluate the hazard index as a ratio of exposure dose to the recommended RfD.
4. Aggregate the hazard for multiple chemical agents and exposure pathways as a hazard index, where appropriate.

The ratio referred to in the third step is utilized to quantify risk from noncarcinogens. As the fourth step indicates, the hazard index for individual chemicals may be summed for chemicals affecting a particular target organ or acting by a common mechanism in order to provide a final measure of noncarcinogenic toxic risk. If the sum of hazard indices is less than one, then the risk of adverse health effects may be considered acceptable (negligible).

24 Carcinogens

The definition of a carcinogen was presented in the previous Chapter. Webster defines cancer as:

1. A malignant new growth anywhere in the body of a person or animal; malignant tumor; cancers tend to spread locally and to distant parts of the body; and
2. Anything bad or harmful that spreads and destroys.

This Chapter, in a very real sense, serves as a companion to the previous Chapter. Noncarcinogens were addressed previously; carcinogens are treated here. Since some of the nanochemicals that will arise on the scene shortly will be classified as carcinogens, it behooves the reader to become familiar with procedures and calculations currently employed for cancer-causing agents.

As indicated, this serves as an extension of the previous Chapter that was concerned with noncarcinogens. The earlier dose–response development is revisited in order to introduce the "slope factor," which serves to define the ratio of the response to the dose. The method the U.S. EPA employs to derive this value is included in the development. Several Problems are presented that involve the application of the slope factor. The Chapter concludes with an outline of a procedure to employ in carcinogenic calculations and a problem involving an application of this procedure.

24.1 NONTHRESHOLD CONCEPT

Discuss the concept of nonthreshold effects.

SOLUTION

Carcinogenesis, unlike many noncarcinogenic health effects, is generally thought to be a phenomenon for which risk evaluation is based on the presumption that the concept of a threshold does not apply, i.e., there is no threshold. For carcinogens, the U.S. EPA assumes that a single molecule can produce changes in a single cell that can ultimately lead to uncontrolled cellular proliferation and eventually to a

Nanotechnology: Basic Calculations for Engineers and Scientists, by Louis Theodore
Copyright © 2006 John Wiley & Sons, Inc.

clinical state of disease, i.e., cancer. This hypothesized mechanism for carcinogenesis is referred to as "nonthreshold" because there is believed to be essentially no level of exposure for an agent that does not pose a finite probability, however small, of generating a carcinogenic effect.

A dose–response relationship is again presented in Figure 24.1. The typical relationship between an agent and an effect (or response) is that as the concentration of the agent increases, the probability of the effect or response in the receptor also increases. A threshold would exist if there was a level of dose for which no apparent effect would be discerned as illustrated in Figure 24.2. This is a strongly debated topic for carcinogens. But, as also indicated above, the regulatory community, in the interest of protecting the health of the public, usually assumes that no thresholds exist for carcinogens.

The U.S. EPA uses a two part approach to determine dose–response information. The agent or chemical is first assigned a weight-of-evidence classification, and secondly, a slope factor is calculated. These two topics receive attention in the next Problem.

24.2   WEIGHT OF EVIDENCE AND SLOPE FACTOR

Discuss the evaluation process for carcinogens.

SOLUTION

In the first step of an evaluation, available data are collected to determine the likelihood that the agent is a human carcinogen. The evidence is characterized separately for human studies and animal studies as sufficient, limited, inadequate, no data, or evidence of no effect. The characterizations of these two types of data are combined, and based on the extent to which the agent has been shown to be a carcinogen in humans or experimental animals, or both, the agent is given a provisional weight-of-evidence classification. EPA scientists then adjust the provisional

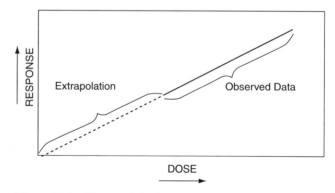

Figure 24.1   Characteristic dose–response curve; nonthreshold.

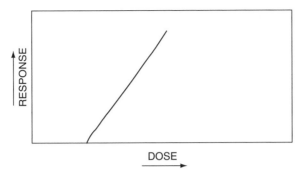

Figure 24.2 Dose–response curve containing a threshold.

classification upward or downward, based on other or new supporting evidence of carcinogenicitv. U.S. EPA's classification system for weight of evidence is provided in Table 24.1 This system is adapted from the approach taken by the international Agency for Research on Cancer (IARC).

A slope factor (see next Problem) is generated in the second part of the evaluation. Based on the valuation that the chemical is a known or probable human carcinogen, a toxicity value that defines quantitatively the relationship between dose and response (i.e., the slope factor) is calculated. Slope factors are typically calculated for potential carcinogens in classes A, B1, and B2. Quantitative estimation of slope factors for the chemicals in class C proceeds on a case-by-case basis [5].

Generally, the slope factor is a plausible upper bound estimate of the probability of a response per unit intake of a chemical over a lifetime. The slope factor is used in risk assessments to estimate an upper-bound lifetime probability of an individual developing cancer as a result of exposure to a particular level of a potential carcinogen. Slope factors should always be accompanied by the weight-of-evidence classification to indicate the strength of the evidence that the agent is a human carcinogen. Calculational details are presented later in the Chapter [5].

TABLE 24.1 Weight-of-Evidence Classification System for Carcinogenicity

Group	Description
A	Human carcinogen
B1 or B2	Probable human carcinogen
	B 1 indicates that limited human data are available
	B2 indicates sufficient evidence in animals and inadequate or no evidence in humans
C	Possible human carcinogen
D	Not classifiable as to human carcinogenicity
E	Evidence of noncarcinogenicity for humans

24.3   CARCINOGENIC TOXICITY VALUES

Discuss the various expressions for toxicity values.

SOLUTION

Toxicity values for carcinogenic effects can be expressed in several ways. The slope factor may be simply viewed as the slope of the dose–response curve and expressed as kg-day/mg or $(mg/kg\text{-}day)^{-1}$ and is occasionally referred to as the q1. Thus:

$$\text{Slope factor} = SF = \text{risk per unit dose}$$
$$= \text{risk per mg/kg-day}$$

Once again (as described in the previous Chapter), the dose should be viewed as the agent rate in mg/day administered to the receptor body in kg, that is, mg/kg-day. The reader should also note that the inverse of the dose, that is, kg-day/mg, is occasionally defined as the potency factor. The term dose–response relationship is a simple ratio of response in terms of risk as probability per unit time to the dose, with the usual units of

$$\left(\frac{\text{probability}}{\text{time}}\right) \bigg/ \left(\frac{\text{mg}}{\text{kg-day}}\right)$$

or, more simply (with time replaced by lifetime)

$$\frac{\text{kg-day-probability}}{\text{mg-lifetime}}$$

Toxicity values for carcinogenic effects also can be expressed in terms of risk per unit concentration of the agent in the medium where human contact occurs. These values, also called unit risks, are obtained by dividing the slope factor by 70 kg and multiplying by either the inhalation rate (20 m³/day) or the water consumption rate (2 L/day) for the risk associated with air or water, respectively. Additional conversion factors may be necessary in the calculation of unit risk; more detailed information is available in the literature [5]. The standardized duration assumption for these unit risks is assumed to be continuous lifetime exposure. The unit risks for these two options are:

$$\text{Air unit risk} = \text{risk per } \mu g/m^3$$
$$= \text{slope factor} \, (1/70 \, kg)(2 \, L/day)(10^{-3})$$
$$\text{Water unit risk} = \text{risk per } \mu g/L$$
$$= \text{slope factor}(1/70 \, kg)(2 \, L/day)(10^{-3})$$

The conversion factor 10^{-3} is required to convert from mg to μg since the unit risk is given in $(\mu g/m^3)^{-1}$ or $(\mu g/L)^{-1}$.

24.4   BENZENE IN WATER APPLICATION

The odor perception threshold for benzene in water is 2 mg/L. The benzene drinking water unit risk is 8.3×10^{-7} L/μg. Calculate the potential benzene intake rate (mg benzene/kg-day) and the cumulative cancer risk from drinking water with benzene concentrations at half of its odor threshold for a 50 year exposure duration. Use the following equation for estimating the benzene ingestion rate:

$$\text{Ingestion rate } [\text{mg}/(\text{kg} \cdot \text{day})] = \frac{(C)(I)(EF)(ED)}{(BW)(AT)}$$

where C = concentration, mg/L

$\quad I$ = water ingestion rate, 2L/day

$\quad EF$ = exposure frequency, 350 day/yr

$\quad ED$ = exposure duration, yr

$\quad BW$ = body weight, 70 kg

$\quad AT$ = averaging time, 70 yr

SOLUTION

The following benzene intake rate is determined using the ingestion rate equation given in the Problem statement.

$$\text{Ingestion rate} = \frac{\left(1\frac{\text{mg}}{\text{L}}\right)\left(2\frac{\text{L}}{\text{day}}\right)(50\,\text{years})}{(70\,\text{kg})(70\,\text{years})}$$

$$= 0.02 \frac{\text{mg}}{\text{kg} \cdot \text{day}}$$

The cancer risk from this ingestion at half of the odor threshold of 1 mg/L is calculated based on the benzene unit risk of 8.3×10^{-7} L/μg or:

$$\text{Cancer risk} = (\text{Water unit risk})(\text{concentration}, \mu g/L)$$

$$\text{Cancer risk} = \left(8.3 \times 10^{-7} \frac{\text{L}}{\mu g}\right)\left(1000 \frac{\mu g}{\text{L}}\right)$$

$$= 8.3 \times 10^{-4}$$

This risk is somewhat high relative to the widely accepted standard range of the environmental lifetime risk of 1×10^{-6} to $\times 10^{-4}$.

24.5   EXCESS LIFETIME CANCER CASES

What are the maximum number of excess lifetime cancer cases expected for a population of 10,000 adults with a daily intake of 0.40 mg of a carcinogenic nanoagent? The slope factor for the nanoagent may be assumed to be 0.029 $(\text{mg/kg-day})^{-1}$.

SOLUTION

Assume a usual adult weight of 70 kg. The lifetime cancer risk is calculated as:

$$\text{Individual cancer risk} = \left(0.40 \, \frac{\text{mg}}{\text{day}}\right)\left(\frac{1}{70 \, \text{kg}}\right)\left(0.029 \, \frac{\text{kg} \cdot \text{day}}{\text{mg}}\right)$$

$$= 1.66 \times 10^{-4}$$

$$= 0.0166\%$$

$$\text{Maximum cases} = (\text{risk})(\text{exposed population})$$

$$= (1.66 \times 10^{-4})(10,000)$$

$$= 1.66 \text{ lifetime cancer cases}$$

24.6   ACTION LEVEL

Determine the action level in $\mu\text{g/m}^3$ for a 60 kg person with a life expectancy of 70 years exposed to benzene over a 30 year period. The acceptable risk is one incident of cancer per one million persons, or 10^{-6}. Assume a breathing rate of 15 $\text{m}^3/$ day and an absorption factor of 75% (0.75). The potency factor for benzene is 1.80 $(\text{mg/kg-day})^{-1}$. The following equation has been used in health risk assessment studies for carcinogens.

$$C = \frac{RWL}{PIA(ED)}; \quad \text{consistent units}$$

where C = action level, this is, the concentration of carcinogen above which remedial action should be taken; R = acceptable risk or probability of contracting cancer; W = body weight; L = assumed lifetime; P = potency factor; I = intake rate; A = absorption factor, the fraction of carcinogen absorbed by the human body; and ED = exposure duration.

SOLUTION

Using the equation provided above, the action level for the carcinogen benzene for an 60 kg person with a life expectancy of 70 years and an exposure duration of 30 years is

$$C = \frac{(10^{-6})(60)(70)}{(1.80)(15)(0.75)(30)\left(\dfrac{1}{1000}\right)}$$

$$= 0.0069 \ \mu g/m^3$$

The risk for this person would be classified as unacceptable if the exposure exceeds $0.0069 \ \mu g/m^3$ in a 30 year period.

24.7 ACCIDENTAL SPILL

A reagent bottle containing 1 kg of benzene falls to the floor of a small enclosed storage area measuring $10 \ m^3$ in volume. In time, the spilled liquid benzene completely evaporates. If 10% of the vapor remains trapped in the storage room, what is the benzene concentration in mg/m^3? Assuming this concentration to be the action level, what exposure time would produce a risk of 5×10^{-4} over a 30 year period for the individual described in Problem 24.6.

SOLUTION

The benzene concentration in the storage room is

$$C = (0.1)\left(\frac{1 \ kg}{10 \ m^3}\right)\left(1\,000\,000 \ \frac{mg}{kg}\right) = 10^4 \ \frac{mg}{m^3}$$

Substituting this value into the risk assessment equation given in Problem 24.6 leads to:

$$10^4 \ \frac{mg}{m^3} = \frac{(5 \times 10^{-4})(60 \ kg)(70 \ year)}{(1.80 \ kg - day/mg)(15 \ m^3/day)(0.75)(ED)}$$

Solving for ED gives:

$$ED = 1.037 \times 10^{-5} \ year$$

$$= 3785 \times 10^{-6} \ days$$

$$= 0.09084 \ hr$$

$$= 5.451 \ min$$

$$= 327 \ s$$

24.8   UNCERTAINTIES AND LIMITATIONS

Discuss some of the uncertainties and limitations in developing cause and effect relationships for carcinogenics.

SOLUTION

This general subject was reviewed earlier for noncarcinogens and the RfD. The reader is referred to Problem 24.8. Several additional points are presented in this solution.

For a variety of reasons it is often difficult to accurately evaluate responses caused by exposures (particularly acute ones) to carcinogens and hazardous chemicals. Five of these concerns are briefly discussed below [5, 6]:

1. Humans experience a wide range of acute adverse health effects, including irritation, narcosis, asphyxiation, sensitization, blindness, organ system damage, and death. In addition, the severity of many of these effects varies with intensity and duration of exposure. For example, exposure to a substance at an intensity that is sufficient to cause only mild throat irritation is of less concern than one that causes severe eye irritation, lacrimation, or dizziness.

2. There is a high degree of variation in response among individuals in a typical population. Generally, sensitive populations include the elderly, pregnant women, children, and individuals with diseases that compromise the respiratory or cardiovascular system.

3. For the overwhelming majority of substances encountered in industry, there are not enough data on responses of humans to permit an accurate or precise assessment of the substance's health potential. Frequently, the only data available are from controlled experiments conducted with laboratory animals. In such cases, it is necessary to extrapolate from effects observed in animals to effects likely to occur in humans. Thus, extrapolation requires the professional judgment of a toxicologist.

4. Many releases involve multicomponents. There are presently no "rules" or "guidelines" on how these types of releases should be evaluated. Are they additive, synergistic, or antagonistic in their effect on the population? Unfortunately, even response data of humans to single component exposures are woefully inadequate for a large number of chemical agents.

5. There are no toxicology testing protocols that exist for studying episodic releases on animals. This has been in general a neglected aspect of toxicology research. There are experimental problems associated with the testing of chemicals at high concentrations for very short durations in establishing the concentration/time profile. In testing involving fatal concentration/time exposures, a further question exists of how to incorporate early and delayed fatalities into the study results.

These five points can be further extended and rephrased (as presented earlier) to dose–response informations. Additional sources of uncertainty associated with these toxicity values may include [5, 6]:

1. Using dose–response information from effects observed at high doses to predict the adverse health effects that may occur following exposure to the low levels expected from human contact with the agent in the environment.
2. Using dose–response information from short-term exposure studies to predict the effects of long-term exposures, and vice versa.
3. Using dose–response information from animal studies to predict effects on humans.
4. Using dose–response information from homogeneous animal populations or healthy human populations to predict the effects likely to be observed in the general population consisting of individuals with a wide range of sensitivities.

24.9 MULTIPLE CHEMICAL AGENTS AND EXPOSURE PATHWAYS

Discuss the effect of multiple chemical agents on risk. Also discuss the effect of multiple exposure pathways.

SOLUTION

The cancer risk equation for N multiple agents is given by:

$$\text{Risk} = \sum_{i=1}^{N} (Risk)_i = Risk_1 + Risk_2 + \cdots + Risk_i + \cdots + Risk_N$$

where Risk = the total cancer risk usually expressed as a unitless probability and $Risk_i$ = the risk estimate for the ith substance.

The cancer risk equation described above estimates the incremental individual lifetime cancer risk for simultaneous exposure to several carcinogens and is based on the U.S. EPA's risk assessment guidelines. This simple additive equation is appropriate for most risk assessments.

The reader should note that the comments on multiple agents and pathways in the previous Chapter on noncarcinogens applies to carcinogens as well. The cancer risks from various exposure pathways are also assumed to be additive. The describing equation is:

$$\text{Total cancer risk (exposure)} = \sum_{i=1}^{N} (Risk, pathway)_i$$

$$= (Risk, pathway)_1 + (Risk, pathway)_2 + \cdots$$
$$+ (Risk, pathway)_i + \cdots + (Risk, pathway)_N$$

24.10   EXPONENTIAL RISK MODEL

Describe the applicability of EPA's exponential risk model.

SOLUTION

The slope factor (SF) converts estimated daily intakes or dose – averaged over a lifetime of exposure – directly to incremental risk of an individual developing cancer. Because relatively low doses (compared to those experienced by test animals) are most likely for environmental exposures, it generally can be assumed that the dose–response relationship will be linear in the low dose range of the dose–response curve. Under this assumption, the slope factor is constant, and the risk will be linearly related to dose. Thus, the linear form of the carcinogenic risk equation described earlier is usually applicable for estimating risks. That equation is:

$$\text{Risk} = (Dose)(SF)$$

where Risk = a probability (e.g., 1×10^{-6}) of an individual developing cancer, dimensionless; $Dose$ = chronic daily intake averaged over 70 years, mg/kg-day; and SF = slope factor, expressed in $(\text{mg/kg-day})^{-1}$. However, the above linear equation is valid only at a low risk level – usually below risks of 0.01. When chemical intakes are higher, an alternate equation is recommended by the U.S. EPA. This is exponential in form and given by:

$$\text{Risk} = 1.0 - \exp[-(Dose)(SF)]$$

24.11   RISK ALGORITHM

Employing typical units, develop a simple algorithm to calculate the following terms for carcinogens.

1. Exposure
2. Dose
3. Lifetime individual risk
4. Risk to exposed population

SOLUTION

The calculations to characterize the risk to a population exposed to an event or chemical agent can be divided into the four steps described below. The units correspond to a chemical emission to the atmosphere posing a risk to a population.

1. Exposure (E).

$$E\left(\frac{\text{mg} \cdot \text{day}}{\text{m}^3}\right) = \left[\text{Agent concentration}\left(\frac{\text{mg}}{\text{m}^3}\right)\right][\text{Exposure duration (day)}]$$

or

$$E = (AC)(ED)$$

This represents the total exposure with units of concentration days. The days can often be the lifetime of an individual (25 550 days).

2. Dose (D).

$$D\left(\frac{\text{mg}}{\text{kg} \cdot \text{day}}\right) = \left[E\left(\frac{\text{mg} \cdot \text{day}}{\text{m}^3}\right)\right]\left[\text{Dose factor}\left(\frac{\text{m}^3}{\text{kg} \cdot \text{day}^2}\right)\right]$$

This dose (D) represents the average mass of pollutant intake per unit mass of receptor on a daily basis. The dose factor (DF) is given by

$$DF = \frac{\left[\text{"Contact rate"}\left(\frac{\text{m}^3}{\text{day}}\right)\right]\left[\text{Intake fraction}\left(\frac{\%}{100}\right)\right]}{[\text{Average receptor weight (kg)}][\text{Exposure duration (day)}]}$$

Thus,

$$D = E(DF)$$

3. Lifetime individual risk (LIR).

$$LIR\left(\frac{\text{probability}}{\text{lifetime}}\right) = \left[D\left(\frac{\text{mg}}{\text{kg} \cdot \text{day}}\right)\right]$$
$$\times \left[\text{Dose–response relationship}\left(\frac{\text{kg} \cdot \text{day} \cdot \text{probability}}{\text{mg} \cdot \text{lifetime}}\right)\right]$$

This represents an individual's risk over a lifetime. Thus,

$$LIR = (D)(DRR)$$

4. Risk to exposed population (*REP*).

$$REP\left(\frac{\text{individuals}}{\text{year}}\right) = \frac{\left[\text{LIR}\left(\frac{\text{probability}}{\text{lifetime}}\right)\right][\text{Exposed population (individuals)}]}{\text{Years per lifetime}\left(\frac{\text{year}}{\text{lifetime}}\right)}$$

This provides a reasonable estimate of the number of individuals or cases in an exposed population per year. Thus,

$$REP = (LIR)(EP)/(YPL)$$

with *YPL* is usually set at 70 years/lifetime.

24.12   RISK ALGORITHM APPLICATION FOR BENZENE

Theodore Associates have been contracted by the U.S. EPA to provide benzene lifetime risk estimates for the population in the United States based on an ambient benzene concentration (ABC) of 9.5×10^{-3} mg/m^3. Employ the algorithm provided in Problem 24.11. The following data are provided by the EPA.

Breathing rate $= 20$ m^3/day
Percent absorption (ABC) $= 50\%$

$$\text{Dose–response relationship} = 0.028\left(\frac{\text{probability}}{\text{lifetime}}\right) \Big/ \left(\frac{\text{mg}}{\text{kg-day}}\right)$$
$$= \frac{0.028 \text{ kg} \cdot \text{day} \cdot \text{probability}}{\text{mg} \cdot \text{lifetime}}$$

Exposure duration $= 70$ years
Average weight of individual $= 70$ kg
Exposed population $= 300\,000\,000$
Ambient conditions $= 25°\text{C}$, 1 atm

SOLUTION

Calculate the exposure (E) in mg-day/m^3 for ABC:

$$ED = (365 \text{ day/yr})(70 \text{ yr})$$
$$= 25\,550 \text{ day}$$
$$E = (AC)(ED)$$
$$= (9.5 \times 10^{-3} \text{ mg/m}^3)(25\,550 \text{ day})$$
$$= 243 \text{ mg-day/m}^3$$

Calculate the dose factor (DF) in m^3/(kg $\cdot$ day^2),

$$DF = \frac{(20 \text{ m}^3/\text{day})(0.50)}{(70 \text{ kg})(25\,550 \text{ day})}$$
$$= 5.59 \times 10^{-6}(\text{kg-day}^2)$$

Calculate the dose (D) in mg/kg-day.

$$D = E(DF)$$
$$D = \left(243 \, \frac{\text{mg-day}}{\text{m}^3}\right)\left(5.59 \times 10^{-6} \, \frac{\text{m}^3}{\text{kg-day}^2}\right)$$
$$= 0.00136 \text{ mg/(kg-day)}$$

Estimate the lifetime individual risk (LIR) in units of probability/lifetime to those exposed to benzene.

$$LIR = D(DRR)$$
$$LIR = \left(0.00136 \, \frac{\text{mg}}{\text{kg-day}}\right)\left(0.028 \, \frac{\text{kg-day-probability}}{\text{mg-lifetime}}\right)$$
$$= 3.80 \times 10^{-5} \text{ probability/lifetime}$$

The annual risk to the exposed population in number of individuals per year is given by:

$$REP = (LIR)(EP)/(YPL)$$

$$REP = \frac{\left(3.80 \times 10^{-5} \, \dfrac{\text{probability}}{\text{lifetime}}\right)(3.00 \times 10^8 \, \text{individuals})}{70 \, \dfrac{\text{year}}{\text{lifetime}}}$$

$$= 163 \, \text{individuals/year}$$

Perhaps the most important consideration in these calculations is the value of the dose–response relationship provided by the EPA. Determining a numerical value for this factor is critical to the calculation; however, a significant degree of uncertainty exists with these probability values provided by researchers in the field.

25 Health Risk Assessment

As has been reported in the literature from several sources, "nanotechnology has the potential to change one's comprehension of nature and life, develop unprecedented manufacturing tools and medical procedures, and even change societal and international relations" And, since nanoapplications involve "exact" manufacturing, the consumption of raw material, energy, water, and so forth is conserved while minimizing waste/pollution. Enter Murphy's law (or Theodore's law, according to the author). The dark clouds in this case are the environmental health impacts associated with these new and unknown operations, and the reality is that there is a serious lack of information on these impacts. Risk assessment studies in the future will be the path to both understanding and minimizing these nanotechnology effects [15].

It is once again noted that the effect of particle size on properties will have to be addressed in the future. One needs to consider not only the variability of physical, chemical, and so on, properties but also the variability of "health-related" properties and effects.

Although the potential nanoagents and the means for dealing with them may be different, the methodology and protocols developed for conventional materials will probably be the same. Thus, environmental health risk assessment remains *environmental health risk assessment*, and nanoapplications will probably use the same techniques and procedures presented in this Chapter.

How is it possible to make decisions dealing with environmental risks from a new application, i.e., nanotechnology, in a complex society with competing interests and viewpoints, limited financial resources, and a lay public that is deeply concerned about the risks of cancer and other illness? Risk assessment and risk management, taken together, constitute a decision-making approach that can help the different parties involved avoid conflict and thus enable the larger society to work out its environmental problems rationally and with good results.

Risk assessment and risk management also provides a framework for setting regulatory priorities and for making decisions that cut across different environmental areas. This kind of framework has become increasingly important in recent years for several reasons, one of which is the considerable progress made in environmental control. Thirty-plus years ago, it was not hard to figure out where the first priorities should be. The worst pollution problems were all too obvious; this may not be so with nanotechnology [15].

Nanotechnology: Basic Calculations for Engineers and Scientists, by Louis Theodore
Copyright © 2006 John Wiley & Sons, Inc.

Regarding health risk assessment, concern arises because nanoparticles can possibly elude defense mechanisms on entering the body of humans. As noted earlier, there are various pathways that nanoparticles can follow to enter the human body; these include inhalation, skin absorbtion (absorption), and ingestion (digestion system). However, it is fair to say that the dominant route for nanoparticles is inhalation.

The Problems in this Chapter are intended to introduce the reader to the general subject of risk assessment, focusing on health risk assessment and the major steps involved with the health risk assessment process. Once again, it is noted that there is overlap with Chapter 22 (Toxicology) as well as the next Chapter (Hazard Risk Assessment).

25.1   RISK DEFINITIONS

Define risk and comment on the difference between risk assessment and risk management [14].

SOLUTION

There are many definitions for the word risk. Hazard risk (see Chapter 26) is a combination of uncertainty and damage; a ratio of hazards to safeguards; a triplet combination of event, probability, and consequences; or even a measure of economic loss or human injury in terms of both the incident likelihood and the magnitude of the loss or injury [24]. People face all kinds of health risk every day, some voluntarily and others involuntarily Therefore, risk plays a very important role in today's world. Studies on cancer caused a turning point in the world of risk because it opened the eyes of risk scientists and health professionals to the world of risk assessments.

Risk assessment and risk management are two different processes, but they are intertwined. Risk assessment and risk management give a framework not only for setting regulatory priorities but also for making decisions that cut across different environmental areas. Risk management refers to a decision-making process that involves such considerations as risk analysis, technology feasibility, economic information about costs and benefits, statutory requirements, public concerns, and other factors. Therefore, risk assessment supports risk management in that the choices on whether and how much to control future exposure to the suspected hazards may be determined [13]. Regarding both risk assessment and risk management this Chapter will primarily address this subject from a health perspective; Chapter 26 will primarily address this subject from a safety and accident perspective.

25.2   THE HEALTH RISK EVALUATION PROCESS

Outline the health risk evaluation process.

SOLUTION

Health risk assessments provide an orderly, explicit, and consistent way to deal with scientific issues in evaluating whether a health hazard exists and what the magnitude of the hazard may be. This evaluation typically involves large uncertainties because the available scientific data are limited, and the mechanisms for adverse health impacts or environmental damage are only imperfectly understood.

Most human or environmental health hazards can be evaluated by dissecting the analysis into four parts: hazard identification, dose–response assessment or hazard assessment, exposure assessment, and risk characterization (see Figure 25.1). For some perceived hazards, the risk assessment might stop with the first step, hazard identification, if no adverse effect is identified or if an agency elects to take regulatory action without further analysis [25]. Regarding hazard identification, a hazard is defined as a toxic agent or a set of conditions that has the potential to cause adverse effects to human health or the environment. Hazard identification involves an evaluation of various forms of information in order to identify the different hazards. Dose–response or toxicity assessment is required in an overall assessment: responses/effects can vary widely since all chemicals and contaminants vary in their capacity to cause adverse effects. This step frequently requires that assumptions be made to relate experimental data for animals to humans. Exposure

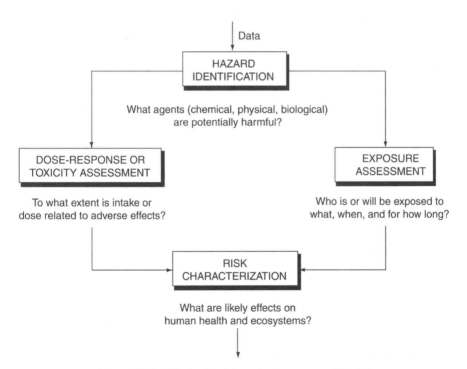

Figure 25.1   The health risk evaluation process [13, 14].

assessment is the determination of the magnitude, frequency, duration, and routes of exposure of human populations and ecosystems. Finally, in risk characterization, toxicology and exposure data/information are combined to obtain a qualitative or quantitative expression of risk. More extensive details are available in the literature [5, 13–15].

Risk assessment thus involves the integration of the information and analysis associated with the above four steps to provide a complete characterization of the nature and magnitude of risk and the degree of confidence associated with this characterization. A critical component of the assessment is a full elucidation of the uncertainties associated with each of the major steps. Under this broad concept of risk assessment are encompassed all of the essential problems of toxicology. Risk assessment takes into account all of the available dose–response data. It should treat uncertainty not by the application of arbitrary safety factors, but by stating them in quantitatively and qualitatively explicit terms, so that they are not hidden from decision makers. Risk assessment, defined in this broad way, forces an assessor to confront all the scientific uncertainties and to set forth in explicit terms the means used in specific cases to deal with these uncertainties [26].

25.3   STANDAND VALUES FOR INDIVIDUALS

To apply risk assessments to large groups of individuals, certain assumptions are usually made about an "average" person's attributes. List the average or standard values used for:

1. Body weight
2. Daily drinking water intake
3. Amount of air breathed per day
4. Expected life span
5. Dermal contact area

SOLUTION

The following are "average" values normally employed for humans.

1. Average body weight is 70 kg for an adult and 10 kg for a child.
2. The average daily drinking water intake is 2 liters for an adult and 1 liter for a child.
3. The average amount of air breathed per day is 20 m^3 for an adult and 10 m^3 for a child.
4. The average expected life span is 70 years.
5. The average dermal contact area is 1000 cm^2 for an adult and 300 cm^2 for a child.

25.4   QUALITATIVE RISK SCENARIOS

Provide examples of qualitative health risk scenarios [5, 6].

SOLUTION

Numerous qualitative approaches to risk assessment have been employed. Some sample categorizations follow.

Health risks may be divided into risk levels as provided below:

Risk Level 1: Does not cause a health hazard

Risk Level 2: Unlikely to cause a health hazard

Risk Level 3: May cause a health hazard

Risk Level 4: May cause a severe health hazard

Risk Level 5: Will cause a severe health hazard

Health risks may also be set in terms of a logarithmic scale of risk levels as seen in Table 25.1. These risk levels and ranges can apply on either an annual or lifetime basis.

Qualitative health risk policies for companies, for example, Theodore and Theodore Enterprises (TATE), could take the following form:

1. TATE will not knowingly pose a greater health risk to the public than it does to its own employees.
2. TATE will not expose its employees or neighbors to health risks that are considered unacceptable, based on industry practice and available technology.
3. TATE will comply with all applicable regulations and industry guidelines related to health risks, and will adopt its own standards where regulations do not exist or are inadequate.
4. TATE will neither undertake nor continue any operations whose associated health risks it does not understand or cannot control at a safe level.

TABLE 25.1   Risk Level and Risk Range

Risk Level	Risk Range	
1	1 in 1 to 1 in 9	$10^0 - 10^{-1}$
2	1 in 10 to 1 in 99	$10^{-1} - 10^{-2}$
3	1 in 100 to 1 in 999	$10^{-2} - 10^{-3}$
4	1 in 1000 to 1 in 9999	$10^{-3} - 10^{-4}$
5	1 in 10 000 to 1 in 99 999	$10^{-4} - 10^{-5}$
6	1 in 100 000 to 1 in 999 999	$10^{-5} - 10^{-6}$
7	1 in 1 000 000 to 1 in 9 999 999	$10^{-6} - 10^{-7}$

One other possible health risk company policy could take the following form:

1. The average individual health risk level for the public should be less than ___.
2. The maximum individual health risk for TATE employees should be less than ___.
3. The probability of one or more public deaths should be less than ___.
4. The probability of 100 or more public deaths should be less than ___.
5. The probability of one or more public illnesses should be less than ___.
6. The probability of 100 or more public illnesses should be less than ___.

Once again, the above can be applied on either an annual or lifetime basis.

25.5   EXAMPLE OF A HEALTH RISK ASSESSMENT

Provide an example of a health risk assessment [27, 28].

SOLUTION

The residents of a small town rely on a community well system for their drinking water. The wells are monitored every three months for the presence of contaminants. The most recent quarterly monitoring has revealed the presence of chlorinated solvents at concentrations near the federal maximum contaminant levels (MCLs). Further investigation reveals that the solvents are coming from the town's landfill, where cleaning solvents from a small agricultural implement company were disposed of over a period of three decades. These solvents have migrated into the drinking water aquifer, and a sizable volume of groundwater has been contaminated. The public becomes extremely concerned and demands that the affected wells either be shut down or treated. The state instructs the water utility and town council to conduct a health risk assessment to determine if the concentration and duration of exposure have been sufficient to impair human health. Lifetime cancer risk becomes an important issue, and the town council decides to request a risk assessment from the state Department of Environmental Quality (DEQ).

The state DEQ must then estimate the lifetime cancer risk that results from drinking groundwater contaminated with trace amounts of chlorinated solvents. After the health risk assessment is completed for the hypothetical situation presented above, it is the responsibility of the participating stakeholders (including state and local elected officials and the public) to decide on a course of action. Action can include doing nothing (if the risk is low, and if "no action" will be politically acceptable), capping the affected wells and providing residents with an alternative water supply, remediating the contaminated aquifer to prevent entry of contaminated water into the distribution system, or treating the contaminated water as it enters the distribution system. A comprehensive assessment might also include evaluating the health risks associated with each possible course of action.

25.6 CHEMICAL EXPOSURE IN A LABORATORY

Over a period of one year, 1 gal of a nanochemical has been spilled in an unventilated 20 ft × 20 ft × 8 ft laboratory room at 68°F and 1 atm pressure on separate occasions. Key information and data are provided as follows:

TLV (ppm): 0.1
IDLH (ppm): 10
MW (g/gmol): 160
Evaporation rate (mL/min): 5.8
Density (g/mL): 3.119

1. Calculate how many milliliters of the chemical would have to evaporate in order to reach the TLV concentration.
2. Determine how long it would take to reach the TLV concentration.

SOLUTION

First, use the ideal gas law to find the number of gmol of gas, n in the laboratory at 1 atm and 68°F = 293 K:

$$V = (20 \text{ ft})(20 \text{ ft})(8 \text{ ft})(28.3 \text{ L/ft}^3)$$
$$= 90\,600 \text{ L}$$

$$n = \frac{(1 \text{ atm})(90\,600 \text{ L})}{\left(0.0821 \dfrac{\text{atm} \cdot \text{L}}{\text{gmol} \cdot \text{K}}\right)(293 \text{ K})}$$

$$= 3766 \text{ gmol}$$

Determine the volume of liquid that must evaporate to bring the concentration to the TLV with the equation:

$$\text{mL to reach TLV} = \frac{[\text{TVL(ppm)}][n(\text{gmol})][\text{MW(g/mol)}]}{\rho(\text{g/mL})}$$

$$= \frac{(0.1/10^6)(3760 \text{ gmol})(160 \text{ g/gmol})}{3.119 \text{ g/mL}}$$

$$= 0.0193 \text{ mL}$$

Determine the time to reach the TLV concentration with the following equation:

$$\text{Time to reach TLV} = \frac{\text{TLV volume evaporated (mL)}}{\text{Rate of evaporation(mL/min)}}$$

$$= (0.0193\,\text{mL})/(5.8\,\text{mL/min})$$
$$= 0.0033\,\text{min} = 0.2\,\text{s}$$

Not surprisingly, the time is low because of the high evaporation rate and low TLV.

25.7 LABORATORY SPILL

A laboratory room has dimensions 10 ft × 10 ft × 10 ft. The IDLH value for the vapor phase of a nanomaterial (defined as x) is 6000 ppm (0.06%). Calculate the volume of the liquid phase that needs to evaporate in order for the concentration inside the lab to meet and exceed the IDLH value. Assume the air in the room is perfectly mixed. Also indicate what measures can be taken to reduce the possibility of a problem [29].
    The following was obtained from an MSDS for the new nanomaterial:

$MW_x = 32.05$ g/gmol
Specific gravity$_x = 0.792$

SOLUTION

Calculate the volume of the room.

$$V = (10)^3 = 1000\,\text{ft}^3$$
$$= (1000)(0.0283)$$
$$= 28.3\,\text{m}^3$$
$$= 28\,300\,\text{L}$$

Find the moles of gas in the room.

$$n = \frac{PV}{RT} = \frac{(1)(28\,300)}{(0.08206)(303)}$$
$$= 1138\,\text{gmol gas}$$

The molar amount needed to reach the IDLH concentration value is given by.

$$\text{Mole fraction } (10)^3, \; x = \frac{\text{gmol}_x}{\text{gmol gas}}$$

$$0.006 = \frac{\text{gmol}_x}{1138}$$

$$\text{gmol}_x = 6.83$$

The liquid volume is calculated using the MW and density ($0.792 \, \text{g}/\text{cm}^3$).

$$V_l = \frac{(6.83)(32.05)}{0.792}$$

$$= 276.4 \, \text{cm}^3$$

One notes that evaporation of a small volume of this liquid could easily exceed the IDLH.

Finally, a ventilation system should be installed in order to reduce or eliminate the possibility of a problem. The reader is referred to Chapter 18 for ventilation details.

25.8 RESPIRATORS

Briefly describe the role respirators can play in health risk management.

SOLUTION

Respirators provide protection against inhaling harmful materials. Different types of respirators may be used depending on the level of protection desired. For example, supplied-air respirators (e.g., a self-contained breathing apparatus) may be required in situations where the presence of highly toxic substances is known or suspected and/or in confined spaces where it is likely that toxic vapors may accumulate. On the other hand, a full-face or half-face air-purifying respirator may be used in situations where measured air concentrations of identified substances will be reduced by the respirator below the substance's threshold limit value (TLV) and the concentration is within the service limit of the respirator (i.e., that provided by the canister).

Air-purifying respirators contain cartridges (or canisters) that contain an adsorbent, such as charcoal, to adsorb the toxic vapor and thus purify the breathing air. Different cartridges can be attached to the respirator depending on the nature of the contaminant. A cartridge for particulates, e.g., nanoparticles, may contain a filter rather than charcoal. The charcoal in a cartridge acts like a fixed-bed adsorber. The performance of any charcoal cartridge may be evaluated by treating it as a fixed-bed adsorber. This is treated in the next Problem.

25.9 PERFORMANCE OF A CARBON CARTRIDGE RESPIRATOR

A respirator cartridge contains 80 g of activated carbon in the form of charcoal, and tests have shown that breakthrough (when organic chemical x starts to be emitted from the cartridge) will occur when 80% of the charcoal is aspirated. How long will this cartridge be effective if the ambient concentration of chemical x is 700 ppm and the temperature is 30°C? Assume that the breathing rate of a normal person is 45 L/min (45 000 cm^3/min).

For a particular blend of carbon cartridge, the adsorption potential (equilibrium concentration) for chemical x can be expressed by the following equation:

$$\log_{10}(C_x) = -0.11\left[\left(\frac{T}{V}\right)\log_{10}\left(\frac{p'_x}{p_x}\right)\right] + 2.076$$

where C_x = amount of x adsorbed in charcoal, cm^3 liq/100 g charcoal at partial pressure, p_x; T = temperature, K; V = molar volume of x (as, liquid) at the normal boiling point, 100 cm^3/gmol for chemical x; p'_x = saturation (vapor pressure) of x; and p_x = partial pressure of x. The vapor pressure of x at 30°C is 60 mmHg and the liquid density is 1.12 g/cm^3. The molecular weight is 110.

SOLUTION

First, calculate the partial pressure of substance x, p_x in units of mmHg:

$$p_x = (700 \times 10^{-6})(760 \text{ mmHg})$$
$$= 0.532 \text{ mmHg}$$

The ratio of vapor pressure to the partial pressure, p'_x/p_x is then

$$p'_x/p_x = (60 \text{ mmHg})/(0.532 \text{ mmHg})$$
$$= 112.8$$

Substitute into the equilibrium relationship provided in the Problem statement.

$$\log_{10}(C_x) = -0.11\left[\left(\frac{T}{V}\right)\log_{10}\left(\frac{p'_x}{p_x}\right)\right] + 2.076$$

$$= -0.11\left[\left(\frac{30 + 273\text{K}}{100 \text{ cm}^3/\text{gmol}}\right)\log_{10}(112.8)\right] + 2.076$$

$$= 1.392$$

$$C_x = 24.65 \text{ cm}^3/100 \text{ g}$$

The mass of x adsorbed in the cartridge in grams at 80% saturation (i.e., at break-through) is

$$m_x = (0.8)\left(\frac{24.65\,\text{cm}^3\text{of }x}{100\,\text{g charcoal}}\right)(80\,\text{g charcoal})(1.12\,\text{g of }x/\text{cm}^3)$$
$$= 17.67\,\text{g of }x$$

Next, calculate the volumetric flowrate of x inhaled through the cartridge from the vapor pressure and breathing rate:

$$q_{air} = (700 \times 10^{-6})(45\,000\,\text{cm}^3/\text{min})$$
$$= 31.5\ \text{cm}^3/\text{min}$$

The intake mass flowrate of x is

$$\dot{m}_x = \frac{pq(MW)}{RT}$$
$$= \frac{(1\,\text{atm})(31.5\,\text{cm}^3/\text{min})(110\,\text{g/gmol})}{\left(82.06\dfrac{\text{cm}^3\cdot\text{atm}}{\text{mol}\cdot\text{K}}\right)(303\,\text{K})}$$
$$= 0.139\,\text{g/min}$$

Finally, the time for breakthrough in minutes is

$$t = (17.67\,\text{g})/(0.139\,\text{g/min})$$
$$= 127\,\text{min}$$

The typical economic concentration limit of organic vapor cartridges is approximately 1000 ppm. If the ambient or local concentration is above 1000 ppm, other methods of personal protection are recommended. It is also important to select the appropriate type of adsorbent since some toxic chemicals are not readily adsorbed by charcoal. For example, hydrogen cyanide is not well adsorbed by charcoal. Since the IDLH level of hydrogen cyanide is 50 ppm and the odor threshold is greater than 50 ppm, by the time the worker smells the vapor, it will be too late to avoid death. Therefore, it is crucial to select a cartridge adsorbent that is specifically designed for the chemical in question. Examples of such special cartridges are the cartridges for chlorine gas and pesticide vapors. Typically, cartridges are color coded so that they are easily distinguished.

Additional details on adsorption are available in the literature [3].

25.10   SAMPLING PROGRAM

It has been proposed to take samples and carry out a mass balance on an industrial operation that is losing small quantities of a toxic nanochemical, identified only as X, to the plant drainage system. The patterns of waste flow and waste concentration in Figures 25.2 and 25.3 are available for the purposes of planning a sampling program.

Based on the information provided, answer the following six questions.

1. Calculate the average flow Q in gal/hr.
2. Calculate the average concentration of X in the discharge stream in Mg/L.
3. Calculate the quantity of X being discharged based on the results from parts (1) and (2).
4. Calculate the mass of X discharged by summing up the quantities $K\,(X_i)\,(Q_i)$, where i identifies the time interval in which $X = X_i$, and $Q = Q_i$ ($i = 1, 2, 3$, and 4), and K is a constant that provides the proper conversion of units.
5. Compare the estimated quantities of X discharged from (3) and (4).
6. Explain why the results of (3) and (4) are different.

SOLUTION

By definition, the average value, $\bar{Y}$, of a dependent variable (say Y) over an interval of an independent variable (say X) is given by

$$\bar{Y} = \frac{\int (\bar{Y})dX}{\int dX}$$

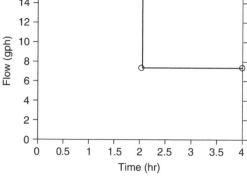

Figure 25.2   Flow versus time.

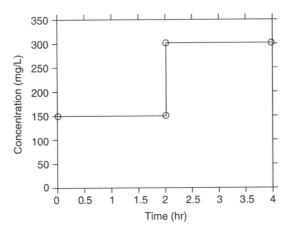

Figure 25.3 Concentration versus time.

If Y is constant over finite intervals of X, the above equation becomes

$$\bar{Y} = \frac{\sum (Y)\Delta X}{\sum \Delta X}$$

Similarly, the average value of the product of two dependent values (say YZ) is

$$\overline{YZ} = \frac{\int (YZ)dX}{\int dX}$$

and YZ is constant

$$\overline{YZ} = \frac{\sum (YZ)\Delta X}{\sum \Delta X}$$

1. Use the flowrate provided in Figure 25.2. Since the flow rates are uniform during the period from 0 to 2 hr and 2 to 4 hr:

$$\overline{Q} = \frac{\text{Total flow volume}}{\text{Total time}} = \frac{\sum Q_i \Delta t}{\sum \Delta t}$$

$$= \frac{(15 \text{ gph})(2 \text{ hr}) + (7.5 \text{ gph})(2 \text{ hr})}{(2 \text{ hr} + 2 \text{ hr})}$$

$$= \frac{30 \text{ gal} + 15 \text{ gal}}{4 \text{ hr}}$$

$$= 11.25 \text{ gph}$$

2. The average concentration of X can be estimated in a similar manner using the second figure.

$$\overline{C} = \frac{\text{Concentration } (\Delta t)}{\text{Total time}} = \frac{\sum C_i \Delta t}{\sum \Delta t}$$

$$= \frac{(150 \ \text{mg/L})(2 \ \text{hr}) + (300 \ \text{mg/L})(2 \ \text{hr})}{(2 \ \text{hr} + 2\text{hr})}$$

$$= \frac{300 \ \text{mg-hr/L} + 600 \ \text{mg-hr/L}}{2 \ \text{hr} + 2 \ \text{hr}} = \frac{900}{4}$$

$$= 225 \ \text{mg/L}$$

3. The total mass rate $(\dot{M}_3)$ of X from these two results is simply the product of the average flow rate and the average concentration.

$$\dot{M}_3 = \overline{Q}\overline{C}$$
$$= (11.25 \ \text{gph})(225 \ \text{mg/L})(3.875\text{L/gal})$$
$$= 9809 \ \text{mg/hr}$$
$$= 9.809 \ \text{g/hr}$$

4. The total quantity of X being discharged is the sum of the products of the flow rate and concentration divided by the appropriate time interval. The appropriate approach is to note that the mass of compound X in a given unit of volume is simply the concentration times the volume. Employing a uniform time interval of 1 hr, one obtains the information in Table 25.2.

The total is 9000 gal-mg/L-hr over the 4 hr period. Converting units, the total mass rate is:

$$\dot{M}_4 = (9000 \ \text{gal-mg/L-hr})(3.875 \ \text{L/gal})/4$$
$$= 8719 \ \text{mg/hr}$$
$$= 8.719 \ \text{g/hr}$$

TABLE 25.2   Data for Part 4 of Problem 25.10

Interval (hr)	Q_i (gph)	C_i (mg/L)	$Q_i C_i$ (gal-mg/L-hr)
0–1	15.0	150	2250
1–2	15.0	150	2250
2–3	7.5	300	2250
3–4	7.5	300	2250
Total			9000

5. The method used in (3) overestimates the discharge in (4) by 12.5%.

6. Simple overall cross-products of the average flow rate and average concentration provide biased results. The correct procedure must follow that provided in the definition of average values, i.e., incremental values need to be calculated over each appropriate interval. Therefore, the result in (4) is correct.

26 Hazard Risk Assessment [15]

As indicated in the previous Chapter, many practitioners and researchers have confused health risk with hazard risk, and vice versa. Although both employ a four-step method of analysis (see Problem 26.2), the procedures are quite different, with each providing different results, information, and conclusions. Both do share a common concern in that when nanoparticles are inhaled (the primary pathway to the human body) they may elude one's defense mechanisms.

The reader is referred to Chapter 25 for details differentiating health risk from hazard risk. As with health risk, there is a serious lack of information on the hazards and associated implications of these hazards with nanoapplications. The unknowns in this risk area may be larger in number and greater in potential consequences. It is the author's judgment that hazard risks have unfortunately received something less than the attention they deserve. However, hazard risk analysis details are available and the traditional approaches successfully applied in the past can be found in this Chapter. Future work will almost definitely be based on this methodology.

Much has been written about Michael Crichton's powerful science-thriller novel titled *Prey*. (The book was not only a best seller, but the movie rights were sold for $5 million.) In it, Crichton provides a frightening scenario in which swarms of nanorobots, equipped with special power generators and unique software, prey on living creatures. To compound the problem, the robots continue to reproduce without any known constraints. This scenario is an example of an accident and represents only one of a near infinite number of potential hazards that can arise in any nanotechnology application. Although the probability of the horror scene portrayed by Crichton, as well as other similar events, is extremely low, steps and procedures need to be put into place to reduce, control, and hopefully eliminate these events from actually happening. This Chapter attempts to provide some of this information.

The previous Chapter defined both "chronic" and "acute" problems. As indicated, when the two terms are applied to emissions, the former usually refers to ordinary, round-the-clock, everyday emissions while the latter term deals with short, out-of-the-norm, accidental emissions. Thus, acute problems normally refer to accidents and/or hazards. The Crichton scenario discussed above is an example of an acute problem, and one whose solution would be addressed/treated by a hazardous risk assessment, as described in this Chapter, rather than the health risk approach provided in the previous Chapter.

As noted in the previous chapter, one of the problems with nanoparticles is that their properties can vary significantly with particle size. In effect, particles of one size possess different properties of the same material of a different size. One size of nanoparticles could pose a hazard, while another size would not be a concern.

The reader is cautioned that this Chapter contains some advanced statistics material. Some introductory probability and statistics principles are presented in Chapter 9, and one can choose to review this first. In addition, much of the material presented has been drawn from several references cited below.

26.1   EXAMPLE OF A HAZARD

Provide an industrial example of a hazard [27, 28].

SOLUTION

In May 1991 an accidental release of chlorine gas from a broken fitting at the Pioneer Chlor-Alkali plant in Henderson, Nevada, released a gas cloud that covered about one square mile. The incident resulted in the evacuation of several thousand people from their homes and their relocation to emergency shelters. Respiratory distress was documented in several hundred people, schools, factories and shopping areas were shut down, and transportation in the area was disrupted for more than 12 hours. The factory that was responsible for the leak paid a large fine and settled numerous civil damage claims caused by the release.

Fortunately, no one was killed as a result of exposure to the cloud. Chlorine leaked out at a fairly slow rate and winds were light, giving people time to either get out of the path of the plume or to implement shelter-in-place procedures. Trained emergency response teams were able to cordon off the affected area and assist in the notification and evacuation of persons from areas in the projected path of the chlorine plume. The accident occurred in the early morning hours, and, once the leak had been fixed, the plume was diluted by atmospheric mixing. The plume had dissipated to safe levels by nightfall.

The reader may have seen televised accounts or read of this incident or similar events that have occurred in their local area. In addition to industrial accidents, transportation accidents, such as train derailments or the overturning of a tanker truck, can also result in the release of hazardous materials into the environment if these accidents result in the rupture of chemical containers.

26.2   RISK EVALUATION PROCESS FOR ACCIDENTS

Provide an outline of a hazard risk assessment.

SOLUTION

There are several steps in evaluating the risk of an accident. These are detailed in Figure 26.1 if the system in question is a chemical plant.

1. A brief description of the equipment and chemicals used in the plant is needed.
2. Any hazard in the system has to be identified. Hazards that may occur in a chemical plant include:

    Fire

    Toxic vapor release

    Slippage

    Corrosion

    Explosions

    Rupture of a pressurized vessel

    Runaway reactions
3. The event or series of events that will initiate an accident has to be identified. An event could be a failure to follow correct safety procedures, improperly repaired equipment or failure of a safety mechanism. This is normally referred to as the hazard identification step. There are several methods used to identify hazards. Two methods are the process checklist and the hazard and operability study (HAZOP), both of which are detailed in the literature [1, 3–5].
4. The probability that the accident will occur has to be determined. For example, if a chemical plant has a 10-year life, what is the probability that the temperature in a reactor will exceed the specified temperature range? The probability can be ranked from low to high. A low probability means that it is unlikely for the event to occur in the life of the plant. A medium probability suggests that there is a possibility that the event will occur. A high probability means that the event will probably occur during the life of the plant.
5. The severity of the consequences of the accident must be determined. This will be described later in detail.
6. If the probability of the accident and the severity of its consequences are low, then the risk is usually deemed acceptable and the plant should be allowed to operate. If the probability of occurrence is too high or the damage to the surroundings is too great, then the risk is usually unacceptable and the system needs to be modified to minimize these effects.

The heart of the hazard risk assessment algorithm provided is enclosed in the dashed box of Figure 26.1. The algorithm allows for reevaluation of the process if the risk is deemed unacceptable. (The process is then repeated, starting with either step one or two.)

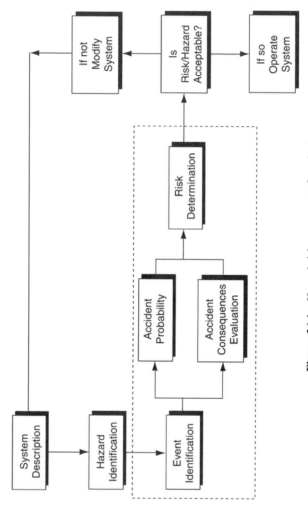

Figure 26.1   Hazard risk assessment flowchart.

26.3 PLANT AND PROCESS SAFETY

Answer the following two questions.

1. List the types of process events that can result in a plant accident and discuss the various kinds of equipment failure that can occur in a process plant. Also list several design approaches that should be employed to minimize accidents in a process plant.

2. List the various pieces of protective equipment that may be found at a chemical plant.

SOLUTION

1. The types of process events that can result in a plant accident are:

Abnormal temperatures

Abnormal pressures

Material flow stoppage

Equipment failures that can occur in a process plant may be described within the major equipment categories of reactors, heat exchangers, vessels, mass transfer operations, pipes and valves, and pumps.

Plant safety and design may be divided into four major areas. These areas are plant equipment and layout, controls, emergency safety devices, and safety factors and backup. These four areas are listed as follows:

Plant and equipment layout

Controls

Emergency safety devices

Safety factors and backup

2. Relief devices and collection systems include:

Safety valves

Rupture disks

Piping networks

Treatment systems include:

Liquid disengagement

Gas quenching

Scrubbing

Disposal systems include:

Atmospheric venting

Flaring (elevated and ground)

Recovery and containment – the recovery and containment of materials includes returning the material to the process, containment of the material external to the process and flare gas recovery.

26.4 SERIES AND PARALLEL SYSTEMS [3, 4]

Determine the reliability of an eight-component system consisting of five operations as shown in Figure 26.2. The reliabilities are provided beneath each component.

SOLUTION

Many systems consisting of several components can be classified as series, parallel, or a combination of both. A series system is one in which the entire system fails to operate if any one of its components fails to operate. If such a system consists of n components that function independently, then the reliability of the system is the product of the reliabilities of the individual components. If R_s denotes the reliability of a series system and R_i denotes the reliability of the ith component. ($i = 1$, $2, \ldots, n$), then the formula for R_s, the reliability of a series system, is as follows:

$$R_s = R_1 R_2 \ldots R_n = \prod_i R_i$$

A parallel system is one that fails to operate only if all of its components fail to operate. If R_i is the reliability of the ith component, then $1 - R_i$ is the probability that the ith component fails. Assuming all n components function independently, the probability that all n components fail is $(1 - R_1)(1 - R_2) \ldots (1 - R_n)$. Subtracting this product from unity yields the following formula for R_p, the fractional reliability of a parallel system,

$$R_p = 1 - (1 - R_1)(1 - R_2) \ldots (1 - R_n) = 1 - \prod_i (1 - R_i)$$

The reliability formulas for series and parallel systems can be used to obtain the reliability of a system that combines features of series and parallel systems.

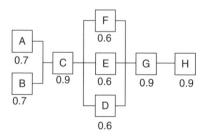

Figure 26.2 Diagram for Problem 26.4.

The reliability of the parallel subsystem consisting of components A and B in the Problem statement is obtained by applying the equation for a parallel system:

$$R_p = 1 - (1 - 0.7)(1 - 0.7)$$
$$= 0.91$$

The reliability of the parallel subsystem consisting of components D, E, and F is

$$R_p = 1 - (1 - 0.6)(1 - 0.6)(1 - 0.6)$$
$$= 0.936$$

The reliability of the entire system is obtained by applying the equation for a series system:

$$R_s = (0.91)(0.9)(0.936)(0.9)(0.9)$$
$$= 0.621$$

26.5 BINOMIAL DISTRIBUTION

A coolant sprinkler system in a highly exothermic reactor producing a new nano-chemical has 20 independent spray components, each of which fails with probability of 0.1. The coolant system is considered to "fail" only if four or more of the sprays fail. What is the probability that the sprinkler, system fails?

SOLUTION

Several probability distributions figure prominently in reliability calculations. The binomial distribution is one of them. Consider n independent performances of a random experiment with mutually exclusive outcomes that can be classified "success" and "failure." These outcomes do not necessarily have the ordinary connotation of success or failure. Assume that P, the probability of success on any performance of the random experiment, is constant. Let $q = 1 - P$ be the probability of failure. The probability distribution of X, the number of successes in n performances of the random experiment, is a binomial distribution with probability distribution function (PDF) specified by

$$f(x) = \frac{P^x q^{n-1} n!}{x!(n-x)!} \qquad x = 0, 1, 2, \ldots, n$$

where $f(x) =$ probability of x successes in n performances, and $n =$ number of independent performances of a random experiment. The binomial distribution can therefore be used to calculate the reliability of a redundant system. A redundant system consisting of n identical components is a system that fails only if more than r components fail. Typical examples include single-usage equipment such as missile engines, short-life batteries, and flash bulbs that are required to operate for one time period and not to be reused.

Assume that the n components of the spray system are independent with respect to failure, and that the reliability of each is $1 - P$. One may associate "success" with the failure of a spray. Then X, the number of failures, has a binomial PDF and the reliability of the random system is

$$P(X \leq r) = \sum_{x=0}^{r} \frac{P^x q^{n-1} n!}{x!(n-x)!}$$

For this problem, X has a binomial distribution with $n = 20$ and $P = 0.10$; the probability that the system fails is given by

$$P(X \geq 4) = \sum_{x=4}^{20} \frac{(0.10^x)(0.90^{20-x})20!}{x!(20-x)!}$$

$$= 1 - P(X \leq 3)$$

This simplifies the problem to:

$$P(X \leq 4) = 1 - \sum_{x-=0}^{3} \frac{(0.10^x)(0.90^{20-x})20!}{x!(20-x)!}$$

$$= 0.13$$

26.6   THE POISSON DISTRIBUTION

The Poisson distribution can also be used to calculate the probability of failure. If λ is the failure rate of each component of a system and λt is the average number of failures per unit of time, the probability of x failures in the specified unit of time is obtained from

$$f(x) = \frac{e^{-\lambda t}(\lambda t)^x}{x!}; \qquad x = 0, 1, 2, \ldots$$

Suppose that the average number of temperature excursions (due to a runaway reaction) during 5000 hours of operation of a nanoreactor is 10. What is the probability of no excursions during an 8-hour work period?

SOLUTION

Substituting $\lambda = (10/5000)$ and $t = 8$ in the Poisson equation yields

$$f(x) = \frac{e^{-0.016}(0.016)^x}{x!}; \qquad x = 0, 1, 2, \ldots$$

as the number of breakdowns in an 8-hour period. The possibility of no breakdowns in an 8-hour working period is therefore

$$P(X = 0) = e^{-0.016} = 0.984$$

In addition to the application cited above, the Poisson distribution can be used to obtain the reliability of a standby redundancy system [29], in which one unit is in the operating mode and n identical units are in standby mode. Unlike a parallel system where all units in the system are active, in the standby redundancy system the standby units are inactive. If all units have the same failure rate in the operating mode, unit failures are independent, standby units have zero failure rate in the standby mode, and there is perfect switchover to a standby when the operating unit fails, the reliability R of the standby redundancy system is given by

$$R = \sum_{x=0}^{n} \frac{e^{-\lambda t}(\lambda t)^x}{x!}$$

This is the probability of n or fewer failures in the time period specified by t. For example, consider the case of a standby redundancy system with one operating unit and one on standby (i.e., a system that can survive one failure). If the failure rate is one unit every five years, then the six-month reliability of the system is obtained by substituting $n = 1$, $\lambda = 0.2$, and $t = 0.5$ above, which gives

$$R = \sum_{x=0}^{1} \frac{e^{-0.1}(0.1)^x}{x!} = 0.9048 + 0.0905 = 0.9953$$

$$= 99.53$$

as the six-month reliability.

26.7 THE WEIBULL DISTRIBUTION

Qualitatively describe the Weibull distribution.

SOLUTION

Frequently, the failure rate of equipment exhibits three stages: a break-in stage with a declining failure rate, a useful life stage characterized by a fairly constant failure rate, and a wearout period characterized by an increasing failure rate. Many industrial parts and components follow this path; mortality rates is another example. A failure rate curve exhibiting these three phases is called a "bathtub" curve (see Figure 26.3) and best characterized by a Weibull distribution.

The Weibull distribution provides a mathematical model of all three stages of the "bathtub" curve. The variety of assumptions about failure rate and the probability

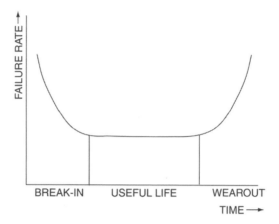

Figure 26.3   Bathtub curve.

distribution of time to failure that can be accommodated by the Weibull distribution make it especially attractive in describing failure time distributions in industrial and process plant applications. Although specific details on this distribution are beyond the scope of this book, extensive information is available in the literature [1, 4, 5].

26.8   THE NORMAL DISTRIBUTION

The time to failure of a new nanotransistor is normally distributed with

$\mu = 100$ months
$\sigma = 2$ months

Determine the probability that one of these new transistors will fail between 98 and 104 months.

SOLUTION

The normal distribution was briefly discussed in Problem 9.9 in Chapter 9. That discussion is now expanded.

SOLUTION

If a variable T is normally distributed with mean μ and standard deviation σ, then the random variable $(T - \mu)/\sigma$ is also normally distributed with mean 0 and standard deviation 1. The term $(T - \mu)/\sigma$ is called a "standard normal variable," and the graph of its distribution is called a "standard normal curve." Table 26.1 is a

TABLE 26.1 Standard Normal, Cumulative Probability in Right-Hand Tail (for Negative Values of z, Areas Are Found by Symmetry)[a]

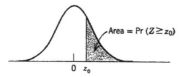

Area $= \Pr(Z \geq z_0)$

				Next Decimal Place of z_0					
z_0	0	1	2	3	4	5	6	7	8	9
00.0	0.500	0.496	0.492	0.488	0.484	0.480	0.476	0.472	0.468	0.464
0.1	0.460	0.456	0.452	0.448	0.444	0.440	0.436	0.433	0.429	0.425
0.2	0.421	0.417	0.413	0.409	0.405	0.401	0.397	0.394	0.390	0.386
0.3	0.382	0.378	0.374	0.371	0.367	0.363	0.359	0.356	0.352	0.348
0.4	0.345	0.341	0.337	0.334	0.330	0.326	0.323	0.319	0.316	0.312
0.5	0.309	0.305	0.302	0.298	0.295	0.291	0.288	0.284	0.281	0.278
0.6	0.274	0.271	0.268	0.264	0.261	0.258	0.255	0.251	0.248	0.245
0.7	0.242	0.239	0.236	0.233	0.230	0.227	0.224	0.221	0.218	0.215
0.8	0.212	0.209	0.206	0.203	0.200	0.198	0.195	0.192	0.189	0.187
0.9	0.184	0.181	0.179	0.176	0.174	0.171	0.169	0.166	0.164	0.161
1.0	0.159	0.156	0.154	0.152	0.149	0.147	0.145	0.142	0.140	0.138
1.1	0.136	0.133	0.131	0.129	0.127	0.125	0.123	0.121	0.119	0.117
1.2	0.115	0.113	0.111	0.109	0.107	0.106	0.104	0.102	0.100	0.099
1.3	0.097	0.095	0.093	0.092	0.090	0.089	0.087	0.085	0.084	0.082
1.4	0.081	0.079	0.078	0.076	0.075	0.074	0.072	0.071	0.069	0.063
1.5	0.057	0.066	0.064	0.063	0.062	0.061	0.059	0.058	0.057	0.056
1.6	0.055	0.054	0.053	0.052	0.051	0.049	0.048	0.047	0.046	0.046
1.7	0.045	0.044	0.043	0.042	0.041	0.040	0.039	0.038	0.038	0.037
1.8	0.036	0.035	0.034	0.034	0.033	0.032	0.031	0.031	0.030	0.029
1.9	0.029	0.028	0.027	0.027	0.026	0.026	0.025	0.024	0.024	0.023
2.0	0.023	0.022	0.022	0.021	0.021	0.020	0.020	0.019	0.019	0.018
2.1	0.018	0.017	0.017	0.017	0.016	0.016	0.015	0.015	0.015	0.014
2.2	0.014	0.014	0.013	0.013	0.013	0.012	0.012	0.012	0.011	0.011
2.3	0.011	0.010	0.010	0.010	0.010	0.009	0.009	0.009	0.009	0.008
2.4	0.008	0.008	0.008	0.008	0.007	0.007	0.007	0.007	0.007	0.006
2.5	0.006	0.006	0.006	0.006	0.006	0.005	0.005	0.005	0.005	0.005
2.6	0.005	0.005	0.004	0.004	0.004	0.004	0.004	0.004	0.004	0.004
2.7	0.003	0.003	0.003	0.003	0.003	0.003	0.003	0.003	0.003	0.003
2.8	0.003	0.002	0.002	0.002	0.002	0.002	0.002	0.002	0.002	0.002
2.9	0.002	0.002	0.002	0.002	0.002	0.002	0.002	0.001	0.001	0.001
z_0			Detail of Tail	$(0._2135,$	For Example,	Means	0.00135)			
2	$0._1228$	$0._1179$	$0._1139$	$0._1107$	$0._2820$	$0._2621$	$0._2466$	$0._2347$	$0._2256$	$0._2187$
3	$0._2135$	$0._2968$	$0._3687$	$0._3483$	$0._3337$	$0._3233$	$0._3159$	$0._3108$	$0._4723$	$0._4481$
4	$0._4317$	$0._4207$	$0._4133$	$0._5854$	$0._5541$	$0._3340$	$0._5211$	$0._5130$	$0._6793$	$0._6479$
5	$0._6287$	$0._6170$	$0._7996$	$0._7579$	$0._7333$	$0._7190$	$0._7107$	$0._8599$	$0._8332$	$0._8182$
	0	1	2	3	4	5	6	7	8	9

[a]From Woonacott and Woonacott [30].

tabulation of areas under a standard normal curve to the right of z_0 for nonnegative values of z_0 [30]. Probabilities about a standard normal variable Z can be determined from this table. For example,

$$P(Z > 1.54) = 0.062$$

is obtained directly from the table as the area to the right of 1.54 (see Figure 26.4). The symmetry of the standard normal curve about zero implies that the area to the right of zero is 0.5 and the area to the left of zero is 0.5. Consequently

$$P(0 < Z < 1.54) = 0.5 - 0.062 - 0.438$$

Also, because of symmetry

$$P(-1.54 < Z < 0) = 0.438 \quad \text{and} \quad P(Z < -1.54) = 0.062$$

Note that the area to the right of 1.54 is 0.062. The following probabilities can also be deduced from Figure 26.4

$$P(-1.54 < Z < 1.54) = 0.876$$
$$P(Z < 1.54) = 0.938$$
$$P(Z > -1.54) = 0.938$$

Table 26.1 also can be used to determine probabilities concerning normal random variables that are not standard normal variables. The required probability is first converted to an equivalent probability about a standard normal variable. As described above, this is accomplished by subtracting the mean from the variable and dividing the resulting term by the standard deviation. Thus, for any random variable X that is normally distributed with mean μ and standard deviation σ,

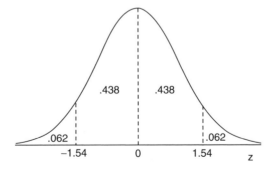

Figure 26.4   Areas under a standard normal curve.

$$P(\mu - \sigma < X < \mu + \sigma) = P\left(-1 < \frac{X - \mu}{\sigma} < 1\right)$$

$$= P(-1 < Z < 1) = 0.68$$

$$P(\mu - 2\sigma < X < \mu + 2\sigma) = P\left(-2 < \frac{X - \mu}{\sigma} < 2\right)$$

$$= P(-2 < Z < 2) = 0.95$$

$$P(\mu - 3\sigma < X < \mu + 3\sigma) = P\left(-3 < \frac{X - \mu}{\sigma} < 3\right)$$

$$= P(-3 < Z < 3) = 0.997$$

With regard to the Problem statement, one notes that the time to failure, T, is normally distributed with $\mu = 100$ months and $\sigma = 2$ months. T may be converted to a standard normal variable in a manner described above.

$$P(98 < T < 104) = P\left(\frac{98 - 100}{2} < \frac{T - 100}{2} < \frac{104 - 100}{2}\right)$$

$$= P(-1 < Z < 2)$$

$$= 0.341 + 0.477$$

$$= 0.818$$

26.9 SOIL CONTAMINATION

An accidental emission from a nanoprocess has contaminated the soil in a local park with a toxic nanochemical. The regulatory specification on this nanoagent in soil calls for a level of 1.0 ppm or less. Numerous sample observations of the concentration (C) of the chemical indicates a normal distribution with a mean of 0.60 ppm and a standard deviation of 0.20 ppm. Estimate the probability that the agent will exceed the regulatory limit.

SOLUTION

This problem requires the calculation of $P(C > 1.0)$. First, normalize the variable C,

$$P\{[(C - 0.6)/0.2] > [(1.0 - 0.6)/0.20]\}$$

$$P(Z > 2.0)$$

See the standard normal table (Table 26.1).

$$P(Z > 2.0) = 0.0228$$
$$= 2.28\%$$

For this situation, the area to the right of 2.0 is 2.28% of the total area. This represents the probability that the nanochemical in the soil will exceed the regulatory limit of 1.0 ppm.

26.10   EVENT TREE ANALYSIS

If a building fire occurs, a smoke alarm sounds with probability 0.9. The sprinkler system functions with probability 0.7 whether or not the smoke alarm sounds. The consequences are minor fire damage (alarm sounds, sprinkler works), moderate fire damage with few injuries (alarm sounds, sprinkler fails), moderate fire damage with many injuries (alarm fails, sprinkler works), and major fire damage with many injuries (alarm fails, sprinkler fails). Construct an event tree and indicate the probabilities for each of the four consequences.

SOLUTION

An event tree provides a diagrammatic representation of event sequences that begin with a so-called initiating event and terminate in one or more undesirable consequences. In contrast to a fault tree (considered in the next Problem), which works backward from an undesirable consequence to possible causes, an event tree works forward from the initiating event to possible undesirable consequences. The initiating event may be equipment failure, human error, power failure, or some other event that has the potential for adversely affecting the environment or an ongoing process and/or equipment.

Note that for each branch in an event tree (moving horizontally), the sum of probabilities must equal 1.0 (see Figures 26.5–26.7). Note again that an event tree includes the following: (1) works forward from the initial event, or an event that has the potential for adversely affecting an ongoing process, and ends at one or more undesirable consequences; (2) is used to represent the possible steps leading to a failure or accident; (3) uses a series of branches that relate the proper operation and/or failure of a system with the ultimate consequences; (4) is a quick identification of the various hazards that could result from a single initial event; (5) is beneficial in examining the possibilities and consequences of a failure; (6) usually does not quantify (although it can) the potential of the event occurring; and (7), can be incomplete if all the initial events are not identified.

Thus, the use of event trees is sometimes limiting for hazard analysis because it usually does not quantify the potential of the event occurring. It may also be incomplete if all the initial occurrences are not identified. Its use is beneficial in examining,

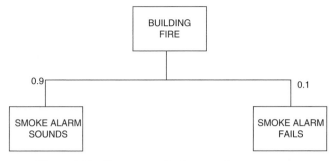

Figure 26.5 Event tree with first set of consequences.

rather than evaluating, the possibilities and consequences of a failure. For this reason, a fault tree analysis should supplement this model to establish the probabilities of the event tree branches. This topic is introduced in the next Problem.

The first consequence(s) of the building fire and the probabilities of the first consequence(s) are shown in Figure 26.5. The second consequence(s) of the building fire and the probabilities of the consequence(s) are shown in Figure 26.6. The final consequences and the probabilities of minor fire damage, moderate fire damage with few injuries, moderate fire damage with many injuries, and major fire damage with many injuries are shown in Figure 26.7.

26.11   FAULT TREE ANALYSIS

In the production of a new nanochemical, a runaway chemical reaction can occur if coolers fail (A) or there is a bad chemical batch (B). Coolers fail only if both cooler

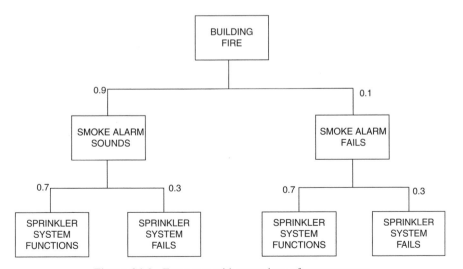

Figure 26.6 Event tree with second set of consequences.

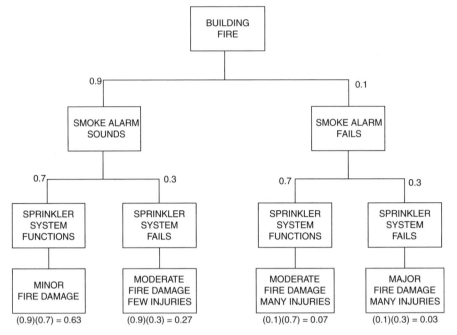

Figure 26.7   Event tree with final set of consequences.

#1 fails (C) and cooler #2 fails (D). A bad chemical batch occurs if there is a wrong mix (E) or there is a process upset (F). A wrong mix occurs only if there is an operator error (G) and instrument failure (H).

1. Construct a fault tree.
2. If the following annual probabilities are provided by the plant engineer, calculate the probability of a runaway chemical reaction occurring in a year's time given the following probabilities:
3. $P(C) = 0.05$
   $P(D) = 0.08$
   $P(F) = 0.06$
   $P(G) = 0.03$
   $P(H) = 0.01$

SOLUTION

Fault tree analysis seeks to relate the occurrence of an undesired event to one or more antecedent events. The undesired event is called the "top event" and the antecedent events are called "basic events." The top event may be, and usually is, related to the basic events via certain intermediate events. The fault tree diagram exhibits

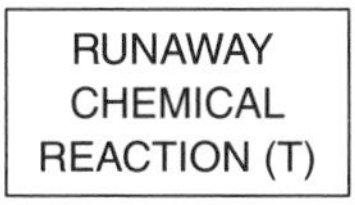

RUNAWAY
CHEMICAL
REACTION (T)

Figure 26.8 Top event of fault tree.

the casual chain linking the basic events to the intermediate events and the latter to the top event. In this chain, the logical connection between events is illustrated by so-called *logic gates*. The principal logic gates are the AND gate, symbolized on the fault tree by ⌓ and the OR gate, symbolized by ⌓.

The reader should note that a fault tree includes the following: (1) works backward from an undesirable event or ultimate consequence to the possible causes and failures, (2) relates the occurrence of an undesired event to one or more preceding events, (3) "chain links" basic events to intermediate events that are in turn connected to the top event, (4) is used in the calculation of the probability of the top event, (5) is based on the most likely or credible events that lead to a particular failure or accident, and (6) analysis includes human error as well as equipment failure.

Begin with the top event for this Problem as shown in Figure 26.8. The first branch of the fault tree can be generated by applying logic gates (see Figure 26.9). The second branch of the fault tree can be obtained by applying the logic gates as shown in Figure 26.10. The third and the last branch of the fault tree, is generated in a similar manner (Figure 26.11).

The probability that a "nano" runaway reaction will occur is then

$$P = (0.05)(0.08) + (0.01)(0.03) + 0.06$$
$$= 0.064$$

Note that the process upset, F, is major contributor to the probability.

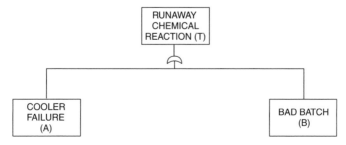

Figure 26.9 Fault tree with first branch.

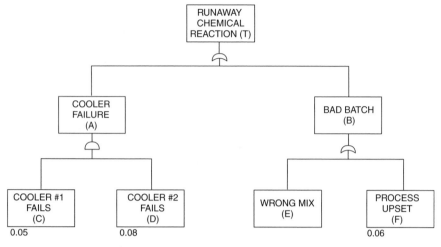

Figure 26.10   Fault tree with second branch.

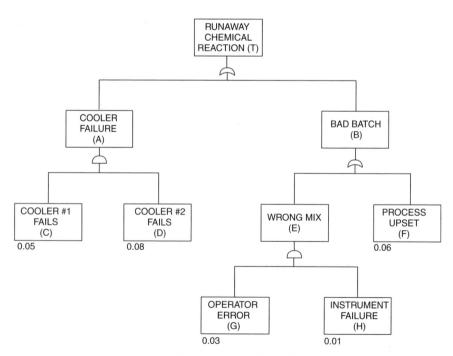

Figure 26.11   Fault tree with third branch.

26.12   UPPER AND LOWER FLAMABILITY LIMITS

1. Calculate the upper flammability limit (UFL) and lower flammability limit (LFL) for a gas mixture consisting of 50% (by volume) methane, 10% ethane, and 40% pentane.
2. Calculate the UFL and LFL of a gas mixture consisting of 40% methane, 5% ethane, 30% pentane and 25% octane.
3. Determine how the UFL and LFL vary when a mixture consisting of 10% ethane is mixed with various volume percents of methane and pentane.
4. Determine how the UFL and LFL vary when a mixture consisting of 10% pentane, is mixed with various volume percents of methane and ethane.

UFL and LFL data are provided in Table 26.2.

SOLUTION

Flammability limits (or explosion limits) for a flammable gas define the concentration range of a gas–air mixture with which an ignition source can start a self-propagating reaction. The minimum and maximum fuel concentrations in air that will produce a self-sustaining reaction under given conditions are called the *lower flammability limit* (LFL) and the *upper flammability limit* (UFL). (The abbreviations LEL and UEL, for *lower* and *upper explosivity limits*, respectively, are sometimes used.) The flammability of a gas mixture can be calculated by using Le Chatelier's law given the flammability limits of the gas components.

$$\text{LFL(mix)} = \frac{1}{y_1/\text{LFL}_1 + y_2/\text{LFL}_2 + \cdots + y_n/\text{LFL}_n}$$

$$\text{UFL(mix)} = \frac{1}{y_1/\text{UFL}_1 + y_2/\text{UFL}_2 + \cdots + y_n/\text{UFL}_n}$$

where $y_1, y_2, \ldots, y_n$ = volume or mole fraction of each of the components; LFL(mix), UFL(mix) = mixture lower and upper flammability limits in volume or mole fraction, respectively; $\text{LFL}_1, \text{LFL}_2, \ldots, \text{LFL}_n$ = component lower flammability limits in volume or mole fraction; and $\text{UFL}_1, \text{UFL}_2, \ldots, \text{UFL}_n$ = component upper flammability limits in volume or mole fraction.

TABLE 26.2   UFL and LFL Data

	LFL	UFL
Methane	4.6	14.2
Ethane	3.5	15.1
Pentane	1.4	7.8
Octane	1.0	6.5

If data are not available for a particular gas mixture, it is possible to estimate the flammability limit (FL) by taking data for a similar material and applying the equation:

$$FL_A = (M_B/M_A)FL_B$$

where M_A and M_B are the molecular weights of components A and B, respectively [1, 4, 5].

1. Applying the UFL and LFL equations gives

$$UFL = \frac{1.0}{\displaystyle\sum_{i=1}^{3}(y_i/UFL_i)}$$

$$= \frac{1}{(0.5/14.2) + (0.1/15.1) + (0.4/7.8)}$$

$$= \frac{1}{0.0352 + 0.0066 + 0.0513}$$

$$= \frac{1}{0.0931}$$

$$= 0.10.74 = 1074\%$$

and

$$LFL = \frac{1}{(0.5/4.6) + (0.1/3.5) + (0.4/1.4)}$$

$$= \frac{1}{0.423}$$

$$= 0.02364 = 2.364\%$$

2. Similarly, for the four compound mixture,

$$UFL = 0.0922 = 9.22\%$$
$$LFL = 0.0177 = 1.77\%$$

3. LFL and UFL variation when a mixture of 10% ethane is mixed with various volume percents of methane and pentane is provided in Table 26.3 and Figure 26.12 [31]. Note that as the volume percent of methane in the mixture increases from 0 to 90, so too does the LVL and UFL. This makes sense because the denominator in the LFL and UFL equations experiences an overall decrease, leading to an increase in LFL and UFL.

TABLE 26.3 **Values of LFL and UFL with Varying Concentrations of Methane and Pentane with 10% Ethane**

% Methane	% Ethane	% Pentane	LFL	UFL
0	10	90	1.49%	8.20%
10	10	80	1.61%	8.60%
20	10	70	1.75%	9.05%
30	10	60	1.91%	9.55%
40	10	50	2.12%	1.11%
50	10	40	2.36%	1.74%
60	10	30	2.68%	11.45%
70	10	20	3.09%	12.26%
80	10	10	3.65%	13.20%
90	10	0	4.46%	14.29%

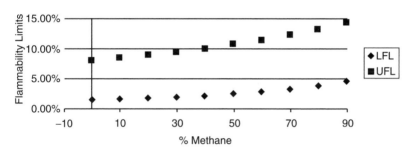

Figure 26.12 LFL and UFL as % methane and % pentane vary (constant % ethane).

TABLE 26.4 **Value of LFL and UFL with Varying Concentrations of Methane and Ethane with 10% Pentane**

% Methane	% Ethane	% Pentane	LFL	UFL
0	90	10	3.04%	13.81%
10	80	10	3.11%	13.73%
20	70	10	3.18%	13.65%
30	60	10	3.25%	13.57%
40	50	10	3.32%	13.49%
50	40	10	3.40%	13.42%
60	30	10	3.48%	13.34%
70	20	10	3.56%	13.27%
80	10	10	3.65%	13.20%
90	0	10	3.74%	13.12%

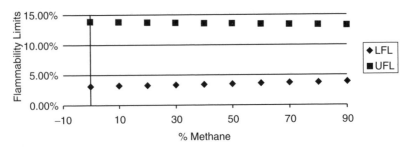

Figure 26.13   LFL and UFL as % methane and % ethane vary (constant % pentane).

4. LFL and UFL variation when a mixture of 10% pentane is mixed with various volume percents of methane and ethane can be found in Table 26.4 and Figure 26.13. As the volume percent of methane in the mixture increases from 0 to 90, the LFL and UFL remain relatively constant. This makes sense when comparing this to the trend observed in (3) because the LFL and UFL values of methane and ethane are more similar than the values of methane and pentane.

27 Epidemiology

As described in the Introduction to this Part, perhaps the greatest danger from nano-materials may be their escape and persistence in the environment, the food chain, and human and animal tissues. Unintended consequences can include health problems, disease, and even death. Epidemiology provides the engineer and scientist with the ability to study and analyze these negative effects.

Epidemiology is a discipline within the health sciences that deals with the study of the occurrence of disease in human populations. Whereas physicians are generally concerned with the single patient, epidemiologists are generally concerned with groups of people who share certain characteristics. A good example would be the interest epidemiologists show in characteristics associated with adverse health effects, e.g., smoking and lung cancer, asbestos exposure and asbestosis, or noise and hearing loss. Epidemiologic data come from many different sources. Acquiring reliable, accurate, and complete data describing health problems is a key concern of the epidemiologist.

Epidemiology also operates within the context of public health with a strong emphasis on the prevention of disease through the reduction of factors that may increase the likelihood that an individual or group will suffer a given disease. In particular, implicit in the practice of epidemiology is the need for different disciplines to participate in studying the influence of certain characteristics, e.g., occupation, or human health.

As one might suppose, there are essentially no epidemiological data available on nanoparticles/nanochemicals. There also appears to be little to no possibility of obtaining meaningful and useful data in the near future. Health effect variation with nanoparticle size will not come easily or quickly. Toxicological data (see Chapter 22) will have to come first.

This last Chapter also serves to introduce the reader to certain simple terms in statistics. This treatment may be viewed as an extension of the material in Chapters 9 and 26.

27.1 HISTORICAL VIEW

Provide some historical background on epidemiology [32].

Nanotechnology: Basic Calculations for Engineers and Scientists, by Louis Theodore
Copyright © 2006 John Wiley & Sons, Inc.

SOLUTION

Epidemiology is a term derived from the Greek words "epic" and "demos" meaning upon and people, respectively. A loose interpretation would be disease upon people. It is a medical science that involves the study of the incidence and distribution of diseases in large populations, and to a lesser extent the conditions influencing the spread and severity of disease.

Initially, epidemiology concerned itself with infectious diseases. The first prominent epidemiological investigation was conducted in 1849 by the English physician John Snow, who observed that the London cholera epidemic occurred chiefly in regions served by the Broad Street pump. Snow is credited with hastening the resolution of the epidemic by removing the pump handle. Modern epidemiology can focus on the effects of age, such as the greater susceptibility of older persons to respiratory death during influenza outbreaks; of sex, such as the higher incidence of heart attacks among men; of nationality, such as the higher rate of the birth defect spina bifida among children born to parents of Irish background; or, of social influences, as exemplified by the greater incidence of tuberculosis among the poor in crowded cities. In addition to searching for the general causes of disease among populations, an epidemiologist could be involved with identifying the source for a specific outbreak of a particular disease.

27.2   OCCUPATIONAL HEALTH

Discuss occupational health.

SOLUTION [5, 6]

Occupational disease is not a new phenomenon. Ample historical evidence exists recounting the effects of lead poisoning, chronic respiratory problems associated with mining, and the hazards of manufacturing (including traumatic injury).

Although it has been known for a long time that occupational exposures can induce human disease, as in the above examples, the fact remains that diseases of occupational origin are underreported. This can be attributed to three major factors. The first is that health professionals generally do not gather enough information concerning the patient's occupational history or the various jobs and duties carried out by the patient to possibly link employment with his/her symptoms [33]. The second is that many of the diseases associated with occupational causes could have been caused by other risk factors. Therefore, the occupationally caused case of lung cancer does not appear with some distinct marker to differentiate it from a lung tumor caused by personal risk factors such as smoking. Exceptions do, of course, exist. Mesothelioma – a relatively rare cancer of the lining of the lung – generally only occurs with exposure to asbestos. A third factor, particularly for chronic diseases, is the long time interval that can exist between initial exposure to an occupational agent and the development of disease. This long time interval can make the recognition of the occupational origin of a disease quite difficult. This is in stark contrast to the relative ease of associating injuries with job-related causes.

27.3 DESCRIPTIVE STUDIES

Briefly discuss descriptive studies, keying on the two types of analytical studies.

SOLUTION [5, 6]

The types of epidemiologic studies that attempt to note the number of cases of specific disease in a specific time period are generally known as descriptive studies. Descriptive studies attempt to provide investigators with information concerning the distribution of the disease in time and space as well as to identify attributes that may increase the chances of an individual contracting the disease. These attributes, called risk factors, include factors subject to change such as physical inactivity as well as those that are immutable, such as gender or age. For example, well-established risk factors for occupationally induced lung cancer include asbestos and coke oven emissions. Descriptive studies are also helpful in the formation of hypotheses regarding exposure and disease. Studies seeking to prove or disprove specific hypotheses are called analytical studies. These are briefly discussed below.

The two basic types of analytical studies are the cohort and the case–control study. Each has strengths and weaknesses as well as different resource and time requirements. The cohort study involves the study of individuals classified by exposure characteristics, e.g., a group of welders. The study then follows the development of disease in the welders' group as well as in an unexposed comparison population. The measure that assesses the magnitude of association between the exposure and disease and indicates the likelihood of developing the disease in the exposed group relative to the unexposed is the relative risk. A relative risk of 3.0 indicates no difference between the disease experience in the two groups. A relative risk of greater than 1.0 indicates a positive association between the exposure and the disease, and an increased risk in those who are subject to the exposure.

In the case–control design, a group with a disease (cases) is compared with a selected group of nondiseased (control) individuals with respect to exposure. The relative risk in control studies can only be estimated as the incidence rate among exposed individuals and cannot be calculated. The estimator used is the odds ratio, which is the ratio of the odds of exposure among the cases to that among the controls.

The main difference between the case–control and the cohort type of study is that in the case–control format, the investigator begins by classifying study subjects as to disease status. With the cohort study, the investigator begins by separating study subjects by exposure status. There are major resource consumption differences between the types of study. Cohort studies generally consume more resources and take longer to complete than do case–control studies.

27.4   PROBABILITY

The term probability appears regularly in both epidemiology studies and risk analysis (see Chapters 25 and 26). Define this in a qualitative sense.

SOLUTION

Probability can be interpreted subjectively as a measure of degree of belief, on a scale from 0 to 1, that the event A occurs. This interpretation is frequently used in ordinary conversation. For example, if someone says, "The probability that I will go to the racetrack tomorrow is 90%," then 90% is a measure of the person's belief that he/she will go to the track. This interpretation is also used when, in the absence of concrete data needed to estimate an unknown probability on the basis of an observed relative frequency, the personal opinion of an expert is sought. For example, a doctor might be asked to estimate the probability that an individual with cancer will die. The estimate would be based on the doctor's familiarity with cancer and the history of the patient.

27.5   PREVALENCE

A pressing challenge for epidemiologists interested in occupational health is to derive an accurate picture of disease frequency. This challenge is met by two broad types of measurement: prevalence and incidence rate. These are introduced in this and the next Problem.

Disease refers to the negative health cases existing in the population. Point-prevalence identifies the prevalence estimated from a survey at a given time, e.g., the number of workers with abnormal heart beats. Prevalence is computed as the number of cases divided by the number of study subjects at that given point in time. The describing equation is therefore

$$\text{Prevalence} = \frac{\text{Number of persons with a disease}}{\text{Total number in the study}}$$

Prevalence is thus not a true rate but simply a proportion, although the term prevalence rate is also used.

TABLE 27.1   Data for Problem 27.5

Plant	Individuals Adversely Affected	Total Population
A	2000	300000
B	121	10000

Studies were recently conducted of populations in the vicinity of a plant producing a new nanochemical that experienced health problems. The data in Table 27.1 was obtained from two plants (A and B), both located in urban areas. Calculate the prevalence for both plants.

SOLUTION

$$\text{Prevalence (A)} = \frac{2000}{300\,000} = 0.0067 \text{ or } 0.67 \text{ per } 100$$

$$\text{Prevalence (B)} = \frac{121}{10\,000} = 0.0121 \text{ or } 1.21 \text{ per } 100$$

The Prevalence number at plant B is nearly twice that of plant A.

The effect of time is described through the term latency. Latency refers to the period of time that elapses between the first contact of a harmful agent and a "host," and the development of identifiable symptoms or disease. Latency may be as short as a few hours, e.g., the time required for photochemical smog to induce watery eyes. Or it may extend to 20–30 years for a chronic condition such as asbestosis or malignant neoplasm of the lung. The association between a given exposure and a disease is all that more difficult because of the passage of time.

27.6   INCIDENCE RATE

Incidence, or incidence rate, refers to the number of new cases of a disease in a defined population in a given period of time. Thus, the incidence rate can be expressed as:

$$\text{Incidence rate} = \frac{\text{Number of new cases of disease during a time period}}{\text{Total number at risk}}$$

Ten new cases of a strange ailment experienced by U.S. women using a face cream containing nanoparticles during the 2005 calendar year were reported. It was previously estimated that 1 325 000 people used the cream. Calculate the incidence ratio (IR).

SOLUTION

Based on the problem statement

$$IR = \frac{10}{1\,325\,000} = 7.55 \times 10^{-6}$$

The reader should note that epidemiology uses ratios to express the health experience of populations. Ratios are important because epidemiology is inherently a comparative discipline. It is common for an epidemiologist to compare the disease experience of a study population with that of a comparison population. A ratio is nothing more than a specialized proportion in which individuals with a particular disease are placed over a denominator that is composed of people who are at risk, that is, who have a chance of developing the disease. Women, for example, would not be included in the denominator used to calculate the incidence (or prevalence for that matter) of prostate cancer.

Ratio increases in reported problems over an extended period of time can potentially lead to misleading results. For example, if the number of problems increased from 10 to 40 over a three-year period, one could be (incorrectly) led to believe the average percent increase per year was 100%. If this were true, there would be 20 problems after year 1, 40 problems after year 2, and 80 problems after year 3. The correct approach for an increase from P to F over an M year period in given by

$$F = P(1 + i)^M$$

where i is the average percent (one fractional basis) increase. For this case, $i = 0.59$ or 59% percent.

27.7 THE MEAN

When one has a set of N observations, the individual can use the values of these observations to estimate certain characteristics of a larger number of observations that have not been made, or are simply not available. The usual concern is not directly in the particular set of observations at hand, but rather in using them to estimate something about a potentially larger set. The observation recorded is defined as a *sample*. The larger set, which is usually available is called a *population*. Thus, the characteristics of a sample of observations can be used in estimating similar characteristics for the population.

The mean can be estimated on the basis of a sample of observations. Let X_1, $X_2, \ldots, X_n$ denote a sample of n observations on X. Then the sample mean $\bar{X}$ is defined by

$$\bar{X} = \sum_{i=1}^{n} \frac{X_i}{n}$$

Refer to Problem 27.6. If incidence rates of this strange ailment in France, England, Germany, and Greece were determined to be 10.61, 10.97, 7.20, and 8.42 (all multiplied by 10^{-6}), respectively, calculate the average (or mean) incidence rate, $\overline{IR}$, for all five (including the United States) countries.

SOLUTION

The reader is referred to Problem 9.5 for additional information on the arithmetic mean, geometric mean, and median. For this Problem, employ the above equation:

$$\overline{X} = \overline{IR} = \frac{\sum\limits_{i=1}^{n} (IR)_i}{n}; \quad n = 5$$

$$= \frac{(7.55 + 10.61 + 10.97 + 7.20 + 8.42)10^{-6}}{5}$$

$$= 8.95 \times 10^{-6}$$

The reader should note that the mean can be viewed as the *center of gravity* of the distribution. If all observations were of equal weight bearing down on a bar, this arrangement would be balanced by a knife edge held under the bar at a value of 8.95.

One may also associate *weighing factors* (w) with each observation. For this situation:

$$\overline{X} = \frac{\sum\limits_{i=1}^{n} w_i X_i}{\sum w_i} = \frac{w_1 X_1 + w_2 X_2 + \cdots + w_n X_n}{w_1 + w_2 + \cdots + w_n}$$

$\overline{X}$ is defined as the *weighted arithmetic mean* for this case.

27.8   THE VARIANCE AND THE STANDARD DEVIATION

Although the mean is a satisfactory measure of the previously described central tendency, it does not in itself provide enough information on the distribution, i.e., the mean does not convey any information with respect to the variability (or variance) of the individual observations.

The variance can be estimated on the basis of a sample of observations. Once again let $X_1, X_2, \ldots, X_n$ denote a sample of n observations on X. Then the sample variance σ^2 is defined by

$$\sigma^2 = \sum_{i=1}^{n} \frac{(X_i - \overline{X})^2}{n-1}$$

The calculation of σ^2 can be facilitated by use of the formula

$$\sigma^2 = \frac{n \sum_{i=1}^{n} X_i^2 - \left(\sum_{i=1}^{n} X_i \right)^2}{n(n-1)}$$

One may divide this sum of squares in the first equation by n, the number of observations in the population, and this will also yield the variance. The division should be by n rather than by $n-1$ when dealing with a population rather than a sample. The notation s^2 instead of the σ^2 is used for the population.

Refer once again to Problem 27.6 and 27.7. Calculate the variance and standard deviation for the five countries.

SOLUTION

The reader is referred to Problem 9.6 for additional details on the standard deviation. For this Problem, employ the first equation provided above:

$$\sigma^2 = \sum_{i=1}^{n} \frac{(X_i - \overline{X})^2}{n-1}; \qquad X = IR$$

Substituting,

$$\sigma^2 = \frac{(10.61 - 8.95)^2 + (10.97 - 8.95)^2 + (7.20 - 8.95)^2 + (8.42 - 8.95)^2 + (7.55 - 8.95)^2}{4}$$

$$= \frac{2.7556 + 4.0804 + 3.0625 + 0.2809 + 1.96}{4}$$

$$= 3.035$$

The standard deviation is

$$\sigma = \sqrt{\sigma^2}$$
$$= \sqrt{3.035}$$
$$= 1.74$$

References: Part 4

1. L. Theodore, J. Reynolds and K. Morris, "Health Safety and Accident Management: Industrial Application," A Theodore Tutorial, East Williston, NY, 1992.

2. L. Theodore and R. Allen, "Air Pollution Control Equipment", A Theodore Tutorial, East Williston, NY, 1993.

3. J. Reynolds, J. Jeris and L. Theodore, *Handbook of Chemical and Environmental Engineering Calculation*, Hoboken NJ: John Wiley & Sons, 2002.

4. L. Theodore, J. Reynolds and F. Taylor, *Accident & Emergency Management*, Hoboken, NJ: John Wiley & Sons, 1989.

5. A. M. Flynn and L. Theodore, *"Health, Safety and Accident Management in the Chemical Process Industries"*, New York City: Marcel Dekker, 2002.

6. Personal notes, L. Theodore, 2000.

7. Author unknown, *Nanotechnology Opportunity Report*, 2nd ed., location unknown, 2003.

8. National Center for Environmental Research, *Nanotechnology and the Environment Applications and Implication, STAR Progress Review Workshop*, Washington, DC: Office of Research and Development, National Center for Environmental Research, 2003.

9. J. Spero, B. Devito and L. Theodore, *Regulatory Chemicals Handbook*, New York City: Marcel Dekker, 2000.

10. P. L. McCarty and C. N. Swayer, *Chemistry for Environmental Engineering*, 3rd ed., New York: McGraw-Hill, 1978, p. 416.

11. Metcalf & Eddy. Inc., *Wastewater Engineering: Treatment, Disposal, and Reuse*, 3rd ed., New York: McGraw-Hill, p. 82.

12. J. Santoleri, J. Reynolds and L.Theodore, *Introduction to Hazardous Waste Incineration*, 2nd ed., Hoboken NJ: John Wiley & Sons, 2000.

13. G. Burke, B. Singh and L. Theodore, *Handbook of Environmental Management and Technology*, Hoboken NJ: John Wiley & Sons, 2000.

14. M. K. Theodore and L. Theodore, "Major Environmental Issues Facing the 21st Century", (originally published by Simon & Schuster), Theodore Tutorials, East Williston, NY, 1996.

15. L. Theodore and R. Kunz, *Nanotechnology: Environmental Implications and Solutions*. Hoboken NJ: John Wiley & Sons, 2005.

16. L. Bergeson, Nanotechnology and TSCA, *Chemical Processing*, (Nov. 2003).

17. L. Bergeson, Nanotechnology Trend Draws Attention of Federal Regulators, *Manufacturing Today* (March/April 2004).

18. L. Bergeson, and B. Auerbach, The environmental regulatory implications of Nanotechnology, *BNA Daily Environment Reporters*, pp. B-1 to B-7 (April 14. 2004).

19. L. Bergeson, Expect a Busy Year at EPA, *Chemical Processing* 17 (Feb. 2004).

20. L. Bergeson, Genetically Engineered Organisms Face Changing Regulations, *Chemical Processing*, (Mar. 2004).

21. D. Gute and N. Hanes, *An Applied Approach to Epidemology and Toxicology for Engineers*. Cincinnati, OH: NIOSH, 1993.

22. Threshold Limit Values and Biological Exposure Indices for 1987. Cincinnati OH: ACGIH, 1987.

23. D. LaGreda, P. Buckingham and J. Evans, *Hazardous Waste Management*. New York City: McGraw-Hill, 1994.

24. AIChE *Guidelines for Chemical Process Quantitative Risk Analysis*. New York: Center for Chemical Process Safety of the American Institute of Chemical Engineers, 1989.

25. D. Pautenbach, *The Risk Assessment of Environmental and Human Health Hazards: A Textbook of Case Studies*. Hoboken NJ: John Wiley & Sons, 1989.

26. J. Rodricks, and R. Tardiff. *Assessment and Management of Chemical Risks*. Washington, DC: American Chemical Society, 1984.

27. S. Lowe and D. Jamea, *NSF Sponsored Workshop on Environmental Workshop*, Utah: Logan, 1996.

28. R. Dupont, T. Baxter and L. Theodore, *Environmental Management: Problems and Solutions*. Boca Raton FL: Lewis Publishers, 1998.

29. B. L. Amstadler, *Reliability Mathematics*. New York City: McGraw-Hill, 1971.

30. R. Woonacott and T. Woonacott, *Introductory Science*, 4th ed., Hoboken NJ: John Wiley & Sons.

31. R. Ventimiglia, Problem and Solution Prepared for L. Theodore, New York City, 2003.

32. Adopted from *Funk & Wagnalls New Encyclopedia*, Volume 9, New York: Rand McNally & Company, 1983.

33. C. Klassen, M. Amdur and J. Doull, ASA *Casarett and Doull's Toxicology*. New York: Macmillan Book Company, 1986.

APPENDIX
Quantum Mechanics

Quantum mechanics is the theory used to describe the behavior of the electrons, atoms, and photons that make up the microscopic world of nanotechnology. It allows fundamental work to be done at atomic levels, work that will be needed to create the new structures and devices that many scientists predict will lead to the next industrial revolution. The theory itself is abstract, counterintuitive, and quite removed from everyday experience. But it has become an essential language for many areas of nanotechnology, and in spite of its mathematical difficulties has come to dominate theoretical work in almost every field of physics and chemistry during this past century.

Quantum mechanics was developed at the beginning of the twentieth century because of the failure of classical physics, the physics of Newton and Maxwell, to produce correct results when applied to the atomic world. Largely based on the work of such eminent scientists as Planck, Einstein, Bohr, Schrödinger, and Feynman, quantum mechanics has triumphantly provided a correct theory of atomic structure and has evolved to become the most successful theory in the history of physics.

Stated as simply as possible, the essential feature of quantum mechanics is that all of the fundamental particles, the electrons and photons and protons that make up the universe have a dual nature. As Richard Feynman observed "things on a very small scale behave like nothing you have direct experience about." These "things" can behave like particles at one time, and at other times they behave like waves. When a beam of electrons is passed through a system consisting of two closely spaced slits, for example, the pattern that forms on the observing screen is not the heaping up of electrons opposite the two slits. This would be the expected result if the beam behaved like a stream of bullets. The actual pattern produced consists of a series of bands spread across the screen with regions made up of large electron clusters, and regions where the electron density is zero. This so called diffraction pattern is identical to the pattern of light and dark regions that would be observed if light were to shine on the slit system. In other words, the electrons in this particular experiment behave like a wave.

A French graduate student named Louis de Broglie was the first to recognize the wave character of electrons and called the waves associated with the electrons

Nanotechnology: Basic Calculations for Engineers and Scientists, by Louis Theodore
Copyright © 2006 John Wiley & Sons, Inc.

"matter waves." On the basis of an analogy with Einstein's work with photons, he demonstrated that the "wavelength" of the electron was given by

$$\lambda = h/p$$

where λ = the wavelength of the electron, p = the electron's momentum, and h = Planck's constant, the basic quantum constant equal to 6.626×10^{-34} Joule-seconds.

In the photoelectric effect, radiation such as x-rays or ultraviolet light shining on a piece of metal can transfer part of their energy to the atomic electrons of the metal, and cause them to escape from the metal. These electrons are called photoelectrons. In spite of the fact that the wave theory of electromagnetic radiation had been well tested for centuries, it gave wrong answers when trying to predict the energy and general behavior of these "liberated" electrons. In 1905, however, Einstein made one of his astonishing breakthroughs, successfully explaining the photoelectric effect by assuming that the energy of the radiation was not distributed along its wave front, but concentrated in a stream of particle-like corpuscles he called photons. It is difficult to fully appreciate the shock this assumption made in the scientific world. It took ten years before scientists were ready to accept this "unreasonable" idea that "seemed to violate everything they knew about the interference of light." Einstein found that he obtained correct answers by assuming that the energy of a photon was given by

$$E = hf$$

where E = energy, h = Planck's constant, and f = the frequency of the light.

The inescapable fact is that, depending on the experiment, electromagnetic energy can behave either like a particle or a wave. Allow light to pass through a diffraction grating and it behaves like a wave. Allow light of sufficient energy to fall on a metal surface so that it causes electrons to be emitted from the plate, and it behaves like a particle.

The present theory of quantum mechanics has largely abandoned almost all the concepts of classical physics. Systems are now described by the information contained in a mathematical function, Ψ (psi), called a wave function or probability amplitude. Ψ contains all the information one can possibly know about a particle: its position, momentum, and energy, now and in the future. There is one drawback; the results of an experiment can never be exactly predicted, but only the probability of a certain outcome. Certainty has been abandoned, and the theory no longer predicts that a particle will be at a certain place at a specific time, but tells only the probability that the particle will be at that point. In keeping with its name as a probability amplitude, a large value of Ψ implies a high probability of finding the particle at that point, while a small value of Ψ implies that the chance of finding the particle at the point being examined is low.

For a given set of initial conditions, the wave function is found by solving, if possible, the time-dependent second-order partial differential equation known as the Schrödinger equation.

$$h^2/8\,m\pi^2\nabla^2\Psi + E_p\Psi = ih/2\pi\delta\Psi/\delta t$$

where h = Planck's constant, m = mass, ∇^2 = Laplace operator, $i = \sqrt{-1}$, and E_p = potential energy of the system.

Solutions of the Schrödinger equation often give values of the wave function that are negative or contain complex numbers. A complex number is a number that contains imaginary numbers such as the square root of -1. In the real world, however, the probability assigned to any event is always positive and always a real number. To make sure, therefore, that answers generated by the Schrödinger equation make physical sense, the final result of any computation is always expressed as the square of the absolute value of Ψ. The quantity

$$|\Psi|^2 V$$

is therefore the probability that the particle being measured will be found in the volume V.

The solution of the Schrödinger equation for a multiparticle system can be a lengthy and laborious task. However, it is the conceptual rather than the computational aspects of the theory that have given many scientists the most trouble. Many physicists, including one as distinguished as Einstein, have found it difficult to accept a world ruled by chance and in which the principle of cause and effect is no longer valid. Given a set of initial conditions, the outcome of any experiment is not precisely determined but it is only possible to know the likelihood of certain experimental outcomes. As powerful, far ranging, and successful as the Schrödinger equation has been, the fact that the information obtained from Ψ is largely statistical, and is expressed in terms of probabilities, has continued to disturb physicists.

To further compromise the ability to predict the outcome of an experiment is the unhappy fact that quantum mechanics comes with built-in uncertainties. There is an interaction between the observer and the observed, which in the quantum world of the very small necessarily introduces uncontrollable disturbances into the system. These uncertainties are not due to faulty experimental technique or design, but are basic and go to the heart of the wave–particle duality. The simultaneous measurement of the position and momentum of an electron, for example, requires a probe such as a beam of light to "see" the electron. The photon character of the light, however, as it strikes and interacts with the electron will inevitably produce a distorted measurement of its momentum. Werner Heisenberg, in his famous Uncertainty Principle, first recognized this fundamental limitation. He stated that because particles have both a wave and particle character, the attempt to simultaneously measure such "complementary variables" as position and momentum, or energy and time, will always produce a distortion of one of the variables. This result is assumed to be a fundamental law of nature. Mathematically this is expressed as:

$$\Delta x \Delta p \geq h/2\pi$$

where Δx is the uncertainty in the measurement of the position and Δp is the uncertainty in the measurement of the momentum. The constant h is the familiar Planck's constant. Any attempt to gain accuracy in position will be defeated by the loss of accuracy in the measurement of momentum.

A similar relationship exists between energy and time:

$$\Delta E \Delta t \geq h/2\pi$$

where ΔE is the uncertainty in the energy of a system, and Δt is the time required to make the measurement.

It is important to realize that since h, Planck's constant, is an extremely small number, the restraints placed on the simultaneous measurements of position and momentum are important chiefly for subatomic systems that include particles with an extremely small mass such as electrons and photons. For the purposes of nanotechnology, measurements of massive clusters of atoms or molecules would not necessarily be subject to the same restraints.

There is another important property of wave functions, called superposition, that lies at the very heart of quantum mechanics and is particularly relevant to nanotechnology. It has already been noted that a beam of electrons passing through two closely spaced slits, produced a typical interference effect – regions of high electron density alternating with regions of zero electron density on the observing screen. However, interference of this kind is only possible if the matter-wave associated with the electron behaved as if it were passing through both slits at the same time. That is, the electron wave that passes through the top slit interferes with the electron wave that simultaneously passes through the bottom slit. The seemingly logical statement that the electron reached the screen by either passing through one or the other slit is wrong.

A more formal and mathematical approach can clarify this counterintuitive result. The quantum description of the electron consists of the linear superposition of two wave functions, one that describes the electron passing through the top slit, and one that describes the electron passing through the bottom slit.

$$\Psi = a_1 \Psi_{\text{top slit}} + a_2 \Psi_{\text{bottom slit}}$$

where a_1, a_2 are constants.

Since the probability of finding the electron at any spot is given by the square of the absolute value of its wave function

$$|\Psi|^2 = a_1{}^2 |\Psi_{\text{top slit}}|^2 + a_2{}^2 |\Psi_{\text{bottom slit}}|^2 + 2a_1 a_2 |\Psi_{\text{top slit}} \Psi_{\text{bottom slit}}|^2$$

It is easy to see that the probability of an electron reaching a certain spot on a screen behind the slits is not simply the sum of the probability of reaching the screen by passing through the top slit plus the probability of reaching the screen by passing through the bottom slit. The sum of the first two terms of the above equation is what would be observed if the system behaved classically. It is the third term, the famous interference term, that expresses the essence of quantum mechanics. The interference term contains contributions from the wave functions describing both states; one state being the electron passing through the top slit, and the other state being the electron passing through the bottom slit. The interference terms make it possible to explain that under certain conditions, contributions from the two states can cancel each other so that the probability of observing the electron at a certain point is zero.

Since the electron can be in a superposition of two states in this experiment, one can no longer speak of it as being either "here" or "there." In the mysterious world of the quantum, the electron must be thought of as being both here *and* there. It seems to go through both slits at once. As difficult as this is to accept, the exciting new field of quantum computing depends on this principle of superposition that permits atoms and electrons to be in two different states at the same time.

Consider the example of a conventional computer where so-called "bits" are used to store and manipulate data. The bits act as switches that register either 1 or 0. That is, they are either on or off. A register of eight bits would be required to store all the possible permutations of three data values such as 0,0,1. Typical values would be 011,010,111... and so on. In a quantum computer, however, quantum bits called "qubits" are used to handle data. By using the principle of superposition, a qubit can be both off and on at the same time. Only three qubits would be required to store all of the eight possible combinations in a quantum superposition. It is easy to see how the possible combinations presented by a string of these qubit switches could be processed in a fraction of the time required by conventional computers. It has been estimated that quantum computers will be able to produce calculations a billion times faster than today's integrated circuits.

In a reversal of roles, quantum mechanics also permits physical processes that would be impossible under conventional analysis. One of the most striking effects described by quantum mechanics is the ability of particles to pass into regions whose potential energy barrier height is greater than the kinetic energy of the particle. It is as though the driver of a car who took his foot off the gas and allowed his car to coast along at 10 mph were able to climb a 1000 ft mountain. Classically, of course, this is impossible, but in the world of quantum mechanics there exists a finite chance that this barrier penetration or "tunneling," as it is called, will occur. Consider the example of an alpha particle whose kinetic energy is 5 MeV (million electron volts) trapped inside a uranium nucleus. The nuclear forces preventing the alpha particle from immediately escaping present an effective barrier of some 20–30 MeV. When the Schrödinger equation is applied to this situation most of the wave function is found inside the nucleus, but a small part of this wave appears outside the nucleus as well, although with an extremely small amplitude. Small as it may be, the existence of any Ψ in any region implies that a finite probability exists of finding the particle there. There is, in other words, a chance of finding the alpha particle outside the nucleus, and the alpha particle is thought of as "tunneling" through the potential barrier of the nucleus. The process being described here is, of course, familiarly known as radioactivity, and the quantum mechanical explanation of this process along with correct predictions of half-lives were one of the early triumphs of quantum mechanics.

An example of tunneling in the nanoworld is the limitation it places on the size of electronic devices such as transistors. Attempts to further miniaturize conventional transistors to smaller and smaller sizes, beyond the currently attainable tens of nanometers, has reached a natural limit, since it has been found that electrons tunneling between wires would disrupt anything smaller. Powerful scanning tunneling microscopes are now one of the important tools used to understand how electrons can tunnel through barriers in nanosized electronic devices.

Another important consequence of the wave nature of particles in the quantum world is that such important parameters of a system as energy, momentum, angular momentum, and spin, are often "quantized." For example, the energies of the electrons that make up a typical bound quantum system in an atom are quantized in that only certain allowable values are possible. Here again, as in so many other quantum systems, the wave nature of the electron accounts for this. Because of the self-interference of the electrons as they "orbit" the nucleus at the center of the atom, quantization can be thought of as demanding that, like the standing waves on a violin string, an integer number of waves fits along one of the electron's stationary orbits. The electrons are therefore restricted to certain noninterfering orbitals that are often called shells. The possible shells are assigned a set of numbers called quantum numbers ($n = 1, 2, 3, \ldots$) that define the energy of the shell. In a similar fashion, the angular momentum and spin of the electron are also quantized and have their own quantum numbers. The angular momentum of electrons can only have certain integral multiples of $h/2$, where h is Planck's constant and the spin can only have values of $(+\frac{1}{2})h/2\pi$ or $(-\frac{1}{2})h/2\pi$. The number of electrons that occupy each shell ($2n^2$) is strictly limited and determined by the Pauli Exclusion Principle, which states that no two electrons can have all their quantum numbers, including spin quantum number, the same.

Quantum dots are an important class of nanoparticles used in nanotechnology. They are typically made up of semiconductors that can vary in size from 20 to about 200 nanometers in diameter. Experimental work has recently been done, however, with quantum dots formed of electrons, rather than a material such as gallium arsenide. These electron "dots" can contain anywhere from one electron to several thousand. They are still generally many times larger than the size of a typical atom, but the quantum confinement they are subject to makes them behave like the electrons that form an atom. They are, in fact, often called an "artificial atom." Like atomic electrons, they arrange themselves in a series of energy levels, and each electron can exist in two possible spin states, "up" or "down". If the up spin is identified with a "1" and the down spin identified with a "0", the dot behaves very much like a conventional binary bit. Quantum dots can be easily connected to electrodes, so that these spins can be detected by the behavior of an electric current passed through the dots. The application of these dots to quantum computers is very much at the forefront of research today.

Much progress has recently been made in the area of atom manipulation and quantum control. Modern laser systems, for example, can capture and cool individual atoms, so that they can be positioned with incredible accuracy. Positioning cool atoms on the surface of a crystal can make nanowires a mere two atoms wide. Lasers can be used like pincers to move and interchange atoms on large organic molecules. The tools used to manipulate atoms are often the same size or slightly greater than the atoms themselves. Physicists call them "sticky fingers," since single atoms are, in general, extremely reactive, and will stick to anything. Interferometers that make use of the wave properties of matter are some of the new tools used to test the properties of these quantum structures.

The number of problems that can be solved exactly by quantum mechanics is not very large. The vast majority of problems more complicated than the single electron hydrogen atom must be treated by approximation methods. The most important of these include perturbation theory and the variation method. There are also large ranges of results that depend only on the symmetry properties of the system being studied and a branch of mathematics known as group theory can handle these.

Albert Stwertka

INDEX

Nanotechnology: Basic Calculations for Engineers and Scientists, by L. Theodore
Copyright © 2006 John Wiley & Sons, Inc.